Liposomes

a practical approach

£22

TITLES PUBLISHED IN
THE
PRACTICAL APPROACH
SERIES

Series editors:
Dr D Rickwood
Department of Biology, University of Essex
Wivenhoe Park, Colchester, Essex CO4 3SQ, UK
Dr B D Hames
Department of Biochemistry, University of Leeds
Leeds LS2 9JT, UK

Affinity chromatography
Animal cell culture
Antibodies I & II
Biochemical toxicology
Biological membranes
Carbohydrate analysis
Cell growth and division
Centrifugation (2nd Edition)
Computers in microbiology
DNA cloning I, II & III
Drosophila
Electron microscopy
in molecular biology
Fermentation
Gel electrophoresis of nucleic acids
Gel electrophoresis of proteins
Genome analysis
HPLC of small molecules
HPLC of macromolecules
Human cytogenetics
Human genetic diseases
Immobilised cells and enzymes
Iodinated density gradient media
Light microscopy in biology
Liposomes
Lymphocytes
Lymphokines and interferons
Mammalian development
Medical bacteriology
Medical mycology
Microcomputers in biology

Microcomputers in physiology
Mitochondria
Mutagenicity testing
Neurochemistry
Nucleic acid and
protein sequence analysis
Nucleic acid hybridisation
Nucleic acids sequencing
Oligonucleotide synthesis
Photosynthesis:
energy transduction
Plant cell culture
Plant molecular biology
Plasmids
Prostaglandins
and related substances
Protein function
Protein purification applications
Protein purification methods
Protein sequencing
Protein structure
Proteolytic enzymes
Solid phase peptide synthesis
Spectrophotometry
and spectrofluorimetry
Steroid hormones
Teratocarcinomas
and embryonic stem cells
Transcription and translation
Virology
Yeast

Liposomes

a practical approach

Edited by
R R C New

Formerly Departments of Parasitology and Tropical Medicine,
Liverpool School of Tropical Medicine, Pembroke Place, Liverpool L3 5QA, UK

Current address: Biocompatibles Ltd, Brunel Science Park, Kingston Lane, Uxbridge, UB8 3PQ, UK

IRL PRESS
—at—
OXFORD UNIVERSITY PRESS
Oxford New York Tokyo

Oxford University Press,
Walton Street, Oxford OX2 6DP

Oxford is a trade mark of Oxford University Press

Published in the United States
by Oxford University Press, New York

© Oxford University Press, 1990

British Library Cataloguing in Publication Data
Liposomes.
  1. Organisms. Liposomes.
  I. New, R.R.C. II. Series 574.87'34

Library of Congress Cataloging in Publication Data
Lipsomes: a practical approach/edited by R.R.C. New.
  (The Practical approach series)
  Includes bibliographical references.
  1. Liposomes. I. New, R. R. C. (Roger R. C.) II. Series.
  RS201.L55L55 1989   574.87'4—dc20   89-22802
  ISBN 0-19-963076-3.
  ISBN 0-19-963077-1 (pbk.)

Previously announced as:
ISBN 1 85221 059 1
ISBN 1 85221 060 5

Typeset and printed by Information Press Ltd, Oxford, England.

Preface

Over the last twenty years the liposome has changed its status from being a novel plaything for the laboratory worker to a powerful tool for the industrialist—with the gap between the ideal desired characteristics of liposomes and what is technically feasible becoming narrower all the time. The properties of membrane preparations have been researched extensively, and ingenious ways have been found of manipulating them to confer behavioural characteristics which stretch the imagination—sensitivity to heat, light, pH, magnetic field, and chemical structure. Few other areas of study can routinely bring into play such a wide range of phenomena.

Liposomes may be defined simply as lipid vesicles enclosing an aqueous space. They were brought to the attention of the scientific world by A.D. Bangham in 1965, and proposed as useful models for cell membranes. Indeed, using the definition above, even cells and organelles themselves may be considered to be just sophisticated types of liposome. Using these artificial membrane vesicles, great insight was brought to many aspects of cell physiology such as permeability, fusion, membrane-bound enzyme properties etc, and will continue to do so. More recently, the potential of liposomes in the medical field is slowly becoming realized, with several clinical trials in progress examining their use as drug delivery agents. Applications in the areas of diagnosis, immuno-modulation, and genetic engineering have been identified and developments will follow.

In spite of numerous books and papers written on the subject, many people are still unclear about what liposomes are, and how work employing them is carried out. The aim of this book is to dispel some of that mystery. It has been written with two groups of people in mind. Firstly, the laboratory worker, who wishes to have at his/her fingertips detailed, tried-and-tested methods which will be accepted by experienced workers in the field as giving results which are reliable and convincing. The methods presented here are not a comprehensive list of everything which can be, or has been done, but are a careful selection of the most useful and most easily applied methods for the general laboratory. The second category of reader is the graduate student who may have had little exposure to membrane techniques, and who will benefit from an understanding of the theory behind membrane processes. It is hoped that this may act as an introductory textbook for the basic principles of liposomology, before embarking on detailed study of more learned treatises.

Liverpool School of Tropical Medicine   Roger R.C.New
August 1989

Contributors

C.D.V.Black
Radiation Oncology Branch, National Cancer Institute, National Institutes of Health (10/B3-B69), Bethesda, MD 20892, USA

P.J.Bugelski
Department of Experimental Pathology, L-60, Smith Kline & French Laboratories, PO Box 7929, Philadelphia, PA 19104, USA

A.R.Cossins
Department of Environmental and Evolutionary Biology, University of Liverpool, PO Box 147, Liverpool L69 3BX, UK

S.Frøkjaer
NOVO Research Institute, Novo Alle, DK-2880 Bagsvaerd, Denmark

H.Hauser
Laboratorium für Biochemie, ETH-Zentrum, CH-8092 Zürich, Switzerland

T.D.Heath
School of Pharmacy, Center for Health Sciences, University of Wisconsin, Madison, WI 53706, USA

G.R.Jones
MRC/SERC, Biology Support Laboratory, Daresbury Laboratory, Warrington, WA4 4AD, UK

R.L.Kirsh
Department of Advanced Drug Delivery, Smith Kline & French Laboratories, Philadelphia, PA 19101, USA

P.I.Lelkes
Winter Research Building, Sinai Samaritan, 836 North 12th Street, Milwaukee, WI 53201, USA

A.Loyter
Department of Biological Chemistry, Institute of Life Sciences, The Hebrew University of Jerusalem, 91904 Jerusalem, Israel

F.J.Martin
Liposomes Technology Incorporated, 1050 Hamilton Court, Menlo Park, CA 04025, USA

R.R.C.New*
Biocompatibles Ltd, Brunel Science Park, Kingston Lane, Uxbridge, UB8 3PQ, UK

R.J.Parker
Division of Cancer Etiology, National Cancer Institute, National Institutes of Health (37/2D-02), Bethesda, MD 20892, USA

*The support of the Leverhulme Trust in the preparation of this book is gratefully acknowledged by R.R.C.New.

N.Payne
Lederle Labs Ltd, Fareham Road, Gosport, Hants PO13 0AS, UK

D.M.Phillips
Lipid Products, Nutfield Nurseries, Crabhill Lane, South Nutfield, near Redhill, Surrey RH1 5PG, UK

A.Puri
Laboratory of Mathematical Biology, National Cancer Institute, National Institutes of Health (10/4B-56), Bethesda, MD 20205, USA

G.L.Scherphof
Laboratory of Physiological Chemistry, University of Groningen, Bloemsingel 10, 9712 KZ Groningen, The Netherlands

J.M.Sowinski
Department of Experimental Pathology, L-60, Smith Kline & French Laboratories, PO Box 7929, Philadelphia, PA 19104, USA

Contents

Abbreviations

ADC	analogue to digital converter
AM	ammonium molybdate
ANS	anilinonaphthalene sulphonate
APSA	N-(p-aminophenyl) stearylamide
α-T	α-tocopherol
α-TS	α-tocopherol succinate
BCECF	2,7-biscarboxyethyl-5(6)-carboxyfluorescein
BCIP	5-bromo, 4-chloro, 3-indolyl phosphate
BHT	butylated hydroxytoluene
BPS	biotinylated phosphatidyl serine
BSA	bovine serum albumin
CDI	carbodiimide
CF	carboxyfluorescein
CHEMS	cholesterol hemisuccinate
CL	cardiolipin
CMC	critical micelle concentration
DCCI	dicyclohexyl carbodiimide
DLPC	dilauroyl phosphatidyl choline
DMPC	dimyristoyl phosphatidyl choline
DMPG	dimyristoyl phosphatidyl glycerol
DODAC	dioctadecyl ammonium chloride
DOPC	dioleyl phosphatidyl choline
DPPA	dipalmitoyl phosphatidic acid
DPPC	dipalmitoyl phosphatidyl choline
DPPG	dipalmitoyl phosphatidyl glycerol
DRV	dried-reconstituted vesicle
DSPC	distearoyl phosphatidyl choline
DTT	dithiothreitol
EDCI	1-ethyl-3-(dimethyl aminopropyl)-carbodiimide
ESR	electron spin resonance
F	fluorescence
FPL	French pressed liposomes
FTS	freeze-thaw sonication
G6PDH	glucose-6-phosphate dehydrogenase
HDL	high density lipoproteins
IUV	intermediate-sized unilamellar vesicle
LDL	low density lipoprotein
LPC	lyso-phosphatidyl choline
LUV	large unilamellar vesicle
MEL	micro-emulsification liposomes
Mes	morpholino ethane sulphonic acid
MLV	multi-lamellar vesicle
Mops	morpholino propane sulphonic aicd
MPS	monocyte phagocyte system
MTT	3-[4,5-Dimethyl thiazol-2-yl]-2,5-diphenyl tetrazolium bromide
MVL	multi-vesicular liposome
NHSIA	N-hydroxysuccinimido-iodoacetate

PA	phosphatidic acid
PC	phosphatidyl choline
PCS	photon correlation spectroscopy
PE	phosphatidyl ethanolamine
PG	phosphatidyl glycerol
PHA	pulse height analysis
PI	phosphatidyl inositol
PMSF	phenyl methyl sulphonyl fluoride
PS	phosphatidyl serine
PTA	phosphotungstic acid
PTR	phase transition release
PVP	polyvinyl pyrrolidone
RES	reticular endothelial system
RET	resonance energy transfer
REV	reverse-phase evaporation vesicle
RSVE	reconstituted Sendai virus envelope
SAMSA	5-acetylmercaptosuccinic anhydride
SATA	succinimidyl-5-acetylthioacetate
SFFF	sedimentation field flow fractionation
SM	sphingomyelin
SMPB	*N*-succinimidyl (4-[*p*-maleimidophenyl])butyrate
SPDP	*N*-succinimidyl pyridyl dithiopropionate
SPLV	stable plurilamellar vesicle
SUV	small unilamellar vesicle
TAC	time to amplitude converter
TBA	thiobarbituric acid
TBS	Tris-buffered saline
TEA	triethylamine
TEP	1,1,3,3, tetraethoxypropane
TMS	trimethyl silane
TNBS	trinitrobenzene sulphonic acid
TNS	(6-[*p*-toluidinyl]naphthalene-2-sulphonate)
TO	triolein
WCOT	wall-coated open tubular

CHAPTER 1

Introduction

ROGER R.C.NEW

1. AIM OF BOOK

This book is intended both as a compilation of methods which may act as a useful reference source for workers already in the field of liposomes, and as a simple guide to workers outside the field who are wondering whether liposomes might have some application to their own speciality, and if so, what sort of liposome is best to use. Consequently, for most of this volume the emphasis is not on the applications liposomes can be put to, since it is assumed the reader will already have his/her own uses in mind— uses which may be entirely original, and could not be anticipated by the authors of this book. Instead, we hope to make readers sufficiently well-informed about liposomes that they may choose for themselves the best methods to adopt for their purpose, and in attempting this, we have concentrated heavily on methodology as the means of classi-fying different areas of the subject.

The first half of the book describes in detail the different ways of making, modifying, purifying, and characterizing liposomes, while the last half discusses ways in which information can be obtained about the behaviour of the finished product in different biological systems. This introductory chapter gives a very simple guide to what liposomes are, and outlines the general principles involved in choosing a given type of liposome for a particular application.

2. STRUCTURE OF LIPOSOMES

Liposomes are simply vesicles in which an aqueous volume is entirely enclosed by a membrane composed of lipid molecules (usually phospholipids). They form spontaneously when these lipids are dispersed in aqueous media, giving rise to a population of vesicles which may range in size from tens of nanometres to tens of microns in diameter. They can be constructed so that they entrap quantities of materials both within their aqueous compartment and within the membrane. The value of liposomes as model membrane systems derives from the fact that liposomes can be constructed of natural constituents such that the liposome membrane forms a bilayer structure which is in principal identical to the lipid portion of natural cell membranes—the 'sea of phospholipids' in the Singer and Nicholson model. The similarity between liposome and natural membranes can be increased by extensive chemical modification of the liposome membrane, and may be exploited in areas such as drug targeting or immune modulation, both *in vivo* and *in vitro*, where the ability to mimic (or to improve upon) the behaviour of natural membranes, and also to be degraded by the same pathways, makes them a very safe and efficacious vehicle for medical applications. Alternatively,

a Chemical structure

Name

1,2 – diacyl–sn–glycerol – 3 – phosphoryl choline

Phosphatidyl choline

"Lecithin"

N–acyl–trans – 4 –sphingenine–1–phosphoryl choline

Sphingomyelin

1,2 – dialkyl – sn – glycerol – 3 – phosphoryl choline

"Ether – linked" phosphatidyl choline

Figure 1(a) Three main classes of choline-containing phospholipids. Choline-containing phospholipids are the most abundant phospholipids in nature. In phosphatidyl cholines (lecithin), a three-carbon glycerol bridge links two long chain fatty acids with a phosphoryl choline moiety; by convention, the fatty acids are said to occupy the 1 and 2 positions of the glycerol bridge while the polar headgroup is in position 3. The bridge carbon in position 2 (the middle carbon of the three-carbon glycerol bridge) is asymmetric—i.e., it displays optical activity because each of the four bonds is joined to a different chemical group. Natural phospholipids can have many different fatty acids conjugated in positions 1 and 2, usually with the longer and/or more unsaturated chain in position 2 as shown here (see also *Table 2*). Sphingomyelin consists of a single fatty acyl chain conjugated via an amide linkage to the nitrogen of sphingosine, which is again linked to phosphoryl choline. The lipid portion without the headgroup is known as ceramide. In addition to sphingomyelin, this ceramide residue is found in molecules known as gangliosides and cerebrosides, which contain polysaccharide headgroups in place of phosphoryl choline. Hydrocarbon chains can also be joined to the glycerol bridge via ether linkages as in the third class of phosphocholine lipids, of which a synthetic analogue useful in liposome formulations is shown here. In nature, diether phospholipids usually have a glycerol headgroup, while choline or ethanolamine headgroups are more usually found in molecules containing one each of the acyl and ether linkages (e.g. in plasmalogens). These lipids are considerably less prevalent in nature than the diacyl phosphatidyl cholines.

b Phosphatidyl moiety

Headgroup	Common name abbreviation

Phosphatidyl

Headgroup	Common name	abbreviation
$O-CH_2-CH_2-\overset{+}{N}-Me$ (Me, Me)	choline	PC
$O-CH_2-CH_2-NH_3^+$	ethanolamine	PE
$O-CH \overset{NH_3^+}{\underset{COO^-}{}}$	serine	PS
$O-CH_2-CH-CH_2$ (OH OH)	glycerol	PG
$O-H$	acid	PA
(inositol ring, OH groups)	inositol	PI

Figure 1(b) Some common naturally-occurring phosphatidyl phospholipids. In the same way as for phosphatidyl choline, the phosphatidyl moiety, when linked to each of the headgroups shown here, can contain a range of different fatty acids (see *Table 2*). In nature, the hydrocarbon chains of PE and PS are often highly unsaturated. With the exception of PC and PE, all the phosphatidyl phospholipids are negatively charged, since the charge on the phosphate is not neutralized by the headgroup. In phosphatidic acid, the phosphate group can easily lose the second proton, giving the molecule a double negative charge.

c   **Chemical structure**   **Name**

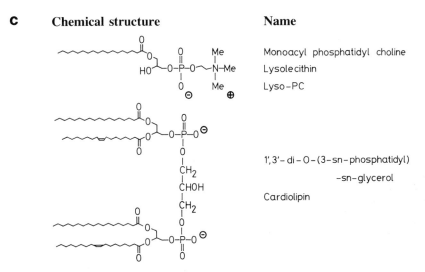

Monoacyl phosphatidyl choline

Lysolecithin

Lyso-PC

1′,3′-di-O-(3-sn-phosphatidyl)

-sn-glycerol

Cardiolipin

Figure 1(c) Phospholipids with unusual fatty acid content. Upon hydrolysis, phosphatidyl phospholipids such as PC can lose one or both of the fatty acid chains, to give lyso- derivatives. Acid hydrolysis gives equal mixtures of lyso- products with the remaining chain in either the 1- or the 2- position, while enzymatic hydrolysis with phospholipase A (e.g. from snake or bee venom, or from pancreatic secretions) exclusively cleaves the ester linkage in the 2- position, giving lyso- derivatives of the type shown here. (Hydrolysis by the enzyme phospholipase C, e.g. from cabbage leaves, does not affect the acyl chain linkages, but removes the headgroup, cleaving between this headgroup and the phosphate moiety.) In mammals, certain tissue membranes such as that of heart muscle mitochondria have relatively high proportions of cardiolipin. This is a negatively charged lipid which has antigen structures in common with the DNA backbone.

liposomes can be composed of entirely artificial components, chosen for their improved chemical properties. In the rest of this book it will be assumed that the liposomes being discussed are made of natural phospholipids, since most information exists relating to vesicles of this composition, but it should be borne in mind that stable bilayer membrane vesicles can be constructed of a wide range of other types of lipid—for example, fatty acids, double chain secondary amines, or cholesterol derivatives.

2.1 Chemical constituents

2.1.1 *Structural components*

Phospholipids are the major structural components of biological membranes, respresentative structures of which are depicted in *Figure 1a* and *b*. Two sorts of phospholipids exist—phosphodiglycerides and sphingolipids, together with their corresponding hydrolysis products. The most common phospholipids are phosphatidyl choline molecules (PC)—amphipathic molecules in which a glycerol bridge links a pair of hydrophobic acyl hydrocarbon chains, with a hydrophilic polar headgroup,

d

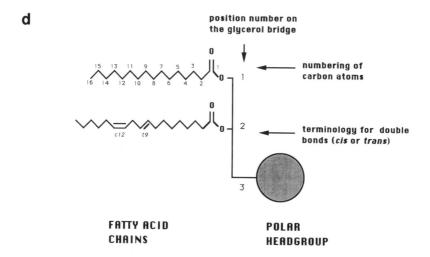

position number on
the glycerol bridge

numbering of
carbon atoms

terminology for double
bonds (*cis* or *trans*)

**FATTY ACID
CHAINS**

**POLAR
HEADGROUP**

Figure 1(d) Numbering of atoms and bonds in phosphodiglycerides. The carbons in a long chain fatty acid are numbered sequentially from the carboxyl end along, as are the $C-C$ single and double bonds. Note that the *trans* double bond shown here is for illustration only, and would be very unusual in a natural fatty acid of mammalian or plant origin.

phosphocholine. Molecules of PC are not soluble in water in the accepted sense, and in aqueous media they align themselves closely in planar bilayer sheets in order to minimize the unfavourable interactions between the bulk aqueous phase and the long hydrocarbon fatty acid chains. Such interactions are completely eliminated when the sheets fold on themselves to form closed sealed vesicles. Phosphatidyl cholines contrast markedly with other amphipathic molecules (detergents, lysolecithin) in that bilayer sheets are formed in preference to micellar structures. This is thought to be because the double fatty acid chain gives the molecule an overall tubular shape, more suitable for aggregation in planar sheets compared with detergents with a polar head and single chain, whose conical shape fits nicely into a spherical micellar structure (see *Figure 2*). Surfactant molecules, such as dioctadecyl ammonium chloride (DODAC), containing two hydrocarbon chains (as well as single chain fatty acids, e.g. oleic) can, under certain circumstances, also form bilayer membrane vesicles.

Phosphatidyl cholines, also known as 'lecithin', can be derived from both natural and synthetic sources. They are readily extracted from egg yolk and soya bean, but less readily from bovine heart and spinal cord. Forming as they do the major phospholipid component of many cell membranes, they are often used as the principal phospholipid in lipsomes for a wide range of applications, both because of their low cost relative to other phospholipids, and their neutral charge and chemical inertness. Lecithin from natural sources is in fact a mixture of phosphatidyl cholines, each with chains of different length and varying degrees of unsaturation. Lecithin from plant sources has a high level of polyunsaturation in the fatty acyl chains, while that from mammalian sources contains a higher porportion of fully saturated chains.

Fatty chains from natural sources have almost exclusively even numbers of carbons in the chain, since biosynthesis involves building up the chain in two-carbon units at the carboxyl end via malonylCoA, from an acetylCoA primer. This process continues

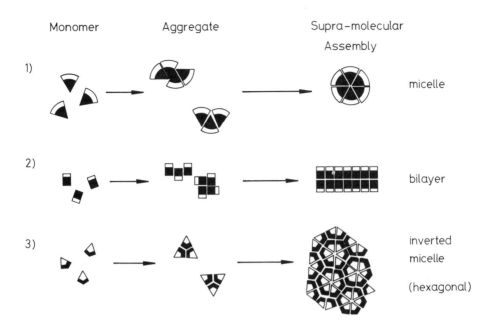

Figure 2. Association patterns of amphiphiles. Relation between molecular shape and structure of aggregates.

uninterrupted until the formation of (C16:0) palmitic acid, which is the fatty acid of shortest chain length usually found in mammalian phospholipids. At this stage, the processes of chain elongation (again in two-carbon units) and desaturation can take place concurrently, The first unsaturation is introduced into the 9-position, giving a family of mono-unsaturated fatty acids, with a *cis*-double bond in the 9, 11, 13, or 15 positions (*Figure 3*). Further unsaturations are added relative to the first and subsequent double bonds, in such a way that each unsaturation is separated by one methylene group, so that none of the olefinic bonds are conjugated with each other. In plants, these unsaturations are inserted in between the last unsaturation and the terminal methyl group, while in mammals they are added on the carboxyl side of the existing unsaturations.

Mammalian fatty acids which contain unsaturations less than seven carbons from the terminal methyl group are derived from plant lipids (e.g., linoleic and linolenic) ingested in the diet (*Figure 4*). Almost all double bonds of natural fatty acids are in the *cis* configuration.

Most natural phospholipids are 'mixed', in that the fatty acids attached in positions 1 and 2 of the same molecule (see legend of *Figure 1*) are usually different from each other; that in the inside position is unsaturated, while the outer chain is usually saturated. Nomenclature and terminology of fatty acids is given in *Table 1* which shows the proportions of different fatty acids making up PC from different sources. *Table 2* shows proportional distributions of saturated and unsaturated acids in positons 1 and 2 of PCs from different sources.

(i) *Phosphatidyl choline membranes*

Phase transitions. At different temperatures, lecithin membranes can exist in different

5

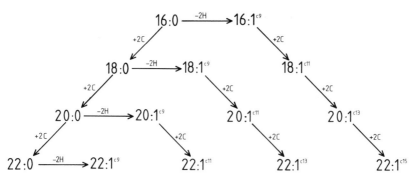

Figure 3. Biosynthetic pathway for some mono-unsaturated fatty acids. This diagram shows the way in which fatty acids of the same chain length can be synthesized with the unsaturation located at different distances along the chain from the carboxyl group, as a result of carrying out the dehydrogenation and two-carbon chain extension steps in different orders. Here, dehydrogenation is represented by horizontal lines, while carbon addition takes place on going from top to bottom. The first unsaturation is introduced in the 9- position, regardless of the chain length. The most common mono-unsaturated acid is oleic acid 18:1 (c9) followed by 16:1 (c9), 18:1 (c11), 20:1 (c9), 20:1 (c11) and 22:1 (c13). Note that the small 'c' here indicates that the double bonds located at the numbered positions along the chain are in the *cis* configuration. *Trans* bonds are referred to with a small 't'.

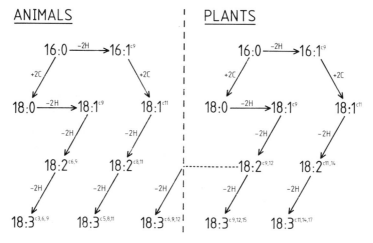

Figure 4. Biosynthetic pathway for some polyunsaturated fatty acids in animals and plants. Fatty acids synthesized by plants and animals differ in the positions of their unsaturations, because plants introduce additional double bonds towards the end of the chain, while animals introduce them towards the carboxyl group at the beginning. Animals also make use of fatty acids obtained from plants in the diet. Thus arachidonic acid (20:4 c5,8,11,14) is derived to a large extent from plant linoleic acid via γ-linolenic acid:

$$18:2 \ (c9,12) \ \rightarrow \ 18:3 \ (c6,9,12) \ \rightarrow \ 20:3 \ (c8,11,14) \ \rightarrow \ 20:4 \ (c5,8,14)$$

Plant linolenic acid 18:3 (c9,12,15) is converted by animals to 20:5 and 22:6 acids.

phases, and transitions from one phase to another can be detected by physical techniques as the temperature is increased. The most consistently observed of these phase transitions is the one occurring at the highest temperature, in which the membrane passes from a tightly ordered 'gel' or 'solid' phase, to a liquid–crystal phase at raised temperatures where the freedom of movement of individual molecules is higher. The most widely

Table 1. Fatty acids of phospholipids in common use.

Abbreviation	Systematic name	Common name	Relative abundance in natural sources of PC (%)		
			Soya PC	Egg PC	Rat
C12:0	dodecanoic	lauric			
C14:0	tetradecanoic	myristic			
C16:0	hexadecanoic	palmitic	17.2	35.3	28.8
C16:1	*cis*-9-hexadecenoic	palmitoleic			
C18:0	octadecanoic	stearic	3.8	13.5	18.0
C18:1	*cis*-9-octadecenoic	oleic	22.6	26.8	8.4
C18:2	*cis,cis*-9,12-octadecadienoic	linoleic	47.8		
C18:2	*cis,cis*-6,9-octadecadienoic			5.7	19.4
C18:3	all *cis*-9,12,15-octadecatrienoic	α-linolenic	8.6		
C18:3	all *cis*-6,9,12-octadecatrienoic	γ-linolenic		0.2	
C20:0	eicosanoic	arachidic			
C20:1	*cis*-9-eicosenoic	gadoleic			
C20:3	all *cis*-8,11,14-eicosatrienoic				1.0
C20:4	all *cis*-5,8,11,14-eicosatetraenoic	arachidonic		1.0	17.0
C20:5	all *cis*-5,8,11,14,17-eicosapentaenoic			3.6	2.4
C22:0	docosanoic	behenic			
C22:5	all *cis*-7,10,13,16,19-docosapentaenoic			1.3	1.5
C22:6	all *cis*-4,7,10,13,16,19-docosahexaenoic			12.6	3.5
C24:0	tetracosanoic	lignoceric			

In the numerical abbreviation for each of the fatty acids (left-hand side), the first figure refers to the total number of carbon atoms in the fatty acid chain, and the second to the number of unsaturations (double bonds) which that chain possesses. Note that fatty acids from mammalian or plant sources all have even numbers of carbon atoms, and that their unsaturations (almost always *cis*) are spaced at three-carbon intervals, each with an intervening methylene group. Data from 'Fatty Acids and Glycerides', Kuksis,A. (ed.). Plenum Press, New York (1978). Handbook of Lipid Research volume 1.

Table 2. Position of different fatty acids in lecithins from various sources.

	Egg yolk PC		Soya bean lecithin		Rat liver	
	(1)	(2)	(1)	(2)	(1)	(2)
16:0	68.8	1.8	34.4	–	23.3	5.6
18:0	25.8	1.2	7.6	–	64.9	3.8
18:1	4.7	48.9	30.2	15.8	7.4	13.3
18:2	0.2	11.1	23.8	71.8	1.3	22.5
18:3	0.5	–	4.0	13.1	–	1
20:4	–	2.1	–	–	0.2	39.4
20:5	–	7.1	–	–	–	–
22:5	–	2.6	–	–	–	–
22:6	–	25.2	–	–	–	7.0

Figures are percentages of total fatty acid in a given position [(1) or (2)] on the glycerol backbone.

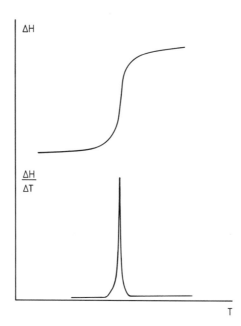

Figure 5. Change in heat absorbed by membrane preparation upon increase of temperature, as a result of passing through a transition from one phase to another. Traces such as this one are obtained using a differential scanning microcalorimeter, in which the heat required by the sample to maintain a steady upward rise in temperature is plotted as a function of temperature. In essence, two small aluminium pans are compared, one empty, and one containing a concentrated sample of liposomal membrane. Thermistors monitor the temperature of each pan as the pans are heated up separately. The heat input of the sample pan is adjusted so that its temperature matches that of the reference pan. At the phase transition temperature, extra heat is required in order to maintain the rise in temperature of the sample pan equal to that of the reference, and this is recorded directly as shown in the upper curve. The lower trace is obtained from the upper by mathematical differentiation, and the area under the peak is the enthalpy of transition.

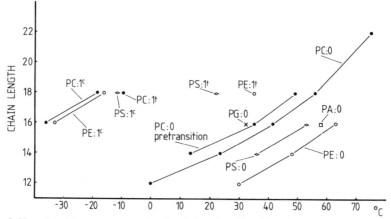

Figure 6. Phase transition temperatures for diacyl phospholipids with different headgroups as a function of chain length. Data from different sources is presented for synthetic phospholipids containing fatty acids of the same chain length and unsaturation in both 1- and 2- positions. Unless otherwise stated, all values quoted are for the main transition. The mono-unsaturated acids all have their unsaturation (either *cis* or *trans*) in the 9-position, which is the position in the chain giving the lowest phase transition temperature (i.e. the position in which the *cis* double bond maximally inhibits close packing of fatty acid chains in the gel phase).

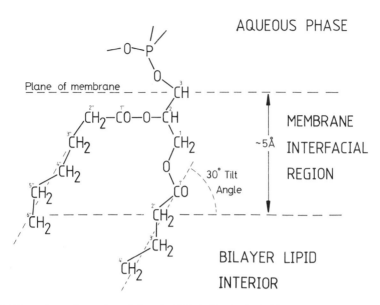

AQUEOUS PHASE

MEMBRANE

INTERFACIAL

REGION

BILAYER LIPID

INTERIOR

Figure 7. Schematic conformation of phosphatidyl choline in the bridge region. This diagram shows the orientation of the glycerol bridge approximately perpendicular to the plane of the membrane, and the first two carbons of the acyl chain in the 2- position roughly horizontal to the plane, as in the X-ray crystal structure. A chain tilt of 30 degrees has been inferred from comparison of the bilayer thickness with the length of the straight chain, but it is not clear in which direction this tilt is manifested relative to the 2-ester linkage. One possibility is shown here.

used method for determining the phase transition temperature (T_c) is microcalorimetry—see *Figure 5*. The influence of hydrocarbon chain length and unsaturation (as well as head group) on the value of T_c for membranes composed of different phospholipids is considerable, and is illustrated in *Figure 6*. In general, increasing the chain length, or increasing the saturation of the chains increases the transition temperature. Membranes made from egg yolk lecithin have a transition temperature from $-15°C$ to $-7°C$, compared with membranes from mammalian sources which are usually in the range zero to $40°C$.

An understanding of phase transitions and fluidity of phospholipid membranes is important both in the manufacture and exploitation of liposomes, since the phase behaviour of a liposome membrane determines such properties as permeability, fusion, aggregation, and protein binding, all of which can markedly affect the stability of liposomes, and their behaviour in biological systems.

Chain tilt. In bilayer membranes, the molecules appear to be aligned with the glycerol backbone approximately perpendicular to the plane of the membrane, and the phosphocholine headgroup in a straight line roughly parallel with the membrane surface. This conformation would be expected to reduce the distance between positive and negative charges within the phospholipid molecule. Because of the three methyl groups attached to the quaternary nitrogen, the headgroup of phosphatidyl choline in this conformation is very bulky, and occupies an area of the membrane (42 $Å^2$) greater than that taken up by the two fatty acids in their straight chain configuration (~ 39 $Å^2$). In order to overcome this, the hydrocarbon chains are thought to tilt relative to

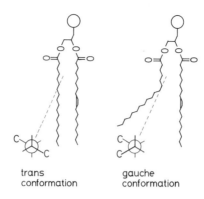

trans
conformation

gauche
conformation

Figure 8. Disposition of phospholipid diacyl chains. This diagram shows the difference in configuration caused by rotation about one C−C single bond from a *trans-* to a *gauche-* conformation. Multiple changes from *trans* to *gauche* conformation increase the total volume occupied by the hydrocarbon chains.

the plane of the membrane at an angle of 58°, in such a way as to fill up the extra space created by the headgroups, and to bring the chains of adjacent molecules into closer proximity, to maximize Van der Waals and other non-covalent interactions (*Figure 7*).

The tilt of the fatty acid chains, together with the fact that the first two carbon atoms of chain 2 are parallel to the membrane surface, before bending down into the bilayer, means that lecithins with chains of equal length will have gaps in the centre of the bilayer where the terminal methyl of chain 2 cannot reach. In natural lecithins this problem is overcome by arranging that chain 2 is usually longer than chain 1. Liposomes composed of lipids whose chains are of very unequal lengths (e.g. chain 2 with 10 and chain 1 with 18 carbon atoms) are thought to form bilayers in which interdigitation of chains from opposite sides occurs end to end. In general though, such behaviour is considered exceptional, and for most natural and synthetic phosphatidyl cholines the bilayer may be regarded as two independent and non-interacting monolayers as far as phase behaviour is concerned.

Phase separation. As the temperature increases, the fatty acid chains tend to adopt conformations other than the all-*trans* straight chain configuration, such as the *gauche* conformation state illustrated in *Figure 8*, and this tends to expand the area occupied by the chains, and hence the membrane, at the same time as it reduces the overall length of the hydrocarbon chains, giving rise to a decrease in bilayer thickness upon transition from a gel to liquid−crystalline phase. In fact, the transition from gel to liquid crystal does not occur in a single step for lecithin, but involves two transitions—the main transition already described, and the pre-transition, about five degrees below the main transition, at which a change in headgroup orientation may occur; the heat of this transition is very low compared with the main transition. In the temperature range between the two transitions, the membrane adopts a ruffled appearance, in which it is transformed from a planar to an undulating surface with a fairly long, regular periodicity (see *Figure 9*). The orientation adopted by the hydrocarbon chains to the plane of the membrane is not certain.

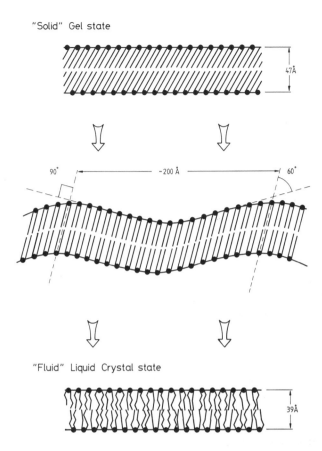

"Solid" Gel state

90° ~200 Å 60°

"Fluid" Liquid Crystal state

Figure 9. Appearance of rippled structure and elimination of chain tilt of membrane on passing through the phase transition. In electron micrographs of membranes in the 'π' phase between the pre- and the main transition the periodicity of the ripples seems to vary considerably from sample to sample. It is thought that the appearance of the rippled structure is related to the switch between a tilted and a perpendicular orientation for the phospholipid molecules, since PE membranes (which have no chain tilt) also have no pre-transition. Nothing is known about the orientation of the chain in the 'π' phase although it seems likely that both tilted and perpendicular configurations will co-exist in the same membrane. One way in which this can come about is shown here. In this configuration, the rippling permits the surface area occupied by the headgroups to remain relatively constant while accommodating differences in the area taken up by chains in different conformations.

Binary mixtures of synthetic lecithins of different chain lengths give a main transition intermediate in temperature between those of the individual components, unless the chain lengths are very different, in which case two separate transitions are observed; in the temperature region between the transitions solid and fluid phases may co-exist, each enriched in one or the other of the components (*Figure 10*). In principle, phase separation could in fact occur over the whole temperature range, giving rise to two co-existing solid phases below the lower transition temperature, and two fluid phases above. The only direct evidence for spatial separation of phase domains in membranes is obtainable in the region close to the pre-transition or main transition, when freeze-

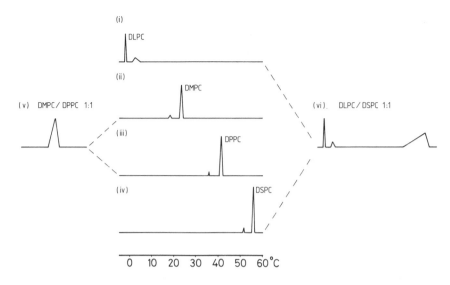

Figure 10. Microcalorimetry curves showing phase transitions of membranes containing single components or mixtures of phospholipids. Abbreviations: DLPC, dilauroyl (12:0) phosphatidyl choline; DMPC, dimyristoyl (14:0) phosphatidyl choline; DPPC, dipalmitoyl (16:0) phosphatidyl choline; DSPC, distearoyl (18:0) phosphatidyl choline. Membranes composed of mixtures of pure phospholipids can display a single phase transition if the T_cs of the individual lipids are close to each other (e.g. DMPC and DPPC). The combined T_c will usually be intermediate between those of the two separate lipids. Note how the combined transition tends to broaden relative to those of the parent compounds. Lipids with T_cs which differ greatly from each other (e.g. DLPC and DSPC) act as isolated components with a degree of mutual immiscibility, undergoing phase transitions independently of each other, and forming membranes composed of two or more separate phases rich in one or other of the individual components.

fracture EM pictures show rippled regions interspersed with clearer patches of membrane corresponding to the different phases (*Figure 11*).

Packing of phospholipids. Even in the upper liquid—crystalline phase, the acyl chains are far from being in the state of free random motion that is found in, for example, a paraffin melt, or the same molecules dissolved in an organic solution. The part of the phospholipid molecule most severely restricted in its motion is the glycerol backbone. The acyl methylene groups closest to the linking carboxyl group display a considerable degree of rigidity, up to about carbon atom number 9, and proceeding further along to the end of the chain, the motion of the carbon atoms becomes progressively less restricted. At the membrane boundary, however, where the lipid compartment interfaces with the bulk aqueous phase, the limited rotational and translational freedom manifests itself in alignment of lipids in a regular two-dimensional array with molecules all adopting a set distance and orientation with respect to each other.

Over a very short time-scale, this matrix can be thought of in the same terms as a solid three-dimensional crystal, in which packing abnormalities occur as a result either of impurities, or of a proportion of the molecules in the array adopting an altered configuration. Packing abnormalities such as point defects, line defects, and grain boundaries (*Figure 12*) give rise to small exposed areas of the membrane which may allow greater access to small molecules to pass through than elsewhere, resulting in a significant permeability of the membrane to such molecules. These packing defects

Figure 11. Electron micrograph of liposome membrane showing phase separation. The sample is a mixture of synthetic dipalmitoyl phosphatidyl choline (DPPC) and natural bovine brain phosphatidyl serine (PS) frozen quenched at 34 °C. Two phases can be seen, a liquid L_α phase with a 'jumbled' surface texture, and a lower temperature intermediate phase with a rippled or 'banded' texture. Micrograph kindly supplied by Dr Sekwen Hui.

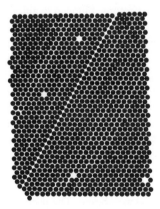

Figure 12. Ball-bearing type model of packing irregularities in two-dimensional matrix: point and line defects.

are quite probably associated with changes in conformation of the acyl chain carbons from *trans* to *gauche*, a transition which has an enthalpy of around 0.4 kcal mol^{-1} and an activation energy of 3 kcal mol^{-1}. This small but significant activation energy means that rotation of adjacent methylene groups in the same chain with respect to each other is not completely free, and that the *gauche* conformation (see *Figure 8*), once achieved, will be maintained for a certain length of time before reverting to the lower energy *trans* configuration. Consequently, packing abnormalities associated with the

13

gauche conformation will have a finite lifetime of the order of milliseconds.

Furthermore, as the temperature increases, the *trans/gauche* transition will become increasingly favourable, and packing defects will increase both in duration and in number, giving rise to a temperature-dependent increase in membrane permeability to small ions and polar molecules. At the phase transition temperature both the gel and liquid − crystalline phases co-exist in different parts of the same membrane, and a consequent large increase in packing defects (such as grain boundaries) is found at the interface of the two phases. As a result, membrane permeability at the phase transition is considerably increased for many small molecules relative to that observed in membranes composed of a single phase, either at higher or lower temperatures than the transition temperature.

Membrane permeability. Liposome membranes are semi-permeable membranes, in that the rate of diffusion of molecules and ions across the membrane varies considerably. For molecules with a high solubility in both organic and aqueous media, a phospholipid membrane clearly constitutes a very tenuous barrier; on the other hand, polar solutes such as glucose, and higher molecular weight compounds pass across the membrane only very slowly. Smaller molecules with neutral charge (e.g. water and urea) can diffuse across quite rapidly, while charged ions differ greatly in their behaviour. Protons and hydroxyl ions cross the membrane fairly quickly, probably as a result of transfer of hydrogen bonds between water molecules (water molecules are found as deep in the bilayer interior as the lower carbonyl group, so the distance to be bridged is considerably reduced). Sodium and potassium ions, on the other hand, traverse the membrane very slowly, not only with respect to protons, but also to anions such as chloride and nitrite.

It would appear that the mechanism for diffusion of metal ions across lipid membranes is quite different from that for other molecule species, as suggested by the fact that permeability to sodium ions decreases with increasing unsaturation of the fatty acyl chains, while permeability to glucose is slightly increased with increasing unsaturation. One explanation may be that metal ions travel along transient kinks in the fatty acid chains formed as a result of single bond rotations. Double bonds cannot rotate freely, and so may hinder the progress of the ions. Glucose, on the other hand, is too large to make use of these conformational changes in individual molecules, and requires the looser packing of membranes containing unsaturated fatty acids for increased transport across the bilayer. However, extending the chain length, thereby increasing the total bilayer thickness, has the same effect with all solutes, charged or otherwise, of decreasing their rate of diffusion through the membrane. The permeability of liposome membranes to calcium and other multivalent ions is lower still than for monovalent ions like sodium, as might be predicted from their increased charge and hydration shell diameter.

That there are several different mechanisms for the passage of molecules across lipid membranes is further illustrated by the difference in behaviour between protons and metal ions at the phase transition temperature (T_c) (*Figure 13*). Proton permeability increases at the T_c, and remains high as the temperature is increased, in contrast to sodium ions (and most other solutes), where the permeability above and below the T_c is lower than that at the T_c itself. Thus protons and water molecules, whose transport is considerably faster than sodium ions, have extra routes across the membrane which are qualitatively different, possibly in the form of water channels.

The above discussion refers to liposomes in which the inner and outer aqueous

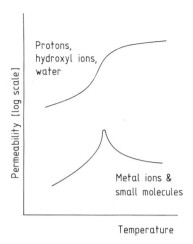

Figure 13. Permeability of membrane below and above T_c. Comparison of protons with sodium ions.

compartments are in equilibrium with each other. Because phospholipid membranes are semi-permeable, however, a concentration difference of solute between one side of the membrane and the other can generate an osmotic pressure which will lead to accumulation of water molecules (being the fastest diffusing species) on one side. In the case of a high concentration of solute entrapped inside liposomes, bathed in low concentration buffer outside, the liposomes will swell up as the internal volume of water increases, to such an extent that the area of the membrane is considerably increased, with the spacing between adjacent phospholipid molecules being correspondingly increased. Under these conditions, an accelerated leakage through the membrane can occur for solutes of molecular size equivalent to or smaller than glucose. Leakage of sucrose is not affected. In some cases, however, the pressure generated may be sufficient to rupture the membrane completely, giving rise to loss of the total contents of the liposome into the bulk aqueous phase, before re-sealing again.

(ii) *Other neutral phospholipids*

In addition to phosphatidyl choline, neutral lipid bilayers may be composed of sphingomyelin, or alkyl ether lecithin analogues which substitute entirely for lecithin in the membranes of halophilic bacteria. Replacement of ester groupings by ether linkages increases the resistance of such lipids to hydrolysis, while not apparently greatly affecting the physical properties of the membranes. The presence in sphingomyelin of the amide linkage and hydroxyl groups, in the region corresponding to the glycerol backbone of lecithin, gives rise to hydrogen bond interactions which may explain the more highly ordered gel phase relative to phosphatidyl choline. As with lecithins, the phase transition temperature increases with increased content of saturated fatty acid, although variation of chain length shows a transition temperature maximum of 52°C for C18 (stearoyl)— greater than for C16 or C24—indicating a more favourable interaction in the gel phase between the stearoyl chain and the invariant sphingosine moiety, presumably due to matching of chain lengths. In nature, however, sphingomyelins exist in forms with chains of very unequal length (e.g. C24 compared with the constant sphingosine chain of 13−15

15

carbons) suggesting the possibility of interdigitation of molecules on opposite sides of the bilayer. Unlike lecithins, mixtures of sphingomyelins of different chain lengths give a combined phase transition temperature below that of the individual components. Natural mammalian sphingomyelin extracts show phase transitions in the region of 37°C, in accord with the paucity of unsaturated fatty acid chains found in these mixtures. In contrast, natural phosphatidyl cholines usually have a much higher content of unsaturations, and give phase transitions well below body temperature. As far as is known, the chains of sphingomyelins are tilted in the gel phase in a similar manner to lecithin. The headgroup probably adopts similar configurations to those in lecithin, although intramolecular hydrogen bonding of the phosphate group may restrict its freedom of motion considerably. In general, sphingomyelin membranes are considered more 'stable' and tightly packed than lecithin bilayers, as shown by decreased values for permeability to solutes, greater resistance to lysis by bile salts, and lower membrane fluidity.

The other neutral phospholipid found commonly in natural membranes is phosphatidyl ethanolamine (PE) (*Figure 1b*). Possessing an unsubstituted quaternary ammonium group, which is protonated at neutral pH, this lipid differs from lecithin in two respects; first, in that its headgroup is smaller than the bulky phosphocholine of lecithin, and secondly, that it is able to take part in hydrogen bonding interactions with its neighbours in the membrane. The small headgroup means that it is not necessary for the chains to tilt in the gel phase in order to approach close enough to maximize Van der Waals interactions of the acyl chains. Consequently the molecules are oriented perpendicular to the plane of the membrane below the gel phase, and no pre-transition is observed. Saturated PEs have transition temperatures approximately 20°C higher than their PC analogues (*Figure 6*); this is attributed to the formation of intermolecular hydrogen bonds which have to be broken before PE can undergo transition to the more expanded liquid−crystalline phase. Unsaturated PEs, on the other hand, in which the *cis* double bonds prevent close approach of adjacent molecules in the bilayer, have almost the same transition temperatures as their PC analogues, presumably because even in the gel phase, the headgroups are too far apart from hydrogen bonding to occur. At low pH, saturated PEs also display PC-like transition temperatures, as the nitrogen becomes protonated, and hydrogen bonding reduced.

Because of its small headgroup, PE can readily adopt a non-bilayer configuration, in which the molecules form long cylindrical structures with the headgroups directed towards the centre (*Figure 2*). This is known as the hexagonal (HII) phase. In all PEs, the transition to hexagonal phase occurs above the main gel−liquid transition, which means that for all common synthetic PEs used, saturated PEs are in the bilayer gel phase at room temperature, while unsaturated PEs are all in the hexagonal (non-liposomal) phase. In mixtures of unsaturated PEs with other phospholipids such as PC, the bilayer phase may be stabilized, and the transition to hexagonal phase prevented. If the concentration of the stabilizing lipid is low, however, a bilayer may be formed in which areas of phase separation exist, where hexagonal or inverted micelle structures appear to occur within the lamellar sheet. In nature, the acyl chains of PE often vary considerably in length, and can be highly unsaturated, suggesting the possibility that PE may be involved in non-bilayer organization of membrane lipids in biological systems.

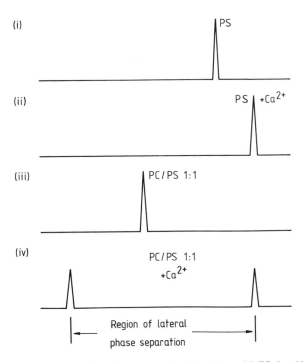

Figure 14. Phase separation arising from interaction of calcium ions with PS. In addition to raising the phase transition temperature of phosphatidyl serine (PS) slightly, calcium can act to separate PS from neutral lipids in the membrane, and create domains of PS-rich gel phase co-existing with a PC-rich fluid phase. Thus at temperatures close to the phase transition temperature, a homogeneous membrane can be converted by calcium to one with numerous boundary regions between solid and liquid phases, giving rise to increased leakage of small molecules, and the opportunity for intermembrane fusion.

(iii) *Negatively-charged phospholipids*

In negatively-charged (acidic) phospholipids (see *Figure 1b*), all three possible forces regulating headgroup interactions (and hence phase transition properties) of bilayer membranes can come into play—namely steric hindrance, hydrogen bonding, and electrostatic charge. Thus for dipalmitoyl phosphatidyl glycerol (DPPG), the bulky glycerol group, in conjunction with electrostatic repulsion of the unprotonated phosphate at pH 7, gives it a main transition temperature of almost 10°C below that of dipalmitoyl phosphatidyl choline (DPPC). In contrast, DPPA (dipalmitoyl phosphatidic acid) has a small headgroup and single proton at neutral pH which, like PE, can undergo hydrogen bonding, resulting in an elevated main transition temperature for pure DPPA bilayers. At high and low pH, the T_c is brought down, particularly at high pH, where electrostatic repulsion can push the headgroups apart.

Membranes composed of acid phospholipids can bind strongly to cations, particularly divalent cations such as calcium and magnesium. The reduction in electrostatic charge of the headgroups as a result of binding causes the bilayer to condense, increasing the packing density in the gel phase, and as expected, raising the transition temperature. Thus, at the appropriate ambient temperature, the addition of cations can induce a phase change from liquid-crystalline to gel phase. In membranes composed of acidic and neutral

17

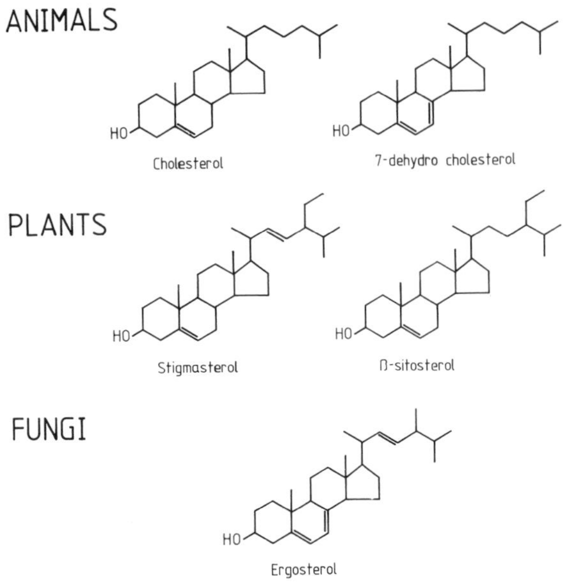

ANIMALS

Cholesterol

7-dehydro cholesterol

PLANTS

Stigmasterol

β-sitosterol

FUNGI

Ergosterol

Figure 15. Major sterols found in natural membranes from different sources.

lipids, displaying a single main transition, addition of cation can induce a phase separation. In the presence of the cations the membrane displays two transitions at different temperatures, corresponding to the individual lipids, with the region in between characterized by a phase separation of the two components (see *Figure 14*).

Lipids such as phosphatidyl serine (PS) show considerable selectivity for calcium compared with magnesium in their binding, the difference perhaps being due to the different coordination number of calcium (7 or 8 in a non-uniform configuration, in contrast to the regular octahedral 6-coordination of magnesium) or the greater readiness of calcium to lose part of its hydration shell, and to displace water upon complex formation. To achieve the same binding, it is often necessary to add a 20-fold greater excess of magnesium relative to calcium. It is not clear in any system whether the binding of calcium to lipid is in a 1:1 or 1:2 molar ratio, or which part of the headgroup the cation interacts with. As a result of neutralization of electrostatic charge, both calcium and magnesium can cause large liposomes composed of phosphatidyl serine to aggregate. Aggregation is a commonly observed phenomenon for all neutral planar membrane preparations, the driving force of which is presumably Van der Waals interaction. For PS vesicles in the presence of calcium the process can go a stage further, with membranes in close apposition fusing with each other. Since the rate of fusion increases with increasing temperature and membrane fluidity, and is especially rapid at the phase transition temperature, it is proposed that the initial step in fusion is very close approach of membranes, followed by alignment of packing defects in the lipid lattices, where mixing of lipids and the formation of non-bilayer structures, such as the hexagonal phase, can take place more easily. Calcium may play a crucial role in bringing the bilayers closer together by partial dehydration of the membrane surface, and cross-linking of opposing molecules of phosphatidyl serine, in a way which is not possible for magnesium.

18

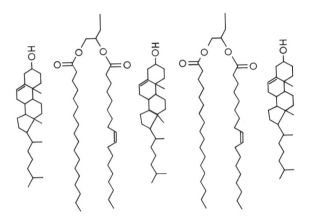

Figure 16. Position occupied by cholesterol in the membrane bilayer. Cholesterol rests in the membrane with the hydroxyl group on a level with the acyl chain carboxyl groups, and the planar steroid nucleus is thus located parallel with the first nine carbon atoms of the acyl chain, which is the portion of the chain where motion is most severely restricted in the absence of cholesterol. In contrast, motion of the carbons later in the chain is relatively free, and is increased in the presence of cholesterol, as insertion of the steroid nucleus at the head of the chain creates more space for the end carbons in the chain to move.

(iv) *Cholesterol*

Sterols are important components of most natural membranes, and incorporation of sterols into liposome bilayers can bring about major changes in the properties of these membranes. In mammals, the predominant sterol is cholesterol, with a significant quantity of 7-dehydro-cholesterol being found in subcellular membranes. Commonly encountered plant sterols are stigmasterol and sitosterol, while ergosterol is an endogenous sterol of yeasts, fungi, and some protozoa (*Figure 15*). Features common to all of these sterols are a 3β-hydroxyl group, a planar steroid nucleus, and an aliphatic side chain, all of which are essential for the characteristic behaviour of sterols in membranes to be displayed.

Cholesterol does not by itself form bilayer structures, but it can be incorporated into phospholipid membranes in very high concentrations—up to 1:1 or even 2:1 molar ratios of cholesterol to PC. In natural membranes, the molar ratio varies from $0.1-1$, depending upon the anatomical and cellular location. Being an amphipathic molecule, cholesterol inserts into the membrane with its hydroxyl group oriented towards the aqueous surface, and the aliphatic chain aligned parallel to the acyl chains in the centre of the bilayer. The 3β-hydroxyl group is positioned level with the carboxyl residues of the ester linkages in the phospholipids, with very little vertical freedom of movement. The presence of the rigid steroid nucleus alongside the first ten or so carbons of the phospholipid chain has the effect of reducing the freedom of motion of these carbons, while at the same time creating space for a wide range of movement for the remaining carbons towards the terminal end of the chain (see *Figure 16*). Above a certain concentration of cholesterol, the membrane area occupied by the acyl chains and sterol combined is greater than or equal to that taken up by the phosphocholine headgroup, so that PC membranes with high levels of cholesterol do not show the chain tilt that is observed in the gel phase of liposomes composed of pure PC (*Figure 17*). Addition

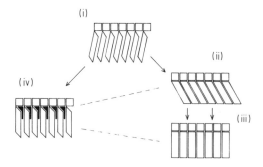

Figure 17. Influence of cholesterol on chain tilt. This diagram illustrates how tilting the chains in DPPC can increase their relative cross-sectional area and enable the chains to pack closely in the presence of a bulky headgroup. The combined area taken up by fatty acyl chains and cholesterol together is equal to that occupied by the headgroup, so when cholesterol is introduced there is no longer any need for chain tilt to maximize lipid chain interactions. The effective surface area of the membrane thus changes only very slightly on addition of cholesterol, since the cholesterol is able to fill the gaps in between the hydrocarbon chains of adjacent PC molecules. Configuration (i) shows the lipids aligned as in the X-ray derived crystal structure. Configurations (ii) and (iii) show the hydrated lipids below and above the phase transition temperature respectively, while the insertion of cholesterol (iv) gives a configuration somewhere in between.

of cholesterol to PC membranes has a marginal effect on the position of the main transition temperature (T_c); in DPPC the T_c changes from 41 °C to 44 °C with 33 mol% cholesterol. With increasing concentration, however, cholesterol is able to eliminate evidence of a phase transition altogether, reducing the enthalpy of phase change to zero at 50 mol% (1:1 ratio), and in so doing altering the fluidity of the membrane both below and above the phase transition temperature (*Figure 18*). Below this temperature, the phospholipids are pushed apart, the packing of the headgroups is weakened, and the fluidity of the ordered gel phase is increased. Above the transition temperature, the reduction in freedom of acyl chains causes the membranes to 'condense', with a reduction in area, closer packing and a decrease in fluidity. These changes in fluidity are paralleled by changes in permeability of the membrane—decreased by high cholesterol at temperatures higher than the T_c but increased at lower temperatures.

The distribution of cholesterol throughout the membrane is not random. At low concentrations of cholesterol (20 mol% or less), a phase separation is thought to occur, as is suggested by resolution of the enthalpy peak of the main transition (for synthetic PCs) into two components, the lower corresponding to pure PC, the upper, cholesterol-rich phase. An unexpectedly high water permeability above the phase transition temperature in egg yolk lecithin at 10 mol% cholesterol is also seen. In the pre-transition temperature range, however, instead of separate domains of smooth and rippled area, a single rippled surface is observed, in which the spacing between the ridges increases from 100 Å (10 nm) upwards with increasing cholesterol until the ridges disappear completely at 20 mol%. It has been proposed that PC membranes with low cholesterol content display microheterogeneity with regard to the distribution of lipids, areas in the troughs of the rippled surface consisting of the cholesterol-rich phase, while the ridges reflect the presence of cholesterol-poor regions. Such behaviour may result from attractive interactions between cholesterol monomers, as are known to occur in

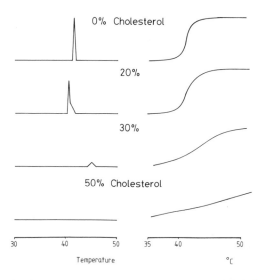

Figure 18. Influence of cholesterol on phase transition. Cholesterol has relatively little effect on the position of the phase transition, but is able to abolish completely the heat of transition. As the concentration of sterol reaches equimolar proportions with phospholipid, the freedom of molecular motion above the phase transition is decreased, while below the phase transition mobility is actually increased.

cholesterol micelles, in which the most favourable arrangement of lipids in the membrane is one in which cholesterol molecules are oriented adjacent to each other. At a ratio of 1:1 of cholesterol to PC, space-filling models show that very efficient packing of components in a two-dimensional lattice is achieved in the form of linear arrays, with rows of cholesterol molecules alternating with rows of phospholipid, such that both cholesterol–cholesterol and cholesterol–lecithin interactions are possible. It may be that concentrations of cholesterol higher than 50 mol% are difficult to achieve because they would disrupt the regular linear structure, and reduce the number of specific intermolecular interactions.

As in the case of single lipids, the addition of cholesterol to membranes composed of mixtures of lipids abolishes the phase transition and alters permeability and fluidity characteristics in the same way. In mixtures giving phase separations, cholesterol interacts preferentially with the component having the lower transition temperature, that is, with the more fluid phase, which is presumably condensed as a result. It is not clear whether cholesterol displays a preference for phospholipids of one headgroup over another, although experiments measuring the rate of uptake of cholesterol into membranes from aqueous phase suggest that sphingomyelin has a particularly high affinity. This might be expected in the light of the possibilities for H-bonding between hydroxyl groups situated at the aqueous lipid interface of the membrane.

While the transfer of phospholipids from one side of the bilayer to the other (or 'flip-flop'), probably via packing irregularities, takes place to a very limited extent in pure lipid bilayers, equilibration of cholesterol between the two sides of the membrane occurs quite rapidly (half-time < 30 min), and may be explained by the smaller size of the polar moiety (the hydroxyl group) possessed by cholesterol relative to phospholipids.

(v) *Small unilamellar vesicles*

The behaviour of bilayer membranes described in this section hold true for planar expanses of bilayer membrane such as comprise large unilamellar liposomes, or large multilamellar liposomes (see Section 2.2). As the curvature of the membrane increases, it develops a degree of asymmetry, which is very marked in small unilamellar vesicles (SUVs) of diameter approximately 250 Å (25 nm), where the restriction in packing geometry dictate that over 70% of the lipids making up the bilayer are in the outer leaflet. Vesicles composed of phospholipids of different headgroups show asymmetry with respect to the distribution of these lipids; those with smaller headgroups (e.g. PI, PA, PS) preferentially occupy the inner bilayer, while PC is located in the outer shell. In contrast to planar lamellae, small vesicles composed of DPPC show no pretransition, and the main transition is much broader, has a lower enthalpy, and occurs about 4°C lower. Evidence suggests that differences in packing may cause intermembrane fusion to occur more readily in small vesicles, and may act as a means of relieving thermodynamically unfavourable packing constraints. This is seen particularly at low temperatures. Below the T_c, the lipid chains are in a more highly disordered state than in planar membranes, forming a sort of glass-like structure; because of the very tight packing (especially in the inner monolayer), however, motion within the bilayer is actually reduced. In membranes composed of mixtures of PCs (e.g. saturated and unsaturated PCs), the effect of bending the lipid bilayer to give a small radius of curvature is to decrease the miscibility of the lipids, giving rise to lateral co-existence of different lipid domains of different composition within the bilayer, even above the gel-to-liquid crystal transition.

2.1.2 *Non-structural components*

For structural purposes, no components other than phospholipids and sterols need be incorporated into the membrane. Because the membrane interior is a very fluid aliphatic medium composed of molecules associated by non-covalent interactions it will readily accept and retain a wide range of lipophilic compounds without the need for any fixed chemical structural specificity. Under normal circumstances, these compounds can probably be accommodated in the membrane to a concentration of about $1-10\%$ by weight without serious disruption of the basic bilayer structure, although the membrane integrity as determined by fluidity or permeability may well be altered. In particular cases, where a specific interaction is known to occur between the compound and other membrane components (e.g. fatty acids, α-tocopherol), concentrations higher than 10% may be achieved. Conversely, relatively low concentrations of certain substances, such as some polyene antibiotics, will completely disrupt the membrane, as a result of specific interactions.

From a practical point of view, the great value of liposomes as carriers of therapeutic and other materials, is the wide spectrum of materials which they can incorporate, ranging from lipophilic agents mentioned above, through amphipathic compounds, located with the phospholipids at the boundary between the aqueous phase and the membrane interior, to water-soluble molecules of any description entrapped in the enclosed aqueous compartment. Because the means of their incorporation is entirely physical, no restrictions are placed on the chemical nature of these agents. Accordingly, only two classes of compound are not good candidates for incorporation into liposomes—first, materials which are insoluble in both aqueous and organic solvents, although with certain methods of preparation even these need not be an obstacle, and secondly, materials whose solubility is high in both media, and for whom a lipid membrane will constitute no barrier to passage from the inside to the outside of a liposome.

2.2 Physical structure of liposomes

Apart from their chemical constituents, which determine such properties as membrane fluidity, charge density, and permeability, liposomes can be characterized by their size and shape. They can range in size from the smallest vesicle obtainable on theoretical grounds (diameter ~25 nm) determined by the maximum possible crowding that headgroups will tolerate as the curvature in the inner leaflet increases with decreasing radius, to liposomes which are visible under a light microscope, with a diameter of 1000 nm (one micron) or greater, equal to the dimensions of living cells. They can also be bound by a single bilayer membrane, or may be composed of multiple concentric membrane lamellae stacked one on top of the other. Partly because it turns out that liposomes of different sizes often require completely different methods of manufacture, and partly because different applications often demand the use of liposomes within particular size ranges, classification of liposomes according to size is the most common index of characterization in current use. It is usual to define the liposomes one works with as belonging to one of several categories:

(i) *Multilamellar vesicles (MLVs)*. These usually consist of a population of vesicles covering a wide range of sizes (100−1000 nm), each vesicle generally consisting of five or more concentric lamellae. Vesicles composed of just a few concentric lamellae are sometimes called oligo-lamellar liposomes, or paucilamellar vesicles (*Figure 19*).

(ii) *Small unilamellar vesicles (SUVs)*. These are defined here as those liposomes at the lowest limit of size possible for phospholipid vesicles. This limit varies slightly according to the ionic strength of the aqueous medium and the lipid composition of the membrane, but is about 15 nm for pure egg lecithin in normal saline, and 25 nm for DPPC liposomes. Since, according to the definition, these liposomes are at or close to the lower size limit, they will be a relatively homogeneous population in terms of size.

(iii) *Large unilamellar vesicles (LUVs)*. These liposomes have diameters of the order of 1000 nm.

(iv) *Intermediate-sized unilamellar vesicles (IUVs)*. This term is not currently found in the literature, but it is introduced here for convenience and to avoid confusion. These have diameters of the order of magnitude of 100 nm (*Figure 20*).

Other terms also used in the literature such as REV (reverse-phase evaporation vesicle), DRV (dried-reconstituted vesicle), MVL (multivesicular liposome) relate to the method of manufacture, and will be discussed in detail in Chapter 2.

For unilamellar vesicles, the phospholipid content is related to the surface area of the vesicles, which is proportional to the square of the radius, while the entrapped volume varies as the cube. In addition, because of the finite thickness of the membrane, as the vesicles become smaller, their aqueous volume is further reduced since the phospholipids occupy more of the internal space. Consequently, for a given quantity

23

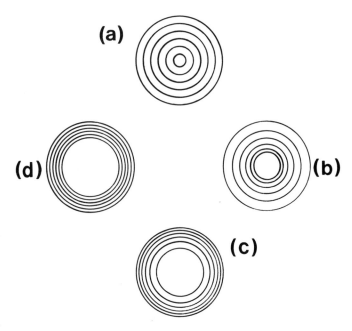

Figure 19. Multilamellar vesicles of different configurations. Although all the multilamellar vesicles (MLVs) in this drawing have the same number of lamellae, and entrap the same volume, they contain markedly different quantities of lipid. In (**a**) the bilayers are evenly spaced, in (**b**) each compartment contains the same volume as the sum of those within it (the outer compartment contains half the total liposome volume); in (**c**) each separate compartment contains the same volume while (**d**) is intended to represent a liposome in which the total contents are within the central core.

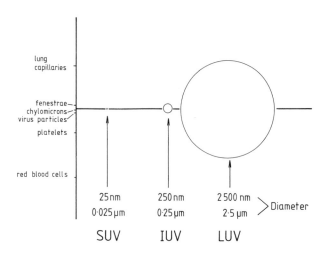

Figure 20. Comparative sizes of liposomes. Comparison of the sizes of small unilamellar vesicles (SUV), intermediate unilamellar vesicles (IUV) and large unilamellar vesicles (LUV) with the sizes of biological structures, relative radii of which are shown along the vertical scale on the left.

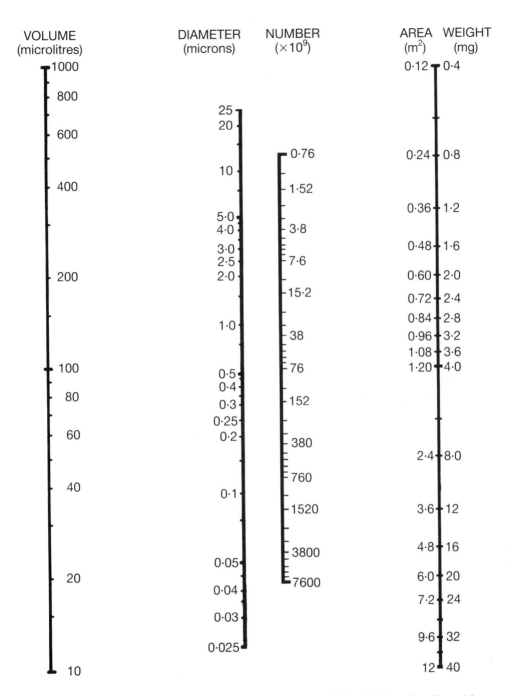

Figure 21. Nomogram relating volume, diameter, number, area and lipid weight for unilamellar vesicles.
To read this nomogram, simply connect two known parameters of your liposome population by a straight
line, and read off the unknown parameter where this line intercepts the third scale. Towards the lower end
of the diameter scale, values have been corrected to allow for a finite membrane thickness and unequal
distribution between inner and outer leaflets of the membrane.

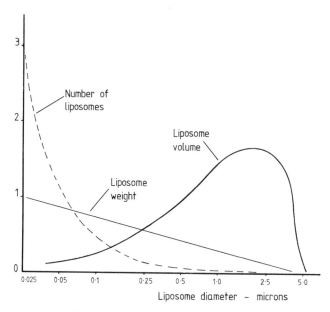

Figure 22. Variation in parameters as a function of size of unilamellar vesicles. In this figure, it has been postulated that a unilamellar vesicle population is being formed by ultrasonication of a multilamellar suspension, so that small vesicles accumulate in an exponential distribution at the lower size limit. Although 80% (by number) of the vesicles are smaller than 0.2 μm radius, 80% of the volume is enclosed in vesicles larger than this size.

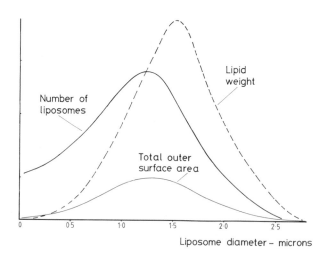

Figure 23. Variation in parameters as a function of size of MLVs. Here it has been assumed that an MLV population follows a standard distribution about a mean with respect to surface area, since the energy required to separate hydrated lipid sheets into individual liposomes should be proportional to the outer vesicle surface area. In this case, no striking differences are seen in distributions measured according to other criteria.

of lipid, large liposomes entrap a far greater aqueous volume than do small liposomes. The theoretical captured volume in litres per gram of lipid is given in *Figure 21* as a function of size.

In practice, most methods of liposome preparation give a fairly heterogeneous population of vesicles with a wide distribution of sizes, particularly towards the lower end of the range. Since the lipid-to-entrapped-volume ratio varies markedly with liposome size, it is a common observation that the size distribution of lipid weight does not correlate with the size distribution of entrapped materials, and it is possible that the behaviour observed for the entrapped material in a biological system is caused by just a small proportion of the total vesicle population (see *Figures 22* and *23*). For this reason it is very important to characterize liposome preparations according to size in order to interpret their *in vitro* and *in vivo* properties correctly (see Chapter 3, Section 4). Even liposomes of the same size and lamellarity can differ in internal distribution of aqueous phase, and quantity of lipid used to construct the vesicles. Thus the liposome type in *Figure 19a* with the lamellae evenly spaced, contains approximately half the lipid found in the liposome type in *Figure 19b*, where all the lipid is exterior to the aqueous compartment.

3. TAILORING LIPOSOMES TO SPECIFIC APPLICATIONS

The choice of what type of liposome to use for a given application depends partly on what one wishes to put inside the liposome, and partly on what one wants to do with the liposome once it is made.

3.1 Considerations related to liposome content

3.1.1 *Chemical composition*

Assuming that one wants to construct liposomes which retain their contents as long as possible, and which will be suitable for interaction with some biological system *in vivo* or *in vitro*, there is no reason why, to begin with, one should not employ the lipid composition mentioned earlier (egg lecithin/cholesterol/negative phospholipid in a molar ratio of 0.9:1:0.1). In many cases, particularly for immunological applications (see Chapter 6), sphingomyelin could be chosen in preference to lecithin, although the much greater cost of sphingomyelin restricts its large scale use considerably. As the system is refined, after the initial trials, a range of lecithins with different levels of unsaturation can be tested, in order to optimize incorporation, leakage, and effector activity. The use of saturated lecithins derived either from hydrogenated natural lecithins, or from partial synthesis, is particularly desirable since this reduces the risk of degradation by oxidation; unfortunately they are not always as easy to work with during manufacture. The inclusion of other membrane components either in quantitative or in trace amounts may be considered useful in view of the specific interactions which may occur which could modify the behaviour of the liposome with regard to either its contents or its mode of action. For example, egg lecithin liposomes retain amphotericin B particularly well when the sterol component is ergosterol rather than cholesterol, because of the formation of a complex between these molecules in the membrane. On the other hand, it appears that ergosterol can be omitted from amphotericin liposomes provided that the egg lecithin is replaced by fully saturated phospholipids. In another system, high uptake by liposomes of the drug potassium antimony tartrate can be achieved by incorporating a positively-charged lipid in the membrane. The electrostatic interaction between drug and membrane results in tight binding to the liposome and reduces leakage.

There are occasions when one wants to construct liposomes which are capable of releasing their contents as rapidly as possible, as for example in systems which require intermembrane transfer of materials. In this case, for reasons discussed earlier, a highly fluid membrane could be chosen, composed of very unsaturated phospholipids (e.g. soya-derived) probably with reduced cholesterol content. Where release is required to take place at a particular temperature, the lipid mix of the components can be adjusted so that this corresponds to the phase transition temperature of the membrane. Membrane components can, of course, also be included in order to alter the interaction of liposomes with their external environment. Thus glycolipids can be incorporated which may bind the liposome to receptors on cells; chemically reactive molecules can be added, which will link proteins to the surface of the liposome, and some proteins can be incorporated directly into the liposome membrane during manufacture. Methods for these are described later.

These examples present a very simplified picture of the ways in which membrane components can modify the characteristics of liposomes. Since individual applications are so diverse, and molecular interactions in and around membranes are unpredictable, it is difficult to give any general rules. In most cases the best approach to adopt is an empirical one, in which a range of possible combinations is tested in an open-ended trial. For a more rigorous discussion of the theory of membrane behaviour, the reader is referred to texts listed in Appendix III at the end of this book.

3.1.2 *Physical size*

As mentioned before, one advantage of liposomes is their singular ability to take up material of all shapes and sizes regardless of their individual chemical peculiarities. In the case of lipid-soluble compounds which are incorporated into the membrane, the maximum amount taken up into liposomes is directly proportional to the quantity of membrane components, and is independent of the size of the liposomes used. In principle, therefore, any type of liposome is suitable for the incorporation of lipid-soluble compounds. However, because both *in vitro* and *in vivo*, this type of compound can readily be removed from the membrane by collision-mediated transfer with solid surfaces, other membranes or large protein molecules, in practice it is desirable to reduce the outer surface area of these liposomes as much as possible in order to minimize loss of material before the liposomes have reached their target area. In other words, multilamellar vesicles are the liposomes of choice. It is also thought that MLVs will give a much more gradual and sustained release of material than unilamellar vesicles, as their concentric membranes are slowly degraded one by one at the target site. Conversely, if rapid intermembrane transfer of lipid-soluble compounds is required, then small unilamellar vesicles (SUVs) are particularly suitable, since abnormalities in packing of lipids in these membranes, because of their high curvature, cause them to be especially susceptible to transfer and degradation in the presence of biological molecules.

For water-soluble compounds where the principle aim is to achieve as high a value as possible for entrapped volume:lipid ratio, large or intermediate-sized unilamellar vesicles (LUVs or IUVs) are the most appropriate types of liposome to employ. One disadvantage of very large unilamellar vesicles is the mechanical weakness of the single membrane which may easily lead to its rupture and partial loss of contents, and which can represent only a tenuous barrier to efflux of water-soluble compounds even when intact. In practice, however, many methods of preparation of LUVs can give rise to a significant proportion of oligo-lamellar liposomes, with a large central compartment enclosed by two of three membrane layers, which may ultimately be an advantage. Alternatively, MLVs can be constructed, in which a small aqueous compartment is contained in between each successive membrane layer. This has been achieved by incorporating a strong negatively-charged lipid into the membrane, which causes adjacent lamellae to repel each other. Even under these circumstances, however, the entrapped volume is not usually as large as for LUVs.

As can be seen from the above, the use of SUVs is not generally recommended from the point of view of efficiency of entrapment, or retention of material. In some cases, however, the disadvantages outlined may not be very important, as for example, when entrapped materials are bound strongly to the membrane, or when uptake by the target site is too rapid for significant loss to occur. Because the preparation of SUVs involves input of high energy (ultrasonication), this is the best method for incorporating materials which for kinetic or thermodynamic reasons do not incorporate readily into the membrane by other means. Furthermore, for some preparations, the conditions may be such that the limit of size is reached, before the curvature of the liposomes markedly alters the membrane characteristics, under which circumstances SUVs can be considered on an equal footing with other types of preparation.

3.2 **Considerations related to the desired behaviour of liposomes**

In recent years, people have become more and more aware of the importance of size in determining the behaviour of liposomes *in vivo*. In *in vitro* systems the parameter under observation is the interaction between the liposomes and a particular cell type, and alternative processes such as fusion, phagocytosis, and cell-to-cell membrane transfer depend largely on chemical composition of the membrane. In contrast, an extra dimension is added in *in vivo* situations, where the sites of action of such liposome – cell interactions in the body are determined to a large extent by size. Thus upon intravenous injection small liposomes can pass through the fenestrae of the liver sinusoids and come rapidly into contact with hepatocytes. Liposomes of intermediate size are retained within the blood compartment and can circulate for a considerable length of time. Larger liposomes pass more slowly through the liver sinusoids, and are rapidly taken up by Kupffer cells. Even larger liposomes are taken out of circulation during first pass through the lung. Superimposed on this picture are variations which take into account receptor-mediated interaction between liposomes and cells, packing-dependent uptake of proteins on liposome membranes, and the size-heterogeneity of lipsome populations generally available.

4. TAILORING APPLICATIONS TO SPECIFIC LIPOSOMES

In concluding this chapter, it may be helpful to offer the reader some guidelines as to the capabilities of liposomes, and what they can reasonably be expected to achieve.

To this end, a bibliography has been appended to this book in which selected key references are cited for different applications, which may act as models to be closely followed, or which may trigger off ideas as to what is possible in totally unrelated areas (see Appendix IV).

While the majority of applications make use of the fact that liposomes are composed of natural, biocompatible components which make them eminently suitable for use in medically-related fields, one should not lose sight of the fact that they can also have applications in physical sciences which may be only marginally related to biological fields of interest. For example, liposomes have been used as a means of maintaining charge separation in the photocatalytic cleavage of water. Recent work in this area uses entirely inorganically-derived photoreceptors and electron-transfer molecules in a phospholipid membrane to mimic the generation of molecular oxygen in nature. In addition to physical methods of examination of liposomes as model membranes, *in vitro* applications include the introduction of genetic material into bacterial and plant cells, the co-presentation of small antigens and class II molecules to lymphocytes in the absence of macrophages, and the introduction of receptors on to cell surfaces to induce susceptibility to virus infection.

4.1 **Natural targeting**

In vivo applications to date can be categorized as follows:

(i)   Diagnosis
(ii)   Chemotherapy
(iii)   Cancer
        Infectious diseases
        Ion overload/metal contamination
        Inborn errors of metabolism
(iii)   Immunotherapy
(iv)   Gene therapy

While in earlier sections of this chapter discussion has centred on tailoring liposomes to the application, it has to be noted that the most successful realization has been achieved when the application has been 'tailored' to the capabilities of liposomes, and when due account has been taken of the fact that *in vivo*, regardless of variations in membrane composition, liposomes will interact almost exclusively with organs of the monocyte phagocyte system (MPS—also known as RES, the reticular endothelial system), namely the liver, spleen, lung, lymph nodes and bone marrow, since it is these organs whose job it is to process, in one way or another, agents of foreign origin which have entered the body. Liposomes are very rapidly taken up by phagocytic cells, and are degraded in lysosomes, entrapped materials being released into the lysosome, and subsequently into the cell cytoplasm or into the external milieu after exocytosis. The markedly altered *in vivo* distribution of drugs entrapped inside liposomes means that systemic toxic side effects after parenteral administration may be considerably reduced, while the concentration of drug in contact with the diseased organ can be increased many-fold over what is seen with the free drug. Liposomes can be of use in therapy of intracellular diseases of macrophages such as leishmaniasis and deep-seated fungal infections, as well as diseases involving liver or lung tissue. They may be effective against tumours either by virtue of the accumulation in large quantities in organs where the tumour is sited (e.g. the liver) albeit within a different cell type, or by the use of macrophage stimulators entrapped in liposomes to activate monocytes to become capable of tumoricidal killing. Liposome-entrapped immune modulators can also be used to stimulate immune responses to antigens presented to macrophages *in vivo* within, or on the surface, of liposomes.

4.2 Directed targeting

The use of molecular determinants attached to the outside of liposomes in an attempt to increase the range and specificity of targets accessible to liposomes has met so far with limited success; the mechanism designed to sweep liposomes into the net of the MPS overrides all refinements which try to modify their behaviour. Even liposomes remaining longest in the bloodstream have no way of leaving the blood compartment, except in small quantities through lesions in blood vasculature, or after uptake by circulating monocytes. Liposomes administered via the subcutaneous or intramuscular route accumulate in draining lymph nodes, while those administered orally are probably almost all degraded, before release of a small proportion of their contents into lymphatics draining the gut. Other sites in the body are accessible to liposomes by local administration and in cases where these compartments are anatomically isolated from the MPS (e.g. the aqueous humour of the eye, intra-articular space of the knee joint) receptor-mediated targeting may be a possibility.

Although no applications have as yet come to light, it should be possible to manipulate the behaviour of liposomes within the blood compartment by modification of surface characteristics. It has been reported that galactosyl residues incorporated into liposome membranes increase the proportion taken up by hepatocytes relative to Kupffer cells in the liver, in line with the observation that hepatocyte plasma membranes bear a lecithin-like receptor protein for terminal galactose-containing determinants. Uptake by the liver as a whole can be partially reduced by procedures which 'blockade' the MPS, in particular the prior administration, intravenously, of a dose of 'cold', empty liposomes. Reduction of liver uptake is accompanied by a reduced clearance rate from the bloodstream, followed by a relative increase in uptake in the spleen, bone marrow, or other tissues. As mentioned earlier, the size of the liposomes can determine the length of time spent in the bloodstream before removal, as can also the composition. Liposomes composed of sphingomyelin and cholesterol have been reported to remain in the bloodstream with a half-life of at least 18 h (compared with a clearance half-life of 1 h in the most unfavourable circumstances). For applications in which targeting to blood cell types (e.g. lymphocytes, red cells) other than phagocytic cells is contemplated, these 'long-lived' liposomes would seem to offer the most promise.

5. GUIDE TO FOLLOWING CHAPTERS

By now it is hoped that the reader will be in a position to decide which of the different classes of liposome will most suit their need, and will be ready to consider the various methods available for making the different types of liposome. As before, the choice between different methods will be determined to a large extent by the material to be entrapped inside liposomes; some compounds will be adversely affected by conditions of manufacture, for example, sonication, or organic solvents, while others may present problems in separation from the unentrapped material.

For all applications the reader will want to refer to Chapters 2 and 3. The fourth chapter contains methods for conjugation of molecules to the membrane of pre-formed liposomes. Thereafter, methods of interest in physical and biological studies are dealt with in Chapters 5 and 6 respectively. Methods for manufacture and purification of starting materials and reagents are given in Appendix I and the use of liposomes in various applications is illustrated by reference works cited in the bibliography (Appendix III and IV).

CHAPTER 2

Preparation of liposomes

ROGER R.C.NEW

1. INTRODUCTION

The key point to grasp in considering the manufacture of liposomes is that phospholipid membranes form spontaneously as a result of unfavourable interactions between phospholipids and water. Thus the emphasis in making liposomes is not towards assembling the membranes (which happens of it own accord), but towards getting the membranes to form vesicles of the right size and structure, and to entrap materials with high efficiency and in such a way that these materials do not leak out of the liposome once formed.

All methods of making liposomes can be said to involve three or four basic stages: drying down of lipids from organic solvents, dispersion of the lipids in aqueous media, purification of the resultant liposomes, and analysis of the final product (*Figure 1*). The main difference between the various methods of manufacture is in the way in which the membrane components are dispersed in aqueous media, before being allowed to coalesce in the form of bilayer sheets. In this chapter, the methods have been classified according to the three basic modes of dispersion; namely, physical dispersion, two-phase dispersion, and detergent solubilization. These will be dealt with in turn in Section 3.

2. HANDLING OF LIPIDS

2.1 Storage

In the methods described in this chapter, it will be assumed that a standard composition of egg lecithin:cholesterol:phosphatidyl glycerol in molar ratio 0.9:1.0:0.1 is being used unless otherwise stated. These lipids can be stored either as solids, or in organic solution at $-20°C$, or at $-70°C$ in order to reduce the chance of oxidation to a minimum. When small quantities of lipids need to be dispersed, organic solutions are more convenient to handle than solids. The solvent most widely used is a mixture of chloroform and methanol in a ratio of 2:1 by volume; compounds which are sparingly soluble in either chloroform or methanol alone will often dissolve readily in this 2:1 solvent mixture. Solvents of the highest purity should be used, particularly since some contaminants may be chemically reactive and may cause the lipids to deteriorate. Impurities which are at low levels in the bulk solvent can be increased in concentration upon drying down. Ether degrades over time to form peroxides, while chloroform gives rise to phosgene on standing. Formation of the latter can be prevented by addition of 1% ethanol to stabilize the chloroform, and most commercial sources of chloroform are sold in this form. Water present in small quantities may be removed by addition of 'molecular sieves'

— microporous clay substrates which absorb any molecules small enough to pass through the holes within the structure of the support, and which bind particularly strongly to water. Methods for purification of solvents are given in Appendix I. All lipid solutions should be stored in the dark, in glass vessels with a securely fastened ground-glass stopper. Polypropylene containers may also be used, although it is difficult to find caps which fasten tightly enough to prevent evaporation of the solvents, which can take place even under refrigeration. Inert rubber (e.g. Neoprene) can be used as a seal, but it does tend to swell in chloroform. It is routine practice to flush the gaseous volume above the liquid in the container with an inert gas prior to storage, in order to reduce the possibility of oxidation of lipids; nitrogen is most commonly used. Since this gas is lighter than air, however, a strong flow of gas is needed in order to ensure complete exchange with air, and one may risk evaporating the solvent in the process. If available, the use of argon is preferable since this is heavier than air, and forms an effective blanket with just a very gentle stream of gas.

2.2 Measurement

Because of the volatility and high density of chloroform, Eppendorf-type pipettors are unsuitable for dispensing the lipids in solution. Graduated glass pipettes and glass capillary micropipettes are strongly recommended for accurate measurement of small volumes of organic solvents. Care must be taken to minimize the evaporation of solvent from stock solutions during handling. Since phospholipids vary in molecular weight, and because it is difficult to be sure that solvent has been removed completely after evaporating an organic solution to dryness, it is preferable to determine the phospholipid concentration of a solution by measuring the phosphate content, rather than by weighing the solid residue after drying down. Since each phospholipid molecule contains a single phosphate group (with the exception of cardiolipin, which contains two phosphates bridging two diglyceride groups) the phosphate content is directly related to the molar concentration of phospholipid, from which a value for concentration can be calculated. The average molecular weight of egg PC is 750; for distearoyl PC — 789, dipalmitoyl PC — 733, and dimyristoyl PC — 677. Methods for the measurement of phosphate are given in Chapter 3.

2.3 Drying down

The starting point for all liposome preparations is an organic solution of membrane lipids. Even if the lipids need to be in a dry form at an early stage, it is good practice to dissolve them together in the same solution beforehand, in order to ensure complete and homogeneous mixing of all the components as they are required in the final membrane preparation. In all the methods to be described, without exception, compounds to be incorporated which are lipid soluble will be added to the organic solution, while compounds to be entrapped in the aqueous compartment of liposomes will be dissolved in the aqueous starting solution.

The drying down of large volumes of organic solutions is most easily carried out in a rotary evaporator fitted with a cooling coil and a thermostatically-controlled water bath (see *Figure 1*). Rapid evaporation of solvents is carried out by gentle warming (20−40°C) under reduced pressure (400−700 mm Hg); rapid rotation of the solvent-containing flask increases the surface area for evaporation. The temperature and shape

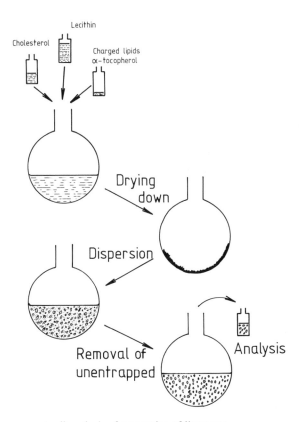

Figure 1. Stages common to all methods of preparation of liposomes.

of the vessel holding the sample are determined by the lipid mix and the particular method to be employed. Vessels varying in capacity from 1 ml to several litres can all be accommodated. The vacuum supplied by a good suction tap is sufficient for normal purposes; a low grade vacuum pump can also be used if a suction tap is not available, but care must be taken not to allow solvents to build up in the pump oil. It is probably wise to run the pump on gas ballast if it is to be used for this application. A vacuum gauge can be fitted to ensure the reproducibility of conditions, and to safeguard against the vacuum rising too high initially which may result in 'bumping' of the solvent if the liquid boils too vigorously, leading to loss of lipid and contamination of the apparatus. The strength of the vacuum may be controlled by introduction of nitrogen via a suitable bleed valve (see *Figure 2*), and the evaporator may also be flushed with nitrogen before and after evacuation to prevent lipid oxidation.

In cases where the vacuum attainable is not sufficient, or if the concentration of lipids is particularly high, it may be difficult to remove the last traces of chloroform from the lipid film. Therefore, it is recommended as a matter of routine that after rotary evaporation, some further means is employed to bring the residue to complete dryness. Attachment of the flask to the manifold of a lyophilizer, and exposure to high vacuum overnight is a good method. In cases where the quantities of lipids and solvents are small – less than one millilitre containing a few milligrams – and where production

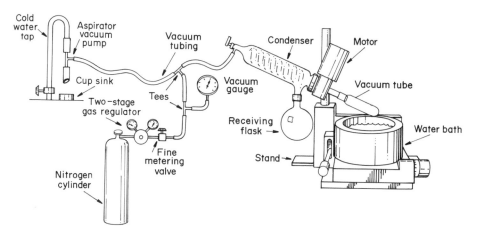

Figure 2. Rotary evaporator apparatus for evaporating off organic solvents. The figure shows the arrangement of rotary evaporator and water bath with a nitrogen bleed to control the vacuum. Drawing kindly provided by Dr F.J.Martin.

of a thin, even film is not of great importance, rotary evaporation may be considered unnecessary, and the whole drying process can be carried out using just a stream of nitrogen from a cylinder. This procedure may result in rather thick deposits of lipid, containing residual chloroform, which may be removed by redissolving the lipids in diethyl ether and repeating the drying down under nitrogen. (Care must be taken to ensure that the ether is peroxide-free − see Appendix I.) In some laboratories, the final drying procedure is carried out by dissolving the lipids in benzene, freezing, and removing the solvent by lyophilization overnight.

3. METHODS OF PREPARATION

3.1 Mechanical dispersion

In this group of methods, essentially the simplest in concept, the lipids are dried down onto a solid support (usually the side of the glass container vessel) and then dispersed by addition of the aqueous medium, followed by shaking. Even before exposure to water, the lipids in the dried-down film are thought to be oriented in such a way as to separate hydrophilic and hydrophobic regions from each other, in a manner not unlike their conformation in the finished membrane preparation. Upon hydration, the lipids are said to 'swell', and peel off the support in sheets, generally to form multilamellar vesicles. In most of the methods described in this section, the aqueous volume enclosed within the lipid membrane is usually only a small proportion of the total volume used for swelling − about 5−10%. The method is, therefore, very wasteful of water-soluble compounds to be entrapped, although the absolute yield of material (the captured volume − ml g^{-1} lipid) may be satisfactory for practical purposes. Lipid-soluble compounds on the other hand, can be encapsulated to 100% efficiency, providing they are not present in quantities which overwhelm the structural components of the membrane. Four of the basic methods of mechanical dispersion are depicted in *Figure 3*.

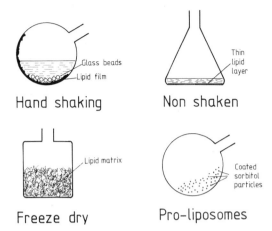

Figure 3. Crucial elements of different mechanical dispersion methods. See text for details of each of these methods.

3.1.1 *Hand-shaken multilamellar vesicles (MLVs)*

The simplest and most widely used method of mechanical dispersion is commonly known as 'hand-shaking', since the lipids are suspended off the sides of a glass vessel into the aqueous medium by gentle manual agitation (1). Experimental details are shown below. In order to increase the entrapment volume, it is advisable to start with a round-sided glass vessel of large volume, so that the lipids will be dried down onto as large a surface area as possible to form a very thin film. Thus, even though the volumes of organic or aqueous starting solutions may be only 1 ml each, it is recommended that one use a 50- or 100-ml vessel for drying down. Care and patience is needed in order to resuspend all the dried lipid with minimum loss of fluid. If the materials to be entrapped are lipid-soluble, however, conditions for drying down are not so critical. The temperature for drying down should be regulated so that it is above the phase transition temperature of the lipids in question. At high temperatures, or under conditions where the dried lipid forms a fairly thick film, it may be difficult to remove it from the glass by manual swirling; also lumps of solid lipid may form which do not disperse easily. In this case, the addition of glass beads, either with the aqueous solution, or dried down with the lipid, is a very effective aid in suspending the lipid. Glass beads of any size from 0.5 to 3 mm in diameter are suitable. After addition of the beads, the flask may be re-attached to the rotary evaporator by means of a clip or tape, and the lipids suspended by rotation of the flask for a period of half an hour at atmospheric pressure. When large volumes of lipid and aqueous solution are used, the suspension can be carried out by vigorous vibratory motion in a mechanical shaker, and complete and homogeneous dispersion ensured by continuing the process for several hours.

Preparation of MLVs by hand-shaking method

Drying down of lipid

(i) Make up the lipid mixture containing 100 mg of egg PC, 40 mg of cholesterol and 10 mg of phosphatidyl glycerol, in 5 ml of chloroform/methanol solvent mixture (2:1 vol/vol).

37

(ii)   Introduce the lipid solution into a 250-ml round-bottomed flask with a ground-glass neck.

(iii)   Attach the flask to a rotary evaporator, evacuate, rotate at about 60 r.p.m. and immerse the flask in a thermostatted water bath, with the temperature set at 30°C (or above the transition temperature of the lipids used).

(iv)   Continue until all the liquid has evaporated from solution, and a dry lipid film has been deposited on the walls of the flask.

(v)   Continue for 15 min after the dry residue first appears.

(vi)   Stop rotation of the flask, and raise it from the water bath.

(vii)   Isolate the evaporator from the vacuum source by closing the tap (not by switching vacuum off, in case suck-back of oil or water occurs).

(viii) Introduce nitrogen from a cylinder into the evaporator via the inlet tap—gradually raise the nitrogen pressure at the cylinder head until there is no pressure differential between the inside and the outside of the flask. The pressure release valve between the cylinder and the evaporator should prevent build-up of pressure inside the apparatus.

(ix)   Remove the flask from the evaporator. Remove residual solvent by wrapping the flask in cling-film, fitting it on to the manifold of a lyophilizer, and subjecting it to high vacuum at room temperature for at least an hour. A gas-line filter (e.g. Millipore FG 50) should be interposed between the flask and lyophilizer, to avoid possible contamination.

Hydration of lipid

(x)   After releasing the vacuum, and removing from the lyophilizer, flush the flask out with nitrogen, add 5 ml of saline (containing solutes to be entrapped) and 0.5 g of glass beads. (Ideally, the saline solution should be purged with nitrogen, or degassed under vacuum to remove traces of oxygen).

(xi)   Attach the flask to the evaporator again (flushed with nitrogen). Secure the flask in position with a clip or sticky tape, and rotate it at room temperature and pressure at the same speed as before (~ 60 r.p.m.).

(xii)   Leave the flask rotating for 30 min, or until all the lipid has been removed from the walls of the flask, and has given a homogeneous milky white suspension free of visible particles.

(xiii) Allow the suspension to stand for a further 2 h at room temperature (or at a temperature above the transition temperature of the lipids) in order to complete the swelling process.

MLVs composed entirely of neutral lipids tend to be very tightly packed multilayer assemblies, with the adjacent bilayers stacked very closely, one upon the other, and with very little aqueous space between them. As discussed later, the distribution of solute throughout the bilayers is often uneven, as a result of sealing of bilayer sheets into vesicles before full hydration of the lipid headgroups. The presence of negatively-charged lipids in the membrane tends to push the bilayers apart from each other, and increase the volume of entrapment significantly. The same effect can be achieved with neutral lipids by freezing and thawing repeatedly the final liposome preparation (2); entrapments of 30% (at 100 mg lipid ml^{-1}) can be achieved, going to higher values at higher lipid concentrations. At the same time the solute inhomogeneities are evened out between adjacent aqueous compartments. The size distribution of the vesicles in the population is the order of several microns. Similar increases in interlamellar spacing and entrapment values can be obtained during the processing procedures (e.g. micro-fluidization or sonication) described later in this section.

3.1.2 *Non-shaken vesicles*

The method described in the previous section is suitable for preparing multilamellar vesicles. Under certain circumstances, large unilamellar vesicles can be formed with much higher entrapment volume by using essentially the same method, but taking more care over the swelling procedure. In its original form developed by Reeves and Dowben (3), using egg or brain phospholipids, only distilled water or solutions of non-electrolytes or zwitterions could be used, presumably because introduction of ions with a net charge in between the membrane bilayers would increase the intermembrane attraction, and prevent their separating from each other during swelling. The hydration and swelling processes are carried out in two separate steps. After drying down the lipid film, hydration is initiated by exposing the film to a stream of water-saturated nitrogen for 15 min followed by swelling in aqueous medium without shaking. The experimental details are as follows.

Preparation of large unilamellar vesicles (LUVs): Reeves and Dowben Method (3)

(i)   Prepare a solution of 4 mg of egg phosphatidyl choline ($\sim 5~\mu M$) in 0.5 ml of 1:2 chloroform:methanol solvent mixture.

(ii)   Spread the solution evenly over the flat bottom of a 2-litre conical flask.

(iii)   Evaporate the solution at room temperature by allowing a flow of nitrogen to circulate through the flask, without disturbing the solution itself.

(iv)   After drying, pass water-saturated nitrogen through the flask, until the opacity of the dried lipid film disappears ($15-20$ min).

(v)   After hydration, swell the lipid by addition of bulk fluid. Incline the flask to one side, introduce $10-20$ ml of 0.2 M sucrose in distilled water (degassed) down the side of the flask, and slowly return the flask to the upright orientation, allowing the fluid to run gently over the lipid layer on the bottom of the flask.

(vi)   Flush the flask with nitrogen, seal, and allow to stand for 2 h at 37°C. Take care not to knock the flask or agitate the medium in any way during this period.

(vii)   After swelling, harvest the vesicles by swirling the contents of the flask gently, to give a milky suspension.

(viii) Spin the suspension at 12 000 g for 10 min in a bench centrifuge at room temperature. Remove the layer of multilamellar vesicles floating on the surface.

(ix)   To the remaining fluid, add an equal volume of iso-osmolar glucose and spin again at 12 000 g.

(x)   Large unilamellar vesicles form a soft pellet which can be resuspended in any required medium of appropriate osmolarity.

The proportion of thin-walled vesicles obtained by this method depends on the nature of the lipid used, lipid mixtures giving higher yields than single phospholipids. Chloroform:methanol in ratio of 1:2 by volume is used because this has a lower contact angle with glass than other solvents, and thus has less tendency to give thick deposits of lipid at the edges when drying down. The method described above for separating

LUVs from MLVs is based on the difference in density between sucrose solution and multiple phospholipid lamellae, and relies on the fact that the liposomes are made up in sucrose solution to start with. However, if preparation of liposomes in other solutions is required, alternative methods of separation (e.g. Ficoll flotation) can be employed (see Section 5.3.1).

A refinement of this method, also for preparing LUVs, has been described by Lasič *et al.* (4), in which the thin film of lipid is dried down on to a finely-etched silicon wafer. Because of the irregularities in the surface of the wafer, its total area is much greater than for an ordinary flat surface. Also, deposition of lipid on to a surface with many different angles and planes encourages the lipid to break up into small sheets upon hydrating, which may help to define the size of the final liposome formed. Hydration from an entirely smooth surface results in large expanses of lipid membrane which easily fold over on each other to give multilamellar vesicles. The same type of principle is employed in the method described in the next section.

3.1.3 *Pro-liposomes*

In order to increase the surface of dry lipid while keeping the aqueous volume down, a method has been devised in which the lipids are dried down on to a finely divided particulate support, such as powdered sodium chloride, or sorbitol or other poly-saccharide (5). Upon adding water to the dried lipid-coated powder (known as 'pro-liposomes'), with mixing on a whirlimixer, the lipids swell, while the support rapidly dissolves to give a suspension of MLVs in aqueous solution. The particle size of the carrier influences the size and heterogeneity of the liposomes finally produced; both of these parameters seem to be smaller in value than for MLVs produced by the conventional hand-shaken method. As with other methods of preparation, efficient dispersion is brought about when carried out above the phase transition temperature of the lipids.

Effective monitoring and control of temperature during the process of drying the lipid onto the support is important in assuring that the components remain in a finely divided form and are evenly dispersed. This has been achieved by using a rotary evaporator, adapted by the introduction of a thermocouple into the evaporation flask via the vacuum line. The shape of the thermocouple is modified so that it lines the inside surface of the flask, for most effective temperature control, and mixing of the dry powder. In order to achieve this, a florentine-shape flask may be more suitable than a conventional full round one.

Of the supports investigated so far, sorbitol has received most attention, both because of its acceptability for clinical use, and because solutions formed from it have lower osmolarity than do those of lower molecular weight compounds. The sorbitol granules used consist of a microporous matrix, so that the actual area for deposition available is considerably greater than that calculated on the basis of the outer surface of the granule alone, and is equal to $33.1 \text{ m}^2/\text{g}^{-1}$. At a loading of approximately 1 g of lipid for 5 g of sorbitol, a coverage of 6 mg/m^{-2} is obtained, which is equivalent to a continous lipid coating of just a few bilayers over the whole surface. Because water has much greater access to lipid in the form of pro-liposomes than when dried on to a glass wall (where the coating ratio may be 10 g m^{-2}), liposomes are formed much more rapidly by this method, and a higher proportion of smaller vesicles is obtained. For example,

MLVs composed of dimyristoyl PC (DMPC) and dimyristoyl phosphatidyl glycerol (DMPG) (7:3 molar ratio) made by the hand-shaking method have a mean diameter of 1.8 μm, compared with pro-liposomes of the same composition which were over 10 times smaller (0.13 μm diameter). The mean size of liposomes produced by hydration of the pro-liposome powder also increases with loading, and decreases with increased negative charge.

Because the pro-liposome method permits lipid deposition under very carefully controlled conditions, it may be a good technique to employ to re-investigate the phenomenon of lipid hydration from a dry film, and examine the parameters which influence it. This method is also a very useful one for commercial applications where large quantities of pro-liposomes can be prepared and stored dry before use in aliquots in sealed vials, then resuspended when required to give batches of liposomes reproducible over a long period of time. This method also overcomes problems encountered when storing liposomes themselves in either liquid, dry or frozen form, and is clearly ideally suited for preparations where the material to be entrapped incorporates into the lipid membrane. In cases where 100% entrapment of aqueous components is not essential, this method is also of value. The procedure for preparation and hydration of pro-liposomes is as follows.

(i) *Preparation of pro-liposomes*

Special equipment

Buchi Rotary Evaporator 'R' with water-cooled condenser coil fitted with a stainless steel covered thermocouple (introduced into the evaporator via vacuum line) connected to a digital thermometer. Modify the end of the glass solvent inlet tube by drawing it out to a fine point, so that the solvent is introduced into the flask as a fine spray.

100- or 250-ml Florentine evaporating flask

Special materials

Sorbitol — Roquett UK, Tonbridge Wells, Kent.

Method

(i)   Dissolve 2 g of lipids in 33 ml of chloroform to give a final concentration of 60 mg ml^{-1}.

(ii) Weigh out 10 g of sorbitol powder, and introduce it into the 100-ml flask.

(iii) Fit the flask onto the evaporator. Make sure that the thermocouple element is shaped so that it lies close to the wall of the flask.

(iv) Seal the evaporator, evacuate, and rotate the flask slowly so that the powder tumbles gently off the wall, to ensure good mixing.

(v)   Lower the flask into a water bath at 30–45°C.

(vi) When a good vacuum has developed (around 100 kPa) introduce an aliquot of 5 ml of methanolic lipid solution into the flask via the solvent inlet tube. The solvent will be absorbed completely by the powder. Monitor the temperature of the bed. As evaporation proceeds, the temperature will decrease.

(vii) When the temperature begins to rise again, slowly introduce a second aliquot of 3 ml. Monitor the temperature. Repeat the process until all the solvent is used up. At no time should the fluid be present in excess, such that a slurry of wet lumps is formed.

(viii) Allow the temperature to rise to 30°C, then release the vacuum, and complete the drying process by connecting the flask containing the powder to a lyophilizer, and leaving it evacuated overnight at room temperature.

(ix) Aliquot the powder into 10 ml glass vials containing 600 mg solid (100 mg lipid and 500 mg sorbitol). Flush with nitrogen, seal well and store.

(x) When ready for use, introduce 10 ml of distilled water into one vial, and mix on a whirlimixer for 30 sec, or in a shaking water bath above the lipid phase transition temperature, to give a 5% w/v solution of sorbitol (isotonic with normal saline) and a lipid concentration of 10 mg ml^{-1}.

Essentially any combination of lipids can be used in this method. If working with lipids with low melting temperature (e.g. egg PC), care must be taken not to allow the temperature to rise too high otherwise agglomeration of powder particles will result. This problem is not so marked with high melting lipids. Any solvent can be used which will dissolve the lipids; in the case of ethanol or methanol, very precise regulation of evaporation conditions is required, in order to ensure that dissolution of the sorbitol does not occur as the temperature rises. It is important to maintain a high vacuum, so that the solvent may be removed rapidly. If one is depositing high melting lipids onto a sorbitol base in chloroform, it may be possible to conduct the operation without temperature control monitoring by the thermocouple. This will considerably reduce the complexity of the set-up for experimental laboratory workers.

Both porous and non-porous supports (e.g. sodium chloride) can be used; in the case of sodium chloride, greater care must be taken not to over-wet the support, but the solvent is easier to remove than for a porous powder, since it is not taken up into the interior of the granules. The surface of powdered sodium chloride is $0.12 \text{ m}^2 \text{ g}^{-1}$. High lipid loadings can only be achieved at the expense of increasing the thickness of lipid layer, so that the liposomes formed by hydrating from an NaCl support will tend to be larger than for sorbitol. Because of the extra lipid exposed on the particle surfaces, agglomeration tends to occur much more readily, and the process can really only be carried out satisfactorily with high-melting lipids.

3.1.4 *Freeze-drying*

An alternative method of dispersing the lipid in a finely-divided form, prior to addition of aqueous media, is to freeze-dry the lipid dissolved in a suitable organic solvent (6). The choice of solvent used is determined by its freezing point, which needs to be above the temperature of the condenser of the freeze-drying, and its inertness with regard to rubber seals which form a part of most commercial lyophilizers. From this point of view tertiary butanol is considered the most suitable solvent. After obtaining the lipid in dry form, in an expanded foam-like structure, water or saline can be added with rapid mixing above the phase transition temperature to give MLVs. Another method making use of freeze-drying of lipids is described later (see Section 3.1.9).

An alternative method of preparing phospholipids in an expanded form has been reported (7), in which a 'hydrated gel' is formed by mixing phospholipids and a charged amphiphile into a paste with a minimal volume of water—sufficient to hydrate the polar headgroups in the bilayer sheets, without giving rise to closed vesicles. The process is facilitated by the presence of a 'hydrating agent', usually a zwitterionic counter ion such as arginine, glutamic acid, or lysine. Liposomes are formed by diluting the gel 10-fold in an aqueous medium.

3.1.5 *Processing of lipids hydrated by physical means*

After preparation of multilamellar vesicles by hydration of dried lipid it is possible to continue processing the liposomes in order to modify their size and other characteristics. For many purposes, MLVs are too large or too heterogeneous a population to work with, so many of the methods have been devised to reduce their size, and in particular to convert liposomes in the large size range into smaller vesicles. These include techniques such as micro-emulsification, extrusion, ultrasonication, and use of a French pressure cell. A second set of methods is designed to increase the entrapment volume of hydrated lipids, and/or reduce the lamellarity of the vesicles formed, and employ procedures such as freeze-drying, freeze-thawing, or induction of vesiculation by ions or pH change.

3.1.6 *Micro-emulsification liposomes (MEL)*

Recently, the use of a 'micro-fluidizer' to prepare small MLVs from concentrated lipid suspensions has been reported (8). The micro-fluidizer is a machine which pumps fluid at very high pressure (10 000 p.s.i., 600−700 bar) through a 5 μm filter, after which it is forced along defined microchannels which then direct the two streams of fluid to collide together at right-angles at a very high velocity, thus effecting a very efficient transfer of energy (*Figure 4*). The lipids can be introduced into the fluidizer, either as a suspension of large MLVs, or as a slurry of unhydrated lipid in an aqueous medium. The fluid collected can be recycled through the pump and interaction chamber until vesicles of the required dimensions are obtained.

After a single pass, the size of vesicles is reduced to between 0.1 and 0.2 μm in

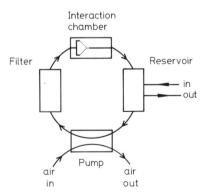

Figure 4. Diagrammatic representation of the use of a micro-fluidizer to prepare liposomes. A suspension of MLVs is introduced, via the reservoir, and pumped at high pressure through the interaction chamber, where the suspension is separated into two streams which collide with each other at high velocity. Large vesicles are broken up into smaller ones and may be collected after a single pass, or recycled several times to produce further reductions in size.

diameter, the exact size distribution depending on the nature of the components of the membrane and of the hydration medium. The presence of negative lipids tends to decrease their size, while increased cholesterol gives larger liposomes. Continuing the cycling time generally brings the size to a steady low value, although in some cases (e.g. liposomes containing doxorubicin) the diameter can increase after prolonged recycling.

The throughput of the machine is 90 ml min^{-1}; working with a minimum volume of 15−20 ml (determined by the dead space of the pump, chamber, etc.), a rate of four cycles per minute can be achieved, so that a finished preparation of liposomes can be obtained in under 10 min. The rapid throughput means that large volumes of liposome suspension can be prepared very easily. In addition to the high rate of production, this method has the advantage of being able to process samples with a very high proportion of lipid (20% or more by weight). Such concentrated suspensions cannot be handled by conventional methods of making MLVs, the resultant liposomes emerging from the micro-fluidizer in the form of a thick paste. Preparing liposomes at this concentration results in a very high proportion of the bulk aqueous phase being captured inside the liposomes (a high percentage entrapment), so that the process is very efficient for encapsulation of water-soluble materials. Capture values of 70% have even been reported, starting with a lipid concentration of approximately 200 mg ml^{-1}.

3.1.7 *Sonicated vesicles (SUVs)*

In order to reduce hydrated lipid to vesicles of the smallest size possible, it is necessary to use a method which imparts energy at a high level to the lipid suspension. This was first achieved by exposure of MLVs to ultrasonic irradiation (9) and it is still the method most widely used for producing small vesicles. The starting point is usually a suspension of multilamellar vesicles. Since these vesicles will be completely broken down in the process, it is not necessary to be too concerned about the initial size of the MLVs, the percentage entrapment or the thickness of the lipid film, so the lipid can be dried down in the same vessel as is used for sonication. There are two methods of sonication, using either a probe or a bath ultrasonic disintegrator (*Figure 5*). The probe is employed for suspensions which require high energy in a small volume (e.g. high concentrations of lipids, or a viscous aqueous phase) while the bath is more suitable for large volumes of dilute lipids, where it may not be necessary to reach the vesicle size limit.

(a) *Probe sonication*

For most efficient transfer of energy from the probe, it is advisable to hold the fluid in a round-bottomed tube with straight sides, having a diameter just slightly greater than that of the probe. The probe is immersed in fluid, approximately 4 mm below the surface, so that the lower face of the probe is just above the centre of curvature of the tube bottom; do not allow the probe to touch the sides of the vessel. Because a lot of heat is generated in the process, the vessel containing the material to be sonicated should be immersed in a cooling bath. In order that heat is conducted away as efficiently as possible, the sonication vessel should be constructed of glass or metal. For preparing volumes of between 5 and 10 ml, a flat tip probe of diameter 19 mm (3/4 inch) is very convenient to use, with the sonication vessel consisting of a tube 6 cm (2.5 inches)

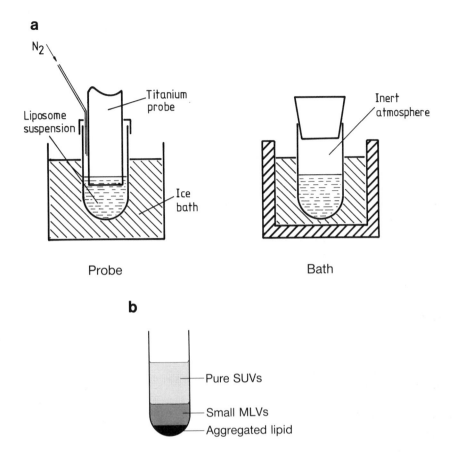

Figure 5. Preparation of liposomes by sonication. (a) Different methods of sonication. (b) Purification of small unilamellar vesicles. A very homogeneous population of small unilamellar vesicles may be obtained by centrifugation at 159 000 *g* for several hours to remove small MLVs after probe sonication of phospholipids in saline.

in length and 2.5 cm (1 inch) in diameter, provided with a B24/29 ground glass neck which will fit directly on to a rotary evaporator. The method outlined below describes the use of an MSE 150 W Probe Ultrasonic Disintegrator, but other models and makes will also give very satisfactory results. A list of manufacturers is given in Appendix II.

Preparation of small vesicles by probe sonication

(i)   Using a Buchi Rotary Evaporator and water bath at 37°C, dry 100 mg of lipid down from chloroform/methanol solution in a wide-bore glass tube with a Quickfit ground glass neck.

(ii) Complete the drying process by attaching the tube to an evacuated lyophilizer (1 −0.1 Torr) at room temperature for several hours.

(iii) Purge the aqueous solution (containing the material to be entrapped, if water-soluble) by bubbling a stream of nitrogen gently through the medium for 5–10 min.

(iv)   When the lipid has dried, remove it from the lyophilizer, introduce 10 ml of aqueous solution into the vessel containing the lipid, flush the gaseous volume with nitrogen, and close the vessel with a ground-glass stopper.

(v)   Vortex the contents rapidly on a whirlimixer for 30 sec, or until all the lipid has been removed from the sides of the vessel.

(vi)   Slide a rubber sealing cap up over the end of the sonicator probe.

(vii)   Remove the stopper and position the sonicator vessel so that the probe (19 mm flat tip) is 4 mm below the surface of the lipid suspension. Clamp it in position. Make sure that the sides of the vessel are not touching the probe at any point.

(viii)   Slide the sealing cap down so that it fits over the rim of the glass vessel.

(ix)   Introduce a Teflon tube (OD. 1 mm) in between the probe and sealing cap so that one end projects into the gaseous space above the fluid.

(x)   Allow a continuous flow of nitrogen to pass gently into the vessel through the Teflon tube. (Displaced gas will escape through gaps around the outside of the tube. Place a swab of cotton wool over this area to reduce contamination.)

(xi)   Surround the vessel with a beaker of ice/water (with magnetic follower) placed on top of a magnetic stirrer.

(xii)   Close the cabinet. With the power knob set to its lowest value (far anticlockwise), switch on the sonicator. Gradually turn the power knob clockwise until the needle on the amplitude scale reads between 8 and 12 (vertically upwards).

(xiii)   Sonicate for 15 sec, then switch off to allow the contents of the vessel to cool down for 30 sec.

(xiv)   Repeat the procedure in Step (xiii) until the liposomes have had a total sonication time of 10–30 min.

(xv)   After sonication has finished, leave the liposomes to stand at a temperature above their transition temperature for approximately 2 h, to allow the 'annealing' process to come to completion.

(xvi)   Spin the liposome suspension at 100 000 g for 20 min to sediment undispersed lipid and MLVs, and to remove particles of titanium released from the probe.

Because of the high input of energy in this method, there is considerable risk of degradation of the lipids resulting from the high temperatures and increased gas exchange associated with operation of the probe. It is essential, therefore, that the sonicator vessel be cooled efficiently at all times. However, it is also desirable that the lipids are sonicated above their transition temperature T_c (breaking and resealing of vesicles does not occur efficiently below this temperature). While this can be achieved very easily for egg lecithin (T_c around $-15\,°C$) with an ice bath, saturated lipids such as DMPC need to be maintained at room temperature or above. In this case, a circulating water bath with a high rate of flow is essential. In the absence of efficient heat transfer, the sonicator probe can raise small volumes of fluid close to boiling point.

During the sonication process, it is very easy to form an aerosol. It is important, therefore, to keep the sonicator vessel sealed at all times, in order to prevent loss of material, and to avoid contamination of the environment with potentially hazardous chemicals or radioisotopes. The probe should be cleaned scrupulously after use by wiping dry, then operating it for 2 min in a large beaker of distilled water.

Probe sonication is a very good method for reducing the size of large liposomes, but it suffers from the drawback that it is impossible to remove completely the risk of lipid degradation by contact with the hot probe, and contamination with titanium from the probe. The time required to bring the liposomes to a given size varies depending on the lipids used, and their initial concentration, as well as the precise positioning of the probe in the sample fluid. Consequently, the process is not a very reproducible one, and the only way to be sure of obtaining liposomes of standard size distribution is to sonicate for an over-long period of time, so that the bulk of the lipid is in the form of vesicles reduced to the minimum size limit (~ 250 Å, i.e. 25 nm). This constitutes a relatively homogeneous population of liposomes in terms of size.

The other drawback of this process is the limitation on the volume of sample which can be handled. As the size of the sample is increased, the distribution of energy throughout the sample becomes much more uneven, and the energy imparted per millilitre is greatly reduced so that it may not be possible to reach the minimum size limit for lipid vesicles. A sonication chamber has been devised (available from Roth Scientific — see Appendix II) which allows for a continuous flow of sample to pass directly under the probe in a controlled manner, so that a large volume could be handled reproducibly. However, it is still difficult to subject the whole of the sample to high energy for a long period of time even with extensive recycling.

For large samples, therefore, there is probably little to be gained by sonicating with a probe, and use of a bath sonicator is much more convenient. Experimental details are as follows:

(b) *Bath sonication of liposomes*

(i)   Prepare a suspension of 20 ml of MLVs from 200 mg of lipid as described earlier, using the hand-shaking method (Section 3.1.1).

(ii)   Transfer the MLVs into a flat-bottomed 100-ml glass conical flask. Flush the flask with nitrogen, and seal it with Parafilm.

(iii) Fill the sonicator bath with water to a level of about an inch (2.5 cm). Add a few drops of detergent to the water.

(iv)   Place the conical flask in the bath. The levels of liquid inside and outside the flask should be roughly equal.

(v)   Switch on the sonicator, and leave for $20-60$ min with occasional shaking.

This method is much milder than probe sonication and runs less risk of degrading the liquid. The volume of sample exposed to sonication is much larger, the field of ultrasonic irradiation more homogeneous, and reproducibility greater, provided the flask is positioned carefully in the same position each time. However, because the energy is dispersed over a much larger area, it may not be possible to reach the minimum size limit for sonicated vesicles. Special precautions for temperature control are not usually necessary since the small amount of heat generated is easily absorbed by the bath. If saturated lipids are used, sonication may be carried out above the transition temperature, by pumping water at the appropriate temperature from a conventional water bath.

Even after prolonged sonication, the population of liposomes obtained will not be homogeneous, and in studies where there is a need for the size range of the vesicles to be narrrow or well characterized, it is necessary to remove the larger liposomes and MLVs from the predominating population of SUVs. This aim was originally achieved by column chromatography on Sepharose 2B or 4B (9), but the method has disadvantages in that

(a)   it is time-consuming (to the extent that physical changes may occur in the SUVs during the timespan of the purification procedure);
(b)   it often results in a low yield of liposomes, since the leading edge of the SUV peak may still be contaminated with larger liposomes;
(c)   up to 10-fold dilution may result, which can only be rectified by ultrafiltration.

   A simpler method has been reported by Barenholtz and co-workers, in which larger vesicles are removed by centrifugation.

(c) *SUV purification by centrifugation*

[This method was devised by Barenholtz *et al.* (10)].

(i)   Prepare a suspension of liposomes in 5 ml of 50 mM KCl (3.775 mg ml^{-1}) by probe sonication of 100 mg of dried egg phosphatidyl choline to give a concentration of 75 μM ml^{-1} (100 mM NaCl may also be used).
(ii)   Place liposomes in a clear plastic-walled 7-ml ultracentrifuge tube.
(iii) Centrifuge at 100 000 *g* for 30 min at 20°C to sediment titanium particles and large MLVs.
(iv)   Follow with higher speed centrifugation at 159 000 *g* for 3−4 h.
(v)   After spinning, remove the tube carefully from the rotor and with a Pasteur pipette, decant off the liquid in the top clear layer, leaving the central opalescent layer (containing small multilamellar vesicles) and the pellet behind (see *Figure 5b*). The top layer will be a pure suspension of SUVs, mean radius 105 Å (10.5 nm).

   The yield of SUVs is about 60% of the total starting lipid. The time required for spinning depends upon the partial specific volume of the liposomes, which will vary according to the lipids used as they pack together to form membranes of different densities, and vesicles of different diameters. Thus while pure egg PC liposomes in 50 mM KCl require centrifugation at 159 000 *g* for at least 3 h, sphingomyelin preparations can be purified in half the time. When using a new lipid combination, it is best to establish the conditions required for centrifugation beforehand by examining the homogeneity of the top layer after centrifugation for different lengths of time, using specific turbidity, or photon correlation spectroscopy. The method is sufficiently reproducible that the same degree of separation will be obtained from one run to another providing the same conditions are maintained. It appears that centrifugation for periods of time longer than the minimum required for separation results in loss of very little further lipids.

   This method is useful when membrane studies on pure SUVs are to be carried out. If, on the other hand, agents are to be entrapped within the vesicles (particularly highly coloured ones) which require separation from liposomes subsequently, then column chromatography may be preferred (see Section 5.1), since liposome separation and fractionation can then be carried out simultaneously using one and the same procedure.

3.1.7 *French pressure cell liposomes* (contributed by P.I.Lelkes)

Because of the problems inherent in subjecting biological materials to ultrasonic irradiation (i.e. degradation not only of lipids, but of macromolecules and other sensitive compounds to be entrapped inside liposomes), methods have been developed which are able to cause fragmentation and restructuring of membranes under very mild conditions.

One of the first and still very useful techniques developed is that of extrusion of pre-formed large liposomes in a French Press under (generally) very high pressure (11,12). This technique yields rather homogeneous uni- or oligo-lamellar liposome preparations of intermediate sizes (30−80 nm in diameter, depending on the pressure used). Besides the more gentle preparation conditions, which permit their use as carriers of sensitive macromolecules, these liposomes are more stable than sonicated ones and can be used advantageously as drug delivery systems *in vitro*. In addition, high pressure extrusion of liposome/protein mixtures appears to be a useful means for reconstitution of solubilized membrane proteins.

The French Press, named after one of its inventors, was originally designed and is still being used for the disruption of plant and bacterial cells. Today it is commercially available from SLM-Aminco Inc., Urbana, Illinois 61801, USA. One of the few drawbacks of this technique is the high initial cost of purchasing an entire new system which consists of an electric hydraulic press and a pressure cell.

The heart of a French Press is the pressure cell. Manufactured in stainless steel, it is designed to resist pressures up to 20 000 or even 40 000 p.s.i. (pounds per square inch; 100 p.s.i. are equivalent to approximately 6.9 atm). For comparison, the pressure in a normal car tyre is of the order of 25−35 p.s.i. Two pressure cells are available, of different sizes. The large one holds up to 40 ml (the operational minimum is approximately 4 ml), while the smaller one is the cell of choice if working with volumes generally used in the laboratory (<4 ml). With some practice the processing of 0.6 ml is quite feasible. An additional feature (at additional cost of course) is an automatic refilling accessory, which is convenient if repetitive, successive extrusions are required. The pressure cell is comprised of the body, containing a precision-bored cylindrical cavity (the pressure chamber) with a small outlet orifice (valve), a bottom seal, a piston, a valve closure and outlet tubing (see *Figure 6a*). The bottom seal and piston are fitted with two different 'O'-rings each, to ensure a tight seal even at very high pressures. The valve closure will close the outlet orifice with the aid of a plastic ball placed in a groove at its tip. The principle of operation of the French Press is essentially as follows:

Manufacture of liposomes using a French Press

(i)   To fill the pressure chamber with the cell or liposome suspension, insert the piston a short distance into the body and turn the cell upside down.

(ii)   Introduce the liquid sample into the cavity. By pressing down on the outer wall of the cell, adjust the height of the piston so that the chamber is completely filled with liquid. Make sure that the chamber is filled right up to the outlet hole. (Observation of this tiny hole, which sometimes might be difficult to see, can be facilitated by slightly tilting the pressure cell.) It is important to fill the chamber up to the hole, in order to prevent compression of trapped air, which will result

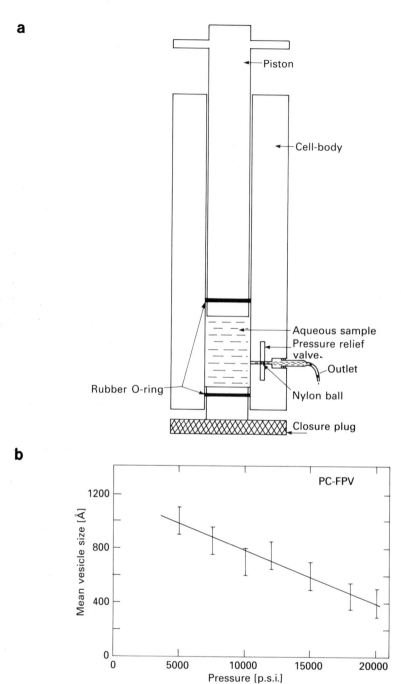

Figure 6. French press technique. (a) Diagram of a French pressure cell. Small vesicles are obtained when phospholipid dipsersions are extruded through the small orifice (lower right) at pressures of 20 000 p.s.i. or greater. (b) Graph showing relationship between pressure and liposome diameter.

in uncontrollable splashing during extrusion (this can be very dangerous if working with radioactive or biohazardous material).

(iii)   Upon filling the chamber, insert the bottom seal, and with the pressure cell still held upside down, close it gently but firmly, and turn it to an upright position. (NB. The plastic ball at the tip of the valve closure gradually wears out during each extrusion; it is thus imperative to check it before each use, to verify that it is still intact. Frequent replacement will ensure proper closing of the valve and easy control of the extrusion process.)

(iv)   Following insertion of the pressure cell into the hydraulic press, select the proper gear, and switch on the power (the order of these operations is important).

(v)   Adjust the pressure inside the chamber (determined by the diameter of the piston) by selecting manually the appropriate setting of the hydraulic press. Once this pressure is reached inside the cell, release the valve closure very very gently, so as to create a pressure gradient across the outlet orifice.

(vi)   Collect the outflow in a suitable container—the flow rate will be about $0.5-1.0$ ml min^{-1}.

The suspension being forced through this outlet experiences tremendous shear forces at the orifice; these forces are sufficient to rip apart intact cells and bacteria, or to disrupt mechanically multilamellar liposomes. The higher the forces the smaller the liposomal fragments. From the theory of hydrophobic interactions we know that these fragments will form liposomes spontaneously. The size of these liposomes is determined by the original size of the membrane fragments created by the shear forces across the orifice. The homogeneity of the resulting liposome suspension is inversely proportional to the flow rate through the orifice. The slower the flow rate, the better the chances are of disrupting the liposomes passing through the orifice. Therefore, it is imperative to open the valve very carefully and let the contents bleed out very slowly drop by drop. This will ensure effective and homogeneous fragmentation of multilamellar liposomes into smaller ones; already after one pass, a significant clarification of the originally milky white suspension is clearly observed. However, it is advisable to repeat the extrusion at least two more times, to ensure transformation of most large liposomes. Even with the most careful extrusion, however, a small fraction of the original liposome population will not be disrupted and/or the procedure will yield some smaller multilamellar vesicles. The fraction generally does not exceed $5-10\%$ of the total lipid. This subpopulation, together with the debris resulting from high pressure effects upon the rubber 'O'-rings, can be removed effectively by centrifuging the resulting liposome preparation for 10 min at approximately 12 000 g in a microcentrifuge.

After use, the pressure chamber should be dismantled and all parts washed thoroughly and if necessary, decontaminated. Particular attention should be paid to cleaning the areas around the 'O'-rings, since experience shows that fluorescent or radioactive material accumulated below them will subsequently interfere with future experiments. After each extrusion, the integrity of the 'O'-rings and of the plastic ball should be checked. If the 'O'-rings need to be replaced, the order of replacement is crucial and should be followed according to the manufacturer's instructions. After thorough cleaning, the pressure cell should be rinsed with distilled water and left to air-dry.

Finally, one word of caution: when using the French Press, one is essentially working with an open system, with respect to radioactive or otherwise hazardous material.

Therefore, all operations should be conducted with the utmost care, to avoid spilling and splashing during the filling and extrusion steps.

The French Press technique was first adapted for liposomology in order to prepare small unilamellar liposomes. A typical preparation contains 20 mg of phospholipid, e.g. egg PC. The lipids are dried under nitrogen or argon in a glass vessel and rehydrated with 2 ml of aqueous buffer containing the proteins, drugs, etc., to be encapsulated. This MLV suspension is then extruded as described above. There are a number of parameters which will determine the characteristics of the resultant liposome preparation.

(i)   To achieve effective fragmentation upon extrusion, the lipid bilayers of the MLVs must be in their fluid state, that is, the preparation must be carried out above the phase transition temperature of the membrane lipids. This is feasible for such lipids as egg PC, DOPC, PS, DMPC, SM, CL or even for DPPC; it is more difficult, however, for DSPC. On the other hand, to avoid impairment of the activity of biological macromolecules and/or oxidation of phospholipids, the temperatures should be as low as possible. Thus, pre-cooling the pressure cell on ice and/or working with ice-cold solutions is advisable. In these circumstances, as with other preparation methods, the composition of the liposomal carrier has to be chosen accordingly. Materials such as cytoskeletal proteins have been incorporated into liposomes (PC:PS:Chol in a 7:2:1 molar ratio) at 20 000 p.s.i. with over 85% retention of biological activity (13).

(ii)   The size of the resulting French pressed liposomes (FPL) is variable, depending on the lipid composition used, the temperature, and, most important, on the pressure (13). As seen in *Figure 6b*, the average diameter of unilamellar French pressed liposomes is roughly inversely proportional to the extrusion pressure. However, below a certain critical pressure, the shear forces do not seem to suffice to completely disrupt the multilamellar parent liposomes. Extrusion at or below approximately 5000 p.s.i. will rather result in small multilamellar vesicles. Lipid mixtures composed of pure phospholipids tend to yield smaller vesicles than mixtures containing cholesterol. Thus inclusion of 30 mol% cholesterol is found to increase the average size of the extruded liposomes by around 50%. Using pure egg PC, the mean size of the liposomes extruded at 0°C is about 80 nm, whereas the same liposomes extruded at 40°C are found to have an average diameter of about 45 nm.

Liposomes prepared by this technique, although still small, are somewhat larger than vesicles prepared by sonication, and are less likely to suffer from the structural defects and instabilities known to arise in sonicated vesicles. Leakage of vesicle contents from liposomes prepared using a French Press has been found to be slower than from sonicated liposomes.

The French Press has also been used to reduce the heterogeneity of populations of proteoliposomes obtained by detergent dialysis techniques (Section 3.3).

3.1.8 *Membrane extrusion liposomes*

An even gentler method of reducing the size of liposomes is to pass them through a membrane filter of defined pore size (14). This can be achieved at much lower pressures (< 100 p.s.i.) than required for the French pressure cell, and can give populations in

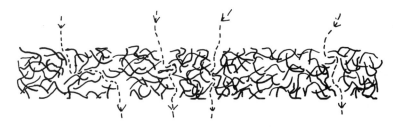

Figure 7. Tortuous path membrane. Compare the difference in size and shape of the pores of this membrane with that of the nucleation track membrane shown in *Figure 8.*

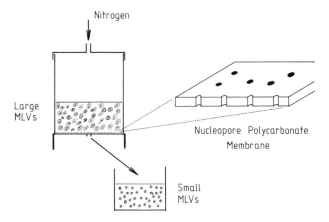

Figure 8. Liposome sizing by extrusion through Nuclepore membranes. MLVs are reduced in size by extrusion through the pores of a Nuclepore polycarbonate membrane. The pores are uniform in diameter, and after several passes through the membrane the liposomes obtained are relatively homogeneous in size, with a diameter equal to or slightly larger than that of the membrane pores.

which one can choose the upper size limit depending on the exact pore size of the filter used.

Membrane filters come in two types. One is the 'tortuous path' type membrane often used for sterile filtration (*Figure 7*), which consists of fibres criss-crossed over each other to give a matrix in which channels are formed from random spaces arising naturally in between the fibres. In three dimensions, the channels wind in and out between the fibres, tracing a tortuous path between one side of the membrane and the other; the average diameter of these channels is controlled by the density of the fibres in the matrix. Because of the convoluted nature of the channels, liposomes which are larger than the channel diameter get stuck when one tries to pass them through such a membrane, and the filter blocks up very easily, with no liposomes reaching the other side.

In contrast, 'nucleation track' membranes will pass liposomes through even if they are larger than the pore diameter. This type of membrane consists of a thin continuous sheet of polymer (usually polycarbonate) in which straight-sided pore holes of exact diameter have been bored through by a combination of laser and chemical etching (see *Figure 8*). Because the pores go straight through from one side to the other, they offer much less resistance to material passing through than do tortuous path membranes. The inherent flexibility of phospholipid lamellae enables liposomes to change their conformation so that they can squeeze through the pores. However, liposomes which are much larger than the pore size are broken up in the process, and emerge from the membrane pore smaller than before. After several passes through the membrane a population of liposomes will be reduced in size to an average diameter somewhat smaller than the membrane pore diameter, but with a small proportion still slightly larger than the pores, having managed to squeeze through without breaking up.

The apparatus used for extrusion is the same as that employed for other membrane pressure filtration applications such as protein concentration or diafiltration. The process can be scaled up simply by increasing the area of the membrane used. Concentration cells in the range from 10 ml to 1 litre capacity manufactured by Amicon are very suitable for use, fitted with polycarbonate membranes of defined pore size from Nuclepore Inc. At present not all cells take membranes in the sizes manufactured by Nuclepore, so it is often necessary to cut one's own. Great care must be taken in cutting to make sure that the membrane is not torn in any way at the edges.

Because of the possibility of the membrane being defective, or of a leak in the cell, it is advisable to test the integrity of the assembled unit by carrying out a 'bubble-point' test. This test relies on the fact that after a membrane has been wetted, the surface tension between water and air is such that air cannot be passed through the membrane until sufficient pressure has been reached to overcome that surface tension. The critical pressure at which air will pass (i.e. the point at which bubbles will appear) is directly related to the size of the pores, 80 p.s.i. (5.5 bar) for a 0.2-μm pore membrane and 44 p.s.i. (3 bar) for a 0.4-μm pore membrane.

Bubble-point test

(i)   Fill assembled unit about one third full of distilled water or saline.

(ii)   Connect the cell to a nitrogen cylinder, with a pressure regulator which reads up to 100 p.s.i. (6.9 bar).

(iii)   Close the inlet valve etc., to seal contents of the cell from the outside.

(iv)   Introduce nitrogen into the system at a pressure of about 10 p.s.i. (0.7 bar), so that water/saline in the cell passes through the membrane and is collected in the outflow.

(v)   When all fluid has passed through, check that the outflow tube is completely full of water still, and that drops have ceased to emerge from the end of the tube. There should be no air bubbles in the tube at this stage. The gas above the membrane should be held back, and the space below the membrane should be completely filled with fluid.

(vi)   Slowly increase the pressure at the cylinder head regulator (at about 2 p.s.i. sec^{-1} – approx 0.15 bar sec^{-1}), at the same time keeping an eye on the fluid in the outlet tube. As the pressure increases, the membrane and support disc may deform slightly, so that a few drops of fluid are expelled through the outlet tube, but this should not constitute a rapid flow.

(vii)   When the bubble point is reached, bubbles of gas will appear in the outflow tube, and the rate of passage of the remaining fluid will increase. At this point, record the pressure attained, allow the last drops of fluid to be forced out, then reduce the pressure to zero at the cylinder head and gradually open the safety valve, so that excess gas is vented to the outside. Take care to release the pressure gradually, as a sudden change may risk unseating the membrane.

If the pressure at which bubbles first appear during this test is markedly below the rated bubble point for a given membrane, the assembled cell should be examined for leaks, the membrane reseated in its bed, and if all else fails, a new membrane should be prepared and inserted into the apparatus.

Once the above procedures have been carried out, the liposome extrusion process itself is very simple. The cell is filled with liposome suspension, and the pressure taken quickly up to 100 p.s.i. (6.9 bar, or the highest pressure for which the cell is rated), in order to force the liposomes through the membrane. The cell can be stirred if desired although this is not always necessary. The rate at which the suspension will pass through a membrane of given pore diameter depends upon the size and concentration of the original liposomes, the cholesterol content of the membrane, and the saturation of the phospholipids, since liposomes with less fluid membranes are more difficult to break up. A typical flow rate through a 0.2-μm membrane for MLVs on the first extrusion at 50 mg ml^{-1} would be $0.1-0.5$ ml min^{-1} cm^{-2} at 70 p.s.i. (5 bar). Very little material is left on the membrane, so that the resultant liposome suspension can be passed through the same membrane a second time to obtain a further reduction in large vesicles. On subsequent passes, the flow rate is often much higher. On occasions the flow can decrease with time during the first pass. Under these circumstances, removing the unextruded suspension and rinsing the membrane with saline can resolve the problem. Alternatively, diluting the remaining liposome suspension in the cell with material that has already passed through may be sufficient to relieve the congestion.

The membrane extrusion technique can be used to process LUVs as well as MLVs. With both these types of liposome, it should be borne in mind that during the extrusion process, vesicle contents are exchanged with the suspending medium during breaking and resealing of the phospholipid bilayers as they pass through the polycarbonate membrane. In order to achieve as high an entrapment as possible of water-soluble compounds, it is important to have these compounds present in the suspending medium during the extrusion process. Material which is not entrapped can be removed subsequently (after the liposomes have been reduced sufficiently in size) by the techniques described in Section 5 of this chapter. Extrusion of liposomes which have been purified beforehand will result in loss of aqueous vesicle contents.

If MLVs are extruded through membranes of pore size 0.1 μm or smaller, then upon repeated extrusions, the liposome suspension becomes progressively more unilamellar in character, with the vesicles still maintaining a size distribution around the pore size of the membrane, but now possessing a considerable internal aqueous volume. An almost completely unilamellar population can be produced after $5-10$ repeated extrusions through two stacked membranes. In order to accomplish this in a short space of time, it is often necessary to increase the pressure up to 500 p.s.i. (35 bar) using a custom-made extrusion cell. The liposomes produced by this high pressure system have been termed 'LUVETs' (Large Unilamellar Vesicles by Extrusion Techniques) by their originators (15), although the size distributions obtained correspond to intermediate-sized submicron liposomes. With neutral phospholipids, a significant proportion of small unilamellar vesicles is obtained, which reduces the expected entrapment volume, but the addition of charged lipids brings the entrapment volume back to calculated values.

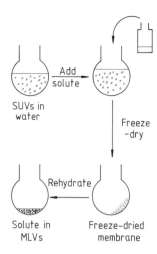

Figure 9. Preparation of dried-reconstituted vesicles. This figure shows schematically the steps involved in entrapment of solutes using the 'DRV' method. The solute is added to a pre-formed suspension of empty SUVs, and the mixture is then freeze-dried and resuspended in a small volume of diluent. In the process the membranes break and reform to give new vesicles enclosing a proportion of the solute which was originally in the extra-liposomal medium. A relatively high entrapment can be achieved, providing that high concentrations of sugars, or other potential cryopreservatives are absent.

Using high lipid concentrations (300 mM PC), capture volumes of up to 30% have been obtained.

In concluding this section on physical methods of dispersion, techniques are described in which liposomes, having been reduced in size by any of the methods described in the preceding paragraphs, can be treated so as to restructure their membranes, leading to a population of liposomes capable of increased entrapment, either through reduction in the volume of bulk aqueous phase, or increase in the captured volume of the liposomes formed. Intermembrane fusion is an important feature of all these methods, and can be brought about by lyophilization, freezing and thawing, introduction of metal ions, or change of pH (*Figures 9* and *10*).

3.1.9 *Dried-reconstituted vesicles (DRVs)*

This method is another one where a way has been sought to disperse the solid lipid in finely divided form before contact with the aqueous fluid which will form the medium for the final suspension. Freeze-drying has again been used, but this time, instead of drying the lipids down from an organic solution, a suspension of empty SUVs is frozen and lyophilized (16). Thus, in contrast to the preparation dried from solvent, where the lipid molecules are in a random matrix, the SUV-dried lipid is already very highly organized into membrane structures (rather like the pro-liposomes without the support) which, on addition of water, can rehydrate, fuse, and reseal to form vesicles with a high capture efficiency. This is partly because the water has very ready access to the lipid in this form, and only a small volume need be added to suspend a large quantity of lipid very rapidly. Liposome suspensions of several hundred milligrams per millilitre can be obtained routinely in this way.

The second novel feature of this method is that the water-soluble materials to be entrapped are added to the suspension of empty SUVs and the two are dried down together, so the material for inclusion is present in the dried precursor lipid before the final step of addition of aqueous medium. This is done because it is found that the intimate contact achieved between lipid and solute in the dry state is essential for optimal entrapment. Although it is possible to add the solute in solution as the final step after lipid has been dried (and some advantage is gained because of the small volumes and higher concentrations which can be used), the best results are obtained when the two are dried down together. A typical procedure is outlined as follows (see *Figure 9*).

Preparation of DRVs

(i)   Dry down a solution of chloroform containing 120 mg of egg PC, 50 mg of cholesterol, 10 mg of phosphatidyl glycerol in a 21-mm diameter glass tube with ground-glass neck, on a rotary evaporator.

(ii)   Remove the last traces of solvent by attaching the tube to the manifold of a lyophilizer for 1 h.

(iii)   Add 5 ml of distilled water, stopper the tube, and vortex to suspend the lipid.

(iv)   Sonicate using a probe for 30−60 min as described in Section 3.1.6.

(v)   Spin the resultant SUV suspension at 2500 *g* in a bench centrifuge for 20 min at room temperature to remove contaminating particles from probe, etc.

(vi)   Transfer the SUV suspension to a 50-ml round-bottomed flask. Add 5 ml (volume equal to liposome suspension) of aqueous solution containing the material to be entrapped. Mix well.

(vii) Connect the flask to a rotary evaporator, and secure it with a clip, or sticky tape.

(viii) Prepare a bath containing acetone and dry ice (solid carbon dioxide).

(ix)   With the flask turning briskly on the rotary evaporator at atmospheric pressure, slowly raise the bath, so that the flask is immersed in it. The liquid inside the flask will be flash-frozen as a thin shell.

(x)   Halt rotation of the flask, remove it from the evaporator and stopper it, keeping it immersed in the dry-ice bath all the while.

(xi)   Take the flask and the bath to the lyophilizer.

(xii) Set the lyophilizer in operation. When the condenser has reached the required temperature, and a good vacuum has been obtained (0.1 Torr), attach the flask to the lyophilizer manifold, and turn the tap to evacuate the flask.

(xiii) Wrap the flask in cling-film and leave it overnight at room temperature.

(xiv) When the material is dry, remove the flask from the condenser, flush with nitrogen, and store until required for use.

(xv)   To obtain a liposome suspension, add 0.5 ml (one-tenth the volume of the original SUVs) of distilled water to the flask and mix gently.

(xvi) Dilute as required in buffer of osmolarity ten times that of the solute added in Step (vi).

Methods for separating unentrapped material are described in Section 5 of this chapter. Liposomes obtained by this method are usually uni- or oligo-lamellar, of the order of one micron or less in diameter. Entrapment yields can vary, but 40% is fairly standard, compared with 2−10% for MLVs prepared by the hand-shaking method. DNA can be entrapped in 70% yield. With low molecular weight solutes, their dilution prior to freezing influences the final percentage entrapment. Increasing the concentration of sodium or potassium chloride increases the efficiency of entrapment, while an increase in sugar concentration decreases the entrapment of these compounds. It is postulated that the latter phenomenon is due to the cryoprotective effect of these sugars, which may here act to prevent the membrane rupture and fusion processes which are essential to effective performance of this method taking place during freeze-drying.

Other methods of drying down can be used with success. The advantage of the DRV technique (apart from its high entrapment of water-soluble compounds) is the mild conditions under which those compounds are treated. The drawback of the process is that it is important for the lipid being lyophilized to be in the form of unilamellar vesicles before drying. MLVs have been tried, but the incorporation rates are not so favourable. Another problem is that the high percentage entrapments by DRVs are achieved by containing the solutes in a low volume of fluid, so that differences in osmolarity between inside and outside of the liposomes on dilution may lead to osmotic rupture and loss of contents. Methods for increasing the stability of membranes in such cases are described in Section 8.2. It is of interest to note that the phenomena occurring in this method for liposome entrapment have been postulated as being responsible for the formation of evolutionary precursors for living cells (17).

3.1.10 *Freeze-thaw sonication (FTS) method*

The DRV method is a variant of an earlier one developed by Pick (18) after Kasahara and Hinckle (19) in which the entrapped material is again introduced into liposomes after they have been formed. In this case, a freezing and thawing process is used to rupture and re-fuse SUVs, during which time the solute equilibrates between the inside and outside, and the liposomes themselves fuse and increase markedly in size, so that their entrapment volume can rise to 30% of the total volume of the suspension (10 μl mg^{-1} phospholipid). Electron microscopy studies show that the liposomes are primarily unilamellar. The starting preparation of empty liposomes is prepared by sonication, and after thawing, the liposomes are subjected to brief sonication again. It is found that this second sonication step considerably reduces the permeability of the liposome membranes, perhaps by accelerating the rate at which packing defects dissipate. Providing sonication is carried out for a short time only (15−30 sec) no reduction in entrapment volume is observed.

In terms of absolute encapsulation efficiency, this method suffers from several disadvantages compared with a DRV technique. It is not possible to prepare neutral liposomes composed of purified phosphatidyl cholines, presumably because the presence of charge is required for the formation of ice crystals to aid in the rupture/fusion process; for similar reasons sucrose (a cryoprotectant), divalent metal ions (which can neutralize surface charge) and high ionic strength salt solutions cannot be entrapped efficiently; concentrations of phospholipid higher than 40 mg ml^{-1} reduce the trapping efficiency, as do high concentrations of polyelectrolytes, perhaps because of a reduction in the chemical activity of water. Nevertheless, the method is very simple, rapid, and mild for entrapped solutes, and results in a high proportion of large unilamellar vesicles which are useful for study of membrane transport phenomena.

(a) *FTS procedure*

(i)   Prepare an organic solution of lipid containing 14 mg of soya lecithin, 6 mg of phosphatidyl serine, and 4 mg of cholesterol, in 1 ml of chloroform.

(ii) Dry down this solution in a 6-ml Pyrex test-tube under a stream of nitrogen.

(iii) When the lipids have dried down, redissolve in 1 ml of diethyl ether (redistilled) and dry down again with a nitrogen stream to remove traces of chloroform.

(iv) Add 1 ml of buffer (30 mM Tricine and 1 mM EDTA at neutral pH purged with nitrogen) to the dried lipids in the tube, flush the gas space with nitrogen, cover the top of the tube with Parafilm, and vortex for about 10 min until a homogeneous milky suspension is obtained.

(v) Transfer the sealed tube to a sonicator bath filled to a depth of about 2−3 cm with water. Hold the tube in a vertical position with a clamp, so that fluid levels inside and outside the tube are equal.

(vi) Sonicate the suspension continuously for 10−15 min at room temperature until partial clarification is observed.

(vii) After sonication, add the materials which are to be entrapped, either in solid form, or as a small aliquot of concentrated solution. Mix well.

(viii) Reflush the tube with nitrogen, seal, then freeze the suspension rapidly by shaking in a bath of liquid nitrogen.

(ix) After complete freezing, remove from the bath, and allow to thaw by standing at room temperature for 15 min.

(x) After thawing, place the tube in the sonicator as before, and sonicate for no longer than 30 sec at room temperature.

Separate unentrapped materials by column chromatography or centrifugation as described in Section 5.

The freeze-thaw technique has been extended by Oku and MacDonald (20) to incorporate a dialysis step against hypo-osmolar buffer, in place of the final sonication. In this case SUVs are first mixed with salt solutions at a concentration of several molar, followed by freeze-thawing several times. During subsequent dialysis, the large vesicles formed by freeze-thawing swell and rupture as a result of osmotic lysis, whereupon they fuse with each other to yield a large number of giant vesicles of diameter between 10 and 50 μm. A wide range of phospholipid compositions is possible, the inclusion of some negatively-charged lipids giving rather higher trapped volumes (20 μl mg^{-1} as compared to 10 μl mg^{-1} for neutral phospholipid). Using a concentration of 1.5 M potassium chloride during the freeze-thawing steps, followed by dialysis against buffer containing no salt, an entrapment percentage of 20% total fluid volume can be achieved, and this is constant over a range of phospholipid concentrations between approximately 10−40 mg ml^{-1}.

It would appear that the factors contributing to the formation of giant vesicles and a high percentage entrapment are more complicated than simple considerations related to the requirements for osmotic rupture. Not all salt solutions perform equally well; lithium chloride yields no large vesicles, caesium chloride requires the presence of calcium, and sodium chloride gives significant yields only at high phospholipid concentrations. In some cases, such observations may be explicable in terms of a cryopreservative effect, since use of high concentrations of glucose or glycerol also give low yields, although osmotic lysis must undoubtedly be taking place. Alternatively, a crucial factor may be the ability of certain salts to permit vesicle aggregation, an essential step in the intermembrane fusion process. The extent of lipid aggregation after freeze-thawing in this technique does appear to correlate with the proportion of large vesicles formed by use of different alkali metal chlorides.

(b) *Preparation of giant liposomes by freeze-thaw-dialysis technique*

(i)   Prepare a suspension of SUVs as described in the previous paragraphs, using a buffer containing 1.5 M KCl, and 1 mM MOPS, pH 7.2. Any additional materials to be entrapped should be added to the suspension after sonication.

(ii)   Freeze by immersing the tube in an acetone-dry ice bath, and then thaw by leaving it to stand at room temperature for 20 min. Repeat this process for at least three cycles.

(iii)   Transfer the suspension to a dialysis bag, and dialyse against three changes of 2 litres of physiological saline for 2 days.

It is important to note that freeze-thawing is an essential step in the production of giant vesicles with high entrapment. Dialysis of sonicated liposomes without freeze-thawing only gives a small increase in size and entrapment. Moreover, performance of repeated cycles of freezing and thawing up to ten times improves the yield of giant vesicles. This is in contrast to preparation of vesicles in 10 mM MOPS buffer without salt (essentially the original FTS method) in which few giant vesicles are seen, and where optimal entrapment values (20–30%) are obtained by using between one and three freeze-thaw cycles at the most.

3.1.11 *pH-induced vesiculation* (contributed by H.Hauser)

MLVs can be induced to reassemble into unilamellar vesicles without the need for sonication or high pressure, simply by changing the pH (21). This process, termed 'pH-induced vesiculation', is an electrostatic phenomenon. The transient change in pH brings about an increase in the surface charge density of the lipid bilayer; provided this exceeds a certain threshold value of around $1-2 \ \mu C \ cm^{-2}$, spontaneous vesiculation will occur. The protocol outlined below will produce small unilamellar liposomes from a starting preparation of MLVs composed of phosphatidic acid.

Formation of liposome by pH-induced vesiculation

(i)   Prepare a solution of 10 mg of phosphatidic acid in 1 ml of chloroform/methanol solution (2:1 by vol).

(ii)   Introduce the solution into a 50-ml round-bottomed flask and dry down in a rotary evaporator.

(iii)   When a dry film has been obtained, complete the drying process under high vacuum, e.g. in a freeze dryer, to remove the last traces of solvent.

(iv)   Disperse the dry phospholipid film in 1 ml of distilled water by hand-shaking at room temperature. Materials to be entrapped inside the vesicles may be introduced at this stage dissolved in the water before addition to the lipid.

(v)   Complete the dispersion by subjecting the suspension to six freeze-thaw cycles between $-15°C$ (ice/methanol) and $5°C$. The pH of the dispersion will be pH 2.5–3.

(vi)   Add 100 μl of 1 M NaOH to the suspension rapidly, with mixing, from a

micropipette, to bring the pH to pH 11.
(vii) Reduce the pH by addition of 0.1 M HCl until a value of pH 7.5 is reached.

The period of exposure of the phospholipids to high pH is less than 2 min, and not long enough to cause detectable degradation of the phospholipid. Phospholipid dispersions with similar properties to those produced by the above procedure are also obtained if concentrated NaOH solution is added directly to the dry lipid film to give a dispersion without freeze-thawing. The resulting suspensions consist of a relatively homogeneous population of small unilamellar vesicles with an outer diameter of 20−60 nm (21).

Variations in the above procedure can bring about changes in the size distribution of the vesicles. The following modifications give rise to a polydisperse population of unilamellar liposomes with an increasing proportion of vesicles greater than 100 nm in diameter.

(i)   Increasing the proportion of neutral phospholipids (e.g. PC−synthetic or natural, cholesterol, mono-olein) mixed with the phosphatidic acid.
(ii)   Reduction in the rate of addition of NaOH when raising the pH.
(iii) Lowering the maximum transient pH attained (from pH 12 downwards).
(iv)   Increasing the concentration of salt in the liquid used for the initial dispersion.

As an electrostatic phenomenon, it is expected that pH-induced vesiculation be independent of the nature of the charged group. Thus lipid dispersions in water (approximately pH 7) consisting of egg PC and quantities of the detergent hexadecyl-amine (up to about 40%) can be induced to vesiculate by transiently lowering the pH, to pH 2, well below the pK of the long chain amine. In this case the vesiculation is brought about by the addition of 0.1 M HCl, producing the fully protonated (positively-charged) form of hexadecylamine. The examples cited above are special cases of a more general vesiculation procedure, where, in any lamellar phase consisting of a lipid and an amphiphile containing an ionizable group, vesiculation may be induced by a transient pH change or any other adjustment leading to full ionization of this group. The principle can also be applied to amphiphilic drug molecules which may be incorporated into lipid membranes, where the agent to be transported with the liposomes can itself be used to modify the size of the vesicles appropriately.

3.1.12 *Calcium-induced fusion to produce large unilamellar vesicles*

This method takes advantage of the fact that small vesicles composed of acid phospholipid aggregate in the presence of calcium, and subsequently fuse. The following method is based on that described by Papahadjopoulos *et al.* (22).

Buffered saline for sonication:
NaCl	100 mM	(0.585 g per 100 ml)
Histidine	2 mM	(31.0 mg per 100 ml)
Tris-base	2 mM	(24.2 mg per 100 ml)
EDTA	0.1 mM	(3.72 mg per 100 ml for disodium dihydrate)
Adjust to pH 7.4.		

Calcium solution:

40 mM $CaCl_2.2H_2O$ − 5.88 mg ml^{-1} in buffered saline.

EDTA solution:

170 mM EDTA $Na_2.2H_2O$ − 63.7 mg ml^{-1} in buffered saline.
Adjust pH to pH 7.4 by the addition of 1 M NaOH.

(i) Dry down 20 mg of phosphatidyl serine, and suspend in 5 ml of sonication buffer.

(ii) Prepare small liposomes either by probe or by bath sonication as described previously.

(iii) Centrifuge at 100 000 *g* for 60 min at 20°C to remove large liposomes and lipid particles.

(iv) To 4 ml of the supernatant add 1 ml of 40 mM $CaCl_2$ solution with rapid mixing; the total calcium and PS are in approximately equimolar proportions. The concentration of free calcium is approximately 2 mM. A white flocculent precipitate will form.

(v) Incubate for 60 min at 37°C.

(vi) Spin down the precipitate by centrifugation at 20 min, 3000 *g*, at room temperature; discard the supernatant.

(vii) Resuspend the pellet in 0.2 ml of buffered saline containing the material to be entrapped.

(viii) Incubate for 10 min at 37°C.

(ix) Add 100 μl of 170 mM EDTA in buffer with mixing. The cloudy precipitate will clear rapidly. Vortex vigorously.

(x) Incubate for 15 min at 37°C, then for a further 15 min at room temperature.

(xi) Dialyse overnight against a litre of buffer to remove the Ca/EDTA complex. The final volume will be about 0.5 ml.

The percentage encapsulation depends on the lipid concentration employed. Under the conditions described here, a 30% capture may be expected. The vesicles obtained range in size from $0.2 - 1$ μm diameter. Larger (perhaps fragile and unstable) liposomes may be produced upon addition of EDTA, but these are broken up deliberately by vortexing. Addition and removal of calcium can be effected alternatively by dialysis (first against 2 mM $CaCl_2$ solution, then against 2 mM EDTA) for entrapped materials of high molecular weight, which will not be removed during dialysis.

The flocculent precipitate obtained forms as a result of aggregation of the negatively-charged vesicles by the calcium cations. After incubation, the membranes fuse to give extended sheets of phospholipid lamellae, which are said to roll up into long cochleate cylinders with a 'swiss-roll' cross-section, presumably again with calcium ions as the driving force, bringing in different parts of the same membrane sheet together, being pulled in upon itself. On addition of EDTA, these lamellae are unravelled and released sequentially, forming LUVs.

This technique has the advantage that it does not expose lipids or entrapped materials to deleterious chemical or physical conditions. Its principle drawback is the requirement for acidic phospholipids, and the inevitable presence of calcium inside the liposomes, even after dialysis.

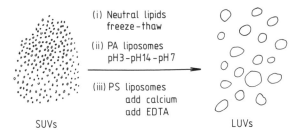

Figure 10. Modification of liposomes by membrane fusion. This drawing summarizes ways in which small liposomes of different compositions can be induced to form large unilamellar vesicles.

The methods of preparation of large liposomes as a result of inducing fusion of smaller vesicles by the techniques described here are summarized in *Figure 10*.

3.2 Solvent dispersion

In this group of methods, the lipids comprising the liposome membrane are first dissolved in an organic solution, which is then brought into contact with the aqueous phase containing materials to be entrapped within the liposome. At the interface between the organic and aqueous media, the phospholipids align themselves into a monolayer which forms the basis for half of the bilayer of the liposome. Methods employing solvent dispersion fall into one of three categories:

(i)   those in which the organic solvent is miscible with the aqueous phase;
(ii)   those in which the organic solvent is immiscible with the aqueous phase, the latter being in a large excess;
(iii)   those in which the organic solvent is in large excess, and is again immiscible with the aqueous phase.

Each of these categories is discussed in turn in the following sections.

3.2.1 *Ethanol injection*

This method was originally reported by Batzri and Korn (23). An ethanol solution of lipids is injected rapidly into an excess of saline or other aqueous medium, through a fine needle (*Figure 11*). The force of the injection is usually sufficient to achieve complete mixing, so that the ethanol is diluted almost instantaneously in water, and phospholipid molecules are dispersed evenly throughout the medium. This procedure can yield a high proportion of small unilamellar vesicles [diameter ~ 250 Å (25 nm)], although lipid aggregates and larger vesicles may form if the mixing is not thorough enough. The method has the advantages of extreme simplicity and a very low risk of bringing about degradative changes in sensitive lipids. Its major shortcoming is the limitation of the solubility of lipids in ethanol (40 mM for PC), and on the volume of ethanol that can be introduced into the medium (7.5% v/v maximum), which in turn limits the quantity of lipid dispersed, so that the resulting liposome suspension is usually rather dilute. The percentage encapsulation is thus extremely low if the materials to be entrapped are dissolved in the aqueous phase. Another drawback is the difficulty with which ethanol can be removed from phospholipid membranes.

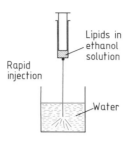

Figure 11. Preparation of SUVs by ethanol injection. This drawing illustrates the salient features of the ethanol injection method for preparation of SUVs. For formation of good vesicles the final ethanol-in-water concentration should not exceed 7.5%.

SUVs by rapid solvent injection [after Batzri and Korn (23)]

(i)   Dissolve 30 mg of PC in 1 ml of ethanol.

(ii)   Introduce 10 ml of nitrogen-purged aqueous medium (water, saline, solute solution etc) into a 25-ml conical flask, and stir rapidly with a magnetic follower.

(iii)   Fit a fine gauge needle to a 1-ml glass syringe and draw up 750 μl of the lipid solution.

(iv)   Position the tip of the needle below the surface of the stirred aqueous solution, and inject the organic solution as rapidly as possible into the medium. Liposomes will be formed immediately. The final concentration is approximately 2 mg of PC ml^{-1}, which is the maximum attainable without further processing.

If required, the liposomes can be concentrated by ultrafiltration at low pressure (10 p.s.i.) in an Amicon concentrator using an XM-100A membrane with rapid stirring. Ethanol may be removed from the bulk medium by dialysis or diafiltration.

3.2.2 *Ether injection*

Although very similar in concept to the method just described, ether injection (24,25) contrasts markedly with ethanol injection in many respects. It involves injecting the immiscible organic solution very slowly into an aqueous phase through a narrow bore needle, at such a temperature that the organic solvent is removed by vaporization during the process (*Figure 12*). The mechanism whereby large vesicles are formed by this method is not clearly understood, but it may be that the slow vaporization of solvent gives rise to an ether:water gradient extending on both sides of the interfacial lipid monolayer, resulting in the eventual formation of a bilayer sheet which folds in on itself to form a sealed vesicle. This process can repeat itself many times.

Ether injection again is a method which treats sensitive lipids very gently, and runs very little risk of causing oxidative degradation, providing the ether has been carefully redistilled to remove peroxides which arise as a result of spontaneous breakdown. Since the solvent is removed at the same rate as it is introduced, there is no limit to the final concentration of lipid which can be achieved, since the process can be run continuously for a long period of time, giving rise to a high percentage of the aqueous medium encapsulated within vesicles. The disadvantages of the technique are the long time taken to produce a batch of liposomes, and the careful control needed for introduction of the lipid solution, requiring a mechanically-operated infusion pump. If working with

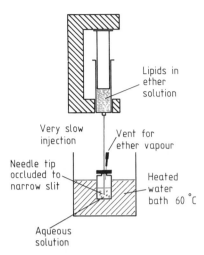

Figure 12. Preparation of LUVs by ether injection. In this method an ether solution of lipids is introduced into an aqueous phase held at a temperature at which the ether vaporizes rapidly. The phospholipids left behind disperse in water forming large bilayer sheets which rapidly seal to give closed LUVs.

substances which might be damaged at elevated temperature (60°C), the method may be adapted to use fluorinated hydrocarbons which vaporize at lower temperatures, instead of ether. The apparatus required for this method can be employed in many different configurations. The use of a simple set-up (*Figure 12*) is described below.

Preparations of liposomes by the ether injection method

(i)   Prepare a solution of lipid in diethyl ether (concentration 4 μmol ml^{-1}) by dissolving 13.5 mg of egg PC, 1.5 mg of PA and 5 mg of cholesterol in 10 ml of ether, to give components in a molar ratio of PC:PA:Chol 9:1:10.

(ii)   Place 5 ml of aqueous solution, containing the material to be entrapped, inside a 10-ml glass vial closed with a silicone rubber injection cap.

(iii)   Prepare the needle for ether injection by taking a long 19 gauge disposable hypodermic needle, and crimping it tightly just behind the tip. Break the tip off by bending, and file the crimped surface to give a narrow slit approximately 0.1 mm across.

(iv)   Introduce the needle into the vial, through the rubber cap, until the tip of the needle (i.e. the newly-formed slit) is well below the surface of the aqueous solution.

(v)   Introduce a second needle (19 gauge or wider) through the cap, projecting 1 cm into the vial, to act as a gas release vent.

(vi)   Fill a 10-ml glass syringe with 2 ml of the ether solution, and fit the syringe into a Harvard infusion pump apparatus.

(vii)   Attach the injection needle to the syringe head, and orient the pump and needle so that the vial hangs vertically below the pump.

(viii) In this orientation place the pump in a clamp so that the vial is suspended in a shaking water bath set at 55°C. Ideally the water bath should be located in a fume hood, in order to remove ether vapour as soon as it is produced.

(ix)   Set the pump working so as to introduce ether solution into the vial at a rate no faster that 0.2 ml min^{-1}.

(x)   After all the ether has been introduced, remove residual vapour by withdrawing the syringe tip from the aqueous phase and passing a stream of nitrogen through the gas space above the liquid in the vial.

(xi)   Remove aggregates by low-speed centrifugation; unentrapped material, as well as the last traces of ether, may be removed either by dialysis or gel permeation chromatography.

The liposomes formed are in the size range 0.1−0.5 μm diameter, with an entrapment volume of 10−15 litres mol^{-1} phospholipid. Under the conditions described above, one would expect a total entrapped volume of approximately 50 μl (i.e. roughly 1% entrapment).

Parameters which are important in the preparation of liposomes with a good entrapment efficiency are as follows:

(i)   The temperature of the medium, which must be between 50−60°C. Injection rates faster than 0.2 ml min^{-1} should not be used, since this can result in cooling of the solution by evaporation.

(ii)   The presence of a negatively-charged phospholipid, to prevent aggregation. There appears to be no advantage in incorporating PA in proportions greater than 10 mol%.

(iii) The shape of the injection needle orifice. The needle is crimped in order to produce a slight back pressure in the syringe, thus preventing pre-evaporation of ether, which can otherwise cause blockage of the needle or the formation of multilamellar vesicles.

Saturated lipids, such as dipalmitoyl PC, which are poorly soluble in pure ether, can be used with this method provided that the ether contains 20% methanol by volume. The concentration of lipid in ether may be increased up to 5-fold without deleterious results, to produce liposomes in higher concentration. In the case of materials which are sensitive to high temperatures, the preparation can be carried out by subjecting the vial to reduced pressure (by attaching a water suction tap to the gas vent needle) and operating at 37°C.

3.2.3 *Water in organic phase*

In this group of methods, the liposome is made up in two steps, first the inner leaflet of the bilayer, then the outer half (*Figure 13*). There are several different variations, the common feature of which is the formation of a 'water-in-oil' emulsion produced by introduction of a small quantity of aqueous medium (containing material to be entrapped) into a large volume of immiscible organic solution of lipids, followed by mechanical agitation to break up the aqueous phase into microscopic water droplets. These droplets are stabilized by the presence of the phospholipid monolayer at the phase interface. The size of the droplets is determined by the intensity of mechanical energy used to form the emulsion, and by the amount of lipid relative to the volume of aqueous phase, since each droplet requires a complete monolayer of phospholipid covering its surface in order to prevent its coalescing with other droplets, or with the solvent−air interface.

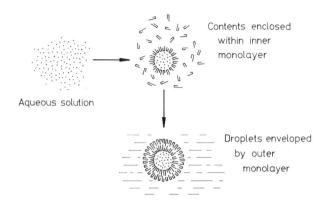

Aqueous solution

Contents enclosed
within inner
monolayer

Droplets enveloped
by outer
monolayer

Figure 13. Basic principle of two-phase systems for the preparation of liposomes. In two-phase preparation methods, the outer bilayer membrane is formed in two steps; first the inner monolayer is assembled enclosing the internal aqueous compartment, then the outer monolayer is wrapped around the droplet, and the intervening solvent is removed.

The aqueous compartment, surrounded by the monolayer of phospholipid, forms the central core of the final liposome. There are several different ways of preparing the water-in-oil droplets, before addition of the outer coating, which under appropriate conditions give droplets of differing sizes. Droplets of diameter 0.1 μm (100 nm) can be formed by probe sonication as outlined below.

(i) *Formation of droplets* (contributed by S.Frøkjær)

(i)   Place 200 mg of egg phosphatidyl choline and 50 mg of cholesterol, dissolved in chloroform/methanol, in a 20-ml glass boiling tube.
(ii) Remove the organic solvent by evaporation in a stream of nitrogen, and redissolve in di-*n*-butyl ether. Dry down again.
(iii) Repeat the previous step once to remove the last traces of methanol, and dissolve the lipids in 3 ml of di-*n*-butyl ether/cyclohexane (5:1 v/v).
(iv) Transfer 1.5 ml of the organic solution to a sonication vessel.
(v)   Add 100 μl of 10 mM Tris$-$HCl pH 7.4 (1.21 g Tris l^{-1}) containing the water-soluble compound to be encapsulated.
(vi) Insert the tip of the sonicator probe 3 mm below the surface of the organic solution. Seal the system with a rubber cap, and surround the vessel with a water bath at 45°C.
(vii) Sonicate the two phase system for 15$-$30 sec. A thick milky white dispersion will be obtained.
(viii) Add 50$-$100 μl of the remaining organic solution to the vessel, mix well, and re-sonicate for 5$-$10 sec.
(ix) Repeat Step (viii) until the system becomes a homogeneous, slightly opalescent dispersion which does not separate (become non-transparent) for at least 10$-$15 min after sonication.

Usually between seven and ten additions of organic solution are required before the suspension becomes clear. In the protocol outlined above, the solvent and lipid

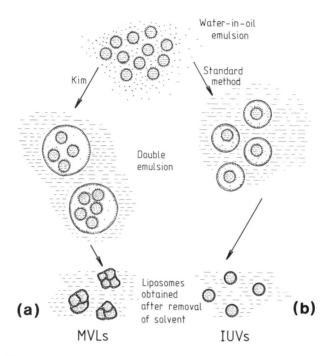

Figure 14. Double emulsion methods. By altering the relative proportions of lipid, water, and organic solvent, either single or multiple water droplets can be enclosed within a single oil droplet of the double emulsion, giving rise to formation of the different types of liposome shown. (**a**) Preparation of multivesicular liposomes (MVLs). (**b**) Preparation of intermediate unilamellar vesicles (IUVs).

concentration have been chosen in order to ensure that a sufficient quantity of lipid remains in solution to provide the outer coating. The dispersion formed in this way can be converted into a liposome suspension by forming a double emulsion.

(b) *Double emulsion vesicles* (contributed by S.Frøkjær)

In this method, the outer leaflet of the liposome membrane is created at a second interface between two phases by emulsification of an organic solution in water (26). If one uses the organic solution which already contains water droplets, and introduces this into excess aqueous medium, followed by mechanical dispersion, a multi-compartment vesicle is obtained, which may be described as a 'water-in-oil-in-water' system (i.e. a double emulsion). These vesicles are suspended in aqueous medium, and have an aqueous core, the two aqueous compartments being separated from each other by a pair of phospholipid monolayers whose hydrophobic surfaces face each other across a thin film of organic solvent (see *Figure 14b*). Removal of this solvent clearly results in intermediate-sized unilamellar vesicles. The theoretical entrapment yield is 100%, and can approach this value on occasions, depending on the nature and concentration of the material to be entrapped.

Preparation of double emulsion

(i)   Introduce 20 ml of 10 mM Tris−HCl pH 7 (1.21 g l^{-1} Tris) and 10 ml of distilled water into a 150-ml two-necked round-bottomed flask.

(ii)   Place the flask in a water bath at 45°C on a magnetic stirrer with a hot plate.

(iii)   Start the magnetic stirrer to give vigorous stirring in the flask.

(iv)   Pass a tube connected to a nitrogen pressure cylinder through the side neck into the flask.

(v)   Rapidly inject the suspension of microdroplets (prepared as above) from a 5-ml glass syringe into the aqueous phase in the flask through a 22 gauge hypodermic needle.

(vi)   Start the evaporation of organic solvent by adjusting a strong jet of nitrogen into the double emulsion system which is still kept under vigorous stirring.

(vii)   Continue to evaporate until the last traces of organic solvent have been removed (60−90 min).

(viii)   Adjust the total volume to 20 ml by adding extra distilled water.

(ix)   Centrifuge the product at 20°C for 30 min at 37 000 g to remove lipid aggregates.

The excess water added to step (a) is the amount of water which is normally lost during the evaporation steps. If working with lipids with a high phase transition temperature, e.g. DSPC, it is necessary to use a solvent composition of di-n-butyl ether/cyclohexane (1:1 v/v) in order to allow for the lower solubility of this type of lipid. It is also recommended that the final liposome preparation be 'annealed' by heating above the transition temperature (to 60°C in this case) for 2 min. The critical point in this process is probably the evaporation step where the organic phase is removed. At this stage some water droplets may tend to disintegrate releasing the entrapped material and/or causing a collapse of the double-emulsion structure, giving rise to marked lipid aggregation. In general, the degree of encapsulation of water-soluble material, expressed as a percentage of the total initial starting amount of this material, depends mainly on the nature of the material itself, and only to a minor extent on the liposome composition. Optimizing the procedure on the basis of the encapsulation of bovine serum albumin (BSA), an entrapment of 50−80% can be obtained routinely. Using the same experimental procedure for evaporation as used for BSA, the amount of insulin which is encapsulated is of the same order, whereas the encapsulation efficiency of compounds like inulin and sucrose is much lower, typically 10%. It has been found, however, that simply by using a more gentle evaporation technique than employed in the standard procedure gives an encapsulation percentage of 40−50% for these compounds.

A method for preparing large unilamellar vesicles by the double emulsion technique has been reported by Kim and Martin (27). This procedure differs in two major respects from that described above; first, in that the preparation of the initial water-in-oil microdroplet emulsion is carried out by mechanical shaking on a vortex mixer (using chloroform as solvent), and secondly in that the aqueous phase used to form the second emulsion consists of a dispersion of ether/lipid solution in water. It is not clear whether the presence of the ether solution is strictly necessary, since only 8% of the lipid in these droplets is eventually incorporated into the final liposome preparation (in contrast with 35% of the lipid from the chloroform solution being incorporated). On the other hand, it may be that the ether itself (after mixing with chloroform) increases the ease with which the second emulsion is obtained, by forming more stable droplets in the aqueous phase than chloroform itself would do.

In all, approximately 25% of the lipid is incorporated into the final liposomes, which are unilamellar spherical vesicles in the 5−10 μm diameter size range, depending on the strength and duration of the initial vortexing. Encapsulation is in the order of 50%. Problems have been reported with incorporation of labile globular proteins such as BSA, which prevent formation of the water-in-chloroform dispersion, perhaps because of denaturing interactions at the organic–aqueous interface. This difficulty is presumably related to the use of pure chloroform, since no such problems are encountered with di-*n*-butyl ether in the previous method.

Preparation of cell-size vesicles [Kim and Martin (27)]

(i)   Prepare a chloroform solution of amphipathic lipids containing approximately 3.33 μmol ml^{-1} by dissolving 10 mg of egg PC, 4 mg of cholesterol, and 5 mg of PG in 10 ml of chloroform.

(ii)   Add 2 mg of triolein (or equivalent amount of other non-polar lipid) to give a molar ratio of components PC:Chol:PG:TO to 4:4:2:1.

(iii)   Prepare another organic solution by dissolving lipids in the same quantities as for Step (ii) in 5 ml of diethyl ether.

(iv)   Take 1 ml of the chloroform solution and introduce into a 1 dram glass vial (1.4 × 4.5 cm) with a screw-cap lined with aluminium foil.

(v)   Dissolve the material to be entrapped in 300 mM aqueous sucrose solution and add 1 ml to the lipid solution in the 1 dram vial. Add the aqueous solution in three equal portions, with manual swirling in between each addition, to prevent formation of a chloroform-in-water emulsion.

(vi)   Close tightly, fix the vial horizontally to the rubber cup of a vortex mixer, and shake mechanically at top speed for 45 sec.

(vii)   Prepare 'mother liquor' by addition of 0.5 ml of ether solution to 2.5 ml of 300 mM sucrose solution (containing no other material) and shaking in a fresh 1 dram vial, as above, for 15 sec.

(viii) Transfer 1 ml of the chloroform-in-water dispersion to the vial containing 3 ml of ether-in-water 'mother liquor'.

(ix)   Shake immediately as before for 10 sec.

(x)   After shaking, introduce the double emulsion dispersion into a 250-ml conical flask (flat bottom diameter 8 cm).

(xi)   Place the flask in a gently shaking water bath set at 37°C, and allow a stream of nitrogen to pass through at a rate of $1-2$ l min^{-1}. After an hour, the organic solvent is virtually all removed and the turbidity of the dispersion is reduced.

(xii)   Mix the suspension with an equal volume of 5% glucose solution, and centrifuge the liposomes at 1000 *g* in a bench centrifuge for 20 min at room temperature. Discard the supernatant (containing unincorporated material and lipid debris) and resuspend the liposome pellet gently in physiological saline.

A wide range of lipids can be used, with good values for entrapment being obtained. It seems to be important to ensure that each of the four different types of lipid is represented in the mixture for best results (i.e. polar uncharged, polar charged, cholesterol, non-polar lipids). Non-polar lipids can include triolein, cholesterol oleate or α-tocopherol. Materials which are not properly dissolved in the solvents used do not seem to give good results.

(c) *Multivesicular liposomes (MVLs)*

This is a variation of the double emulsion method described above, but which gives rise to liposomes of a type which can be produced by no other method. The novelty lies in the fact that the proportions of lipid, solvent composition and duration of shaking are adjusted in such a way that instead of just one single water droplet being contained within the 'oil' droplet of the double emulsion, several water droplets are enclosed (*Figure 14a*). Upon removal of the solvent by evaporation, these water droplets remain intact, and form multiple compartments within a single liposome. As before, the authors of the technique (28) report that a crucial ingredient of the lipid mixture is triolein—a non-polar triglyceride—which may accumulate at the junctions where monolayers from three or more compartments interface, and prevent rupture of these compartments during preparation. It is postulated that the multi-compartment nature of the final liposomes may make them structurally more stable to physical shock than single compartment vesicles of the same size (MVLs are typically $1-2$ μm in diameter). Like ordinary double emulsion vesicles, these have an entrapment yield of 50% or greater. Although no applications have yet been reported for MVLs which could not be performed by normal LUVs, it is conceivable that one may want to deliver to cells or organs of the body a number of different agents simultaneously which normally are not very stable in each other's presence. Entrapping these materials in separate compartments of the same liposome could overcome this problem.

Preparation of MVLs

(i)   Prepare a solution of amphipathic lipids (10 μmol m^{-1}) by dissolving 10 mg of egg PC, 4 mg of cholesterol and 5 mg of egg PG in 3.3 ml of chloroform/diethyl ether (1:1 v/v). Add 2 mg of triolein to give a molar ratio of components PC:Chol:PG:TO 4:4:2:1.

(ii)   Introduce 1 ml of aqueous solution, containing material to be entrapped, into a clean glass 1 dram screw-capped vial.

(iii) Introduce 1 ml of lipid solution into the same vial, flush with nitrogen and cap tightly.

(iv)   Shake the vial vigorously by hand, then tape it in a horizontal position to the cup of a vortex mixer and shake it at the highest setting for 9 min.

(v)   Transfer 1 ml of the water-in-oil emulsion prepared above into a clean 1 dram vial using a Pasteur pipette with a narrowed tip ($0.2-0.4$ mm) and add 2.5 ml of 300 mM sucrose solution.

(vi)   Flush the vial with nitrogen, close tightly, and shake mechanically as above for 10 sec.

(vii) Transfer the water-in-oil-in-water double emulsion to a 250-ml conical flask (8 cm diam bottom), and place the flask in a gently shaking water bath set at 37°C.

(viii) Allow a stream of nitrogen to pass through the flask ($1-2$ l min^{-1}) for up to 1 h until all the solvent has been removed, and the turbidity of the suspension has been reduced.

(ix)   Add an equal volume of 5% glucose solution to the resultant liposome suspension, and spin in a bench centrifuge at 1000 g for 5 min.

(x)   Resuspend the liposome pellet in physiological saline.

The size of the liposomes ranges from 5 to 30 μm in diameter, depending on the duration of shaking employed to produce the second emulsion. Shaking for 10 sec is found to be optimal from the point of view of percentage entrapment, prolonged shaking presumably resulting in rupture and leakage of the internal aqueous compartments to the outside. Under these conditions the greatest reduction in entrapment yield is accounted for by losses during the drying process. A little less than half of the total starting lipid ends up as lipid aggregates in the supernatant after centrifugation; a slightly larger proportion of cholesterol is thus lost than phospholipid, presumably due to differences in partitioning of the two compounds between the bulk organic phase and the monolayer at the interface.

The double emulsion techniques described here have been adapted by Gao and Huang (29) to construct liposomes in which the internal compartment of vesicles is occupied by a 'solid core' composed of agarose and gelatin. The core consists of a matrix of these two components which may be impregnated with colloidal gold, or used as a reservoir for proteins and other large molecules. To make the liposomes, pre-formed solid cores are included in the aqueous phase to be encapsulated during the emulsification procedure. The liposomes employed for this application are typically $1-5$ μm in diameter.

(d) *Reverse-phase evaporation vesicles*

Liposomes made by this method, developed by Szoka and Papahadjopoulos (30), and the first to use 'water-in-oil' emulsions, were so called because the process involved an emulsion which was the reverse of the standard 'oil-in-water' emulsion, and because the key, novel step in the preparation was the removal of solvent from the emulsion by evaporation. Thus, after the droplets have been formed by bath sonication of a mixture of the two phases, the emulsion is dried down to a semi-solid gel in a rotary evaporator, under reduced pressure (as shown schematically in *Figure 15*). At this stage, the monolayers of phospholipid surrounding each water compartment are closely opposed to each other, and in some cases probably already form part of a bilayer membrane separating adjacent compartments (*Figure 16b*). The next step is to subject the gel to vigorous mechanical shaking with a vortex mixer, in order to bring about the collapse of a certain proportion of the water droplets. In these circumstances, the lipid monolayer which enclosed the collapsed vesicle is contributed to adjacent, intact vesicles, to form the outer leaflet of the bilayer of a large unilamellar liposome. The aqueous content of the collapsed droplet provides the medium required for suspension of these newly formed liposomes. After conversion of the gel to a homogeneous free-flowing fluid, the suspension is dialysed in order to remove the last traces of solvent.

Preparation of REVs

(i)   Dissolve 20 mg of egg PC in 10 ml of diethyl ether in a 3-cm diameter glass boiling tube with a B24/29 ground-glass Quickfit neck.

(ii) Introduce 1 ml of phosphate-buffered saline (PBS), containing the material to be entrapped, into the lipid solution by injecting rapidly through a 23 gauge hypodermic needle from a 5 ml syringe.

(iii) Close the system with a glass stopper, and quickly place the vessel in a bath sonicator, filled with a weak detergent solution up to the same level as the mixture in the tube.

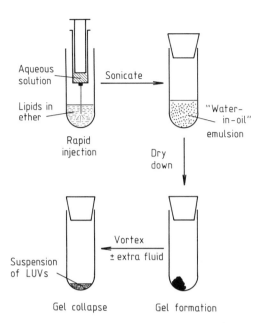

Figure 15. Stages in the preparation of liposomes by reverse phase evaporation. After formation of a water-in-oil emulsion by sonication of the aqueous solute in an organic solution of lipids, the organic solvent is evaporated off to yield a gel. The gel then collapses either naturally, as drying is continued, or as a result of mechanical shaking, to give a free-flowing aqueous suspension of liposomes.

(iv)   Sonicate for approximately 2 min.

(v)   Transfer the sonication vessel directly to a rotary evaporator and dry the contents down gently, at 37°C and low vacuum (approx 450 mm Hg) until a gel is formed.

(vi)   Release the vacuum, remove the tube from the evaporator, replace the stopper, and subject it to vigorous mechanical agitation on a vortex mixer. The gel should collapse and be transformed into a fluid, though rather viscous, suspension of liposomes. If collapse does not occur first time, continue the drying process and then try again. Several attempts may be necessary.

(vii) After the suspension has formed, continue drying for another 5−10 min on the evaporator, then remove the last traces of ether by dialysis against saline overnight.

The vesicles formed are unilamellar, and have a diameter of the order of 0.5 μm. Because it is essential for at least 50% of the vesicles to collapse, in order to provide the outer monolayer for the surviving 50%, values for percentage encapsulation greater than 50% should not be expected. Using the lipid-to-aqueous volume ratio described, the water-in-oil emulsion, in contrast to those described earlier for the double emulsion, has virtually no lipid in free solution after the dispersion has been formed. Almost all the lipid is associated with the monolayer surrounding the water droplets. If, however, the concentration of lipid employed is increased, or the volume of water decreased, then excess lipid will be present which will be dried down with the water droplets, and will coat them with multiple layers of phospholipid membrane. Under these circumstances, it is not necessary for a large number of vesicles to collapse in order

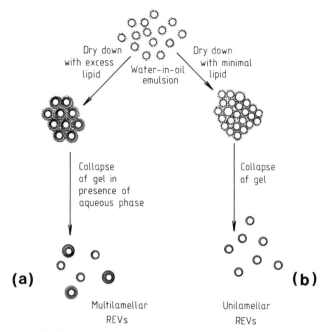

Figure 16. Formation of different types of liposomes using the reverse phase evaporation technique **(a)** MLV-REVs. In the presence of excess phospholipid, each water-in-oil droplet is coated with at least one bilayer of membrane, in addition to the outer monolayer, and in theory, these outer layers can redistribute themselves among vesicles, when the organic solvent is removed, without the need for any of these vesicles to be destroyed. **(b)** LUV-REVs. In the absence of extra lipid, the outer monolayer of the liposome membrane formed is provided by rupture of a proportion of the water droplets, making their envelope of phospholipid available to coat the surviving intact droplets, and releasing their aqueous contents to form the suspension medium for the liposomes which result.

to convert the gel to a liposome suspension, since the outermost lipid monolayer can be transferred from one vesicle to another without destroying the integrity of the donor vesicle (*Figure 16a*). Indeed, in theory, it should be possible to add exogenous water at the gel stage and achieve gel conversion without damaging any of the vesicles. These liposomes, designated MLV-REVs, have a large aqueous core surrounded by ten or more bilayers. The method has been discussed at length by Pidgeon *et al.* (31) and is performed in an identical manner to that for unilamellar REVs, except that 100 mg of lipid and 300 μl of PBS are used. A wide range of lipids and lipid combinations can be employed for both methods. In cases where the lipids are poorly soluble in ether, a 1:1 ether/chloroform solution may be used instead.

(e) 'Stable Plurilamellar Vesicles' (SPLVs)

In this method a water-in-oil dispersion is prepared as described previously, with the lipid in excess, but in which the drying down process, carried out using a stream of nitrogen, is accompanied by continued bath sonication (32), during which time redistribution and equilibration of aqueous solvent and solute occurs in between the various bilayers in each 'plurilamellar vesicle' (see *Figure 17*). The internal structure of the vesicles formed thus differs from that of MLV-REVs in that they lack a large

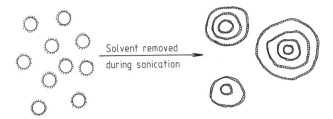

Figure 17. Stable plurilamellar vesicles (SPLVs). Simultaneous sonication and drying down of water-in-oil droplets results in the formation of MLVs with entrapped solute evenly distributed throughout the aqueous compartments.

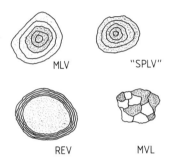

Figure 18. Different types of compartment that can occur in MLVs. Unprocessed MLVs produced by hand-shaking often have non-uniform distributions of the solute entrapped in between concentric bilayers, in contrast to SPLVs (see *Figure 17*), and also to MLV-REVs, in which a single aqueous compartment is enclosed by a thick layer of multiple lamellae closely apposed to each other. In multi-vesicular liposomes (MVLs) the compartments are contiguous but not concentric.

aqueous core, the majority of the entrapped aqueous medium being located in compartments in between adjacent lamellae. Percentage entrapments are normally around 30% (compared with ≥60% for MLV-REVs).

The use of this preparation method has brought to light some anomalies in the properties of unprocessed MLVs prepared by the conventional 'hand-shaking' method. Thus, while SPLVs have solute evenly distributed throughout the different compartments, MLVs prepared in saline from egg PC alone appear to have different concentrations of salt (and other solutes) in adjacent compartments, the outer compartments being solute-depleted (*Figure 18*). This situation is thought to arise in MLVs as a result of sealing of membranes to form vesicles during mechanical dispersion before complete hydration of the phospholipid headgroups has taken place, so that after vesiculation, water continues to be drawn into the liposome across the bilayer, while solute is excluded. The outer compartments thus have a reduced solute concentration relative to both inside and outside the liposome. One consequence of this is that osmotic differences arise between adjacent compartments giving rise to membranes in different parts of the vesicle which are 'stressed'—either under compression or expansion. The spacing between adjacent lamellae (from X-ray diffraction data) is unusually small; penetration of the membranes by molecules such as ascorbate is reduced in MLVs, yet leakage of entrapped solutes is reported to be increased. These differences are manifested relative to SPLVs whose gross physical structure and chemical composition are wholly identical except for the intravesicular distribution of solute (either neutral or ionic). Whether these differences are still seen between SPLVs and MLVs composed of charged lipids, or with MLVs prepared by bath sonication or subjected to subsequent processing, is not clear.

3.3 **Detergent solubilization**

In this third class of methods for the manufacture of liposomes, the phospholipids are brought into intimate contact with the aqueous phase via the intermediary of detergents, which associate with phospholipid molecules and serve to screen the hydrophobic portions of the molecule from water. The structures which form as a result of this association are known as micelles, and can be composed of several hundred component molecules, their shape and size depending on the chemical nature of the detergent, the concentration, other lipids involved, etc. Detergents can form micelles in the absence of any other lipids providing the concentration is high enough. That concentration of detergent in water at which micelles just start to form is known as the critical micelle concentration (CMC). Below the CMC, the detergent molecules exist entirely in free solution. As detergent is dissolved in water in concentrations higher than the CMC, micelles form in larger and larger amounts, while the concentration of detergent in the free form remains essentially the same as the CMC (see *Figure 19*). Micelles containing components in addition to the detergent (or composed of two or more detergents) are known as 'mixed micelles'.

In contrast to phospholipids, detergents are highly soluble in both aqueous and organic media, and there is an equilibrium between the detergent molecules in the water phase, and in the lipid environment of the micelle. The critical micelle concentration can give an indication of the position of this equlibrium, and from that conclusions can be drawn as to the ease with which molecules of detergent can be removed from the mixed micelle, upon lowering the concentration of detergent in the bulk aqueous phase by dialysis. A high CMC indicates that the equilibrium is strongly shifted towards the bulk solution, so that removal from the mixed membrane by dialysis is relatively easy. *Table 1* gives a list of detergents commonly employed for solubilization of phospholipid membranes, together with their CMCs.

As a general rule, membrane-solubilizing detergents have a higher affinity for phospholipid membranes than for the pure detergent micelles. Thus, as detergent is added in increasing amounts to the membrane preparation, more and more detergent will be incorporated into the bilayer, until a point is reached where a transition from the lamellar to the (usually) spherical micellar phase configuration takes place. As the detergent concentration is increased further, the micelles are reduced in size, until they become saturated with detergent, whereupon the concentration of free molecules equal to the CMC is attained, and simple detergent micelles are seen to form. Again, it is generally found that a high CMC is advantageous for solubilizing membrane phospholipids, although one might have expected the converse, since a high affinity for lipid membranes should be reflected by a low CMC. It may be that kinetic factors are important, and that detergents with a higher CMC can incorporate more rapidly into the membrane because of the higher concentration of the free entity. In practice, the

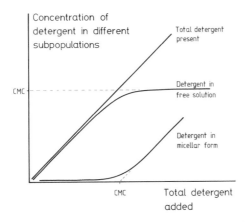

Figure 19. Concentration of detergent (free and micellar) in relation to the critical micelle concentration (CMC). Below the critical micelle concentration (CMC), detergent which is dissolved in aqueous solution exists as individual free molecules. At the CMC, ordered aggregates (micelles) begin to form, and addition of further detergent to the system results in an increase in concentration of these micelles, with no appreciable increase in the concentration of free unassociated detergent molecules.

Table 1. Detergents used in membrane solubilization.

Name	Critical micelle concentration		Molecular wt
	(mM)	*(mg/ml)*	
n-Heptyl glucopyranoside	70	19.5	278
n-Octyl glucopyranoside	23.2	6.8	292
n-Nonyl glucopyranoside	6.5	2.0	306
n-Decyl maltoside	2.19	1.1	499
n-Dodecyl maltotrioside	0.2	0.16	825
Triton X-100			
[PEG(9-10) *p-t*-octylphenol]	0.24	0.15	625
Nonidet P-40			
[PEG(9) *p-t*-octylphenol]	0.029	0.02	603
Tween 20			
[PEG(20) sorbitol monolaurate]	0.033	0.04	1364
Brij 98			
[PEG(29) oleyl alcohol]	0.025	0.04	1527
Sodium deoxycholate	2−6	1.7	415
Sodium taurocholate	10−15	6.7	538
Sodium cholate	14	6.0	431
Sodium dodecyl sulphate	8.3	2.4	289

time taken to achieve a clear mixed micelle suspension depends critically on the way in which the detergent and lipids are presented to each other.

In all the methods which employ detergents in the preparation of liposomes, the basic feature is to remove the detergent from pre-formed mixed micelles containing phospholipids, whereupon unilamellar vesicles form spontaneously. Because removal of detergents is carried out using techniques which inevitably remove other small water-

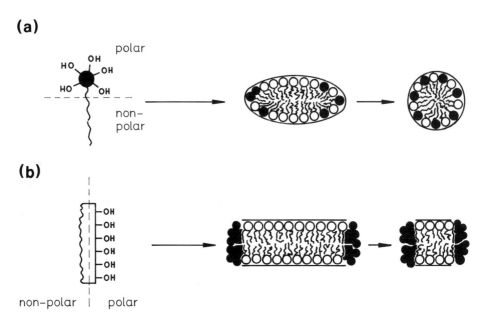

Figure 20. Different structures of mixed micelles. This drawing shows the way in which differences in topological distribution of polar and non-polar moieties in amphiphilic detergents can influence the structure of the mixed micelles they form with phospolipids. (**a**) Longitudinal separation of components within the molecule, as in the alkyl glycosides. (**b**) Latitudinal separation, as in bile salts.

soluble molecules (e.g. dialysis, column chromatography), the detergent methods are not very efficient in terms of percentage entrapment values attainable; on the other hand, they are certainly the best general methods for preparing liposomes with lipophilic proteins inserted into the membranes, since these proteins can be introduced into the mixed micelles in the presence of mild non-denaturing detergents, to achieve 100% incorporation without modification of the general method. Another special feature is the ability to vary the size of the liposomes by precise control of the conditions of detergent removal and to obtain vesicles of very high size homogeneity.

Detergents can differ in the way in which the polar and lipophilic moieties are distributed within the molcule, and these differences lead to differences in the structure of the resultant micelles formed. Thus one may draw distinctions between detergents according to whether the boundary between polar and apolar regions is latitudinal, as in the case of alkyl glycosides (see *Figure 20*a) or longitudinal as for bile salts (*Figure 20b*). The mixed micelles formed are spheroid and discoid respectively. Detergents fall into three classes:

(i)   Ionic — anionic
              — cationic.
(ii) Amphoteric.
(iii) Non-ionic.

Ionic detergents, such as sulphonated hydrocarbons like sodium dodecyl sulphate (SDS), although possessing ideal characteristics as far as solubility and CMC are concerned, are usually considered unsuitable for work with biological membranes since they are highly denaturing for proteins. The long flexible aliphatic chains can fit easily into, and disrupt, the hydrophobic clefts of proteins, and the strong electrostatic charges interfere with the hydrogen bonding interactions that maintain the tertiary structure of the protein. One exception to the rule is the class of ionic detergents including bile salts, for example, sodium cholate or deoxycholate, where the charged carboxyl group is not so bulky, and the rigid steroid nucleus has less access to hydrophobic areas in the protein molecule.

The majority of detergents employed for membrane work are non-ionic, of which two—Triton-X and alkyl glycosides—are very commonly used. Triton-X is a mixture of homologues in which the lipophilic group is an iso-octyl phenol moiety, while the hydrophilic portion is a single long chain consisting of between $8-12$ oxyethylene subunits ($-CH_2-CH_2-O$). Brij, Lubrol, Emulphogen, Nonidet, Tween, and others are all non-ionic detergents comprising a polyoxyethylene hydrophilic chain. In contrast, the alkyl glycosides have hydrophobic chains (short chain alkyl groups), and a sugar residue as the hydrophilic headgroup.

Ionic and non-ionic detergents differ in the way in which the critical micelle concentration, and micelle size, are affected by external conditions. Thus, bile salts are very sensitive to pH, precipitating at pH below neutrality, in contrast to Triton or octyl glucoside. Increasing the ionic strength reduces repulsive interaction between charged groups, and leads to formation of micelles at much lower concentrations. In addition, the size of micelles is increased. For non-ionic detergents, increasing the temperature results in a reduction of hydration of polar moieties (e.g. the oxygen atoms in the oxyethylene subunits of Triton-X) leading to an increase in the size of the micelle.

Many different detergents and variations in technique have been used for liposome manufacture, of which only a handful will be presented here as illustrations. The most common detergents employed are bilt salts, alkyl glycosides, and Triton-X, and these will be dealt with in turn. Detergents may be removed either by dialysis, column chromatography, or by dilution, and the relative merits of these methods will be discussed in the following sections.

3.3.1 *Bile salt preparations*

Under conditions of constant pH, ionic strength and temperature, factors affecting the size and homogeneity of liposomes formed upon careful removal of detergent from mixed micelles are:

(i)   rate of removal of detergent;
(ii) initial ratio of detergent to phospholipid;
(iii) cholesterol concentration;
(iv) presence of charged lipids.

In order for the liposomes formed to be as homogeneous in size composition and lamellarity as possible, it is important for the detergent to be removed at an even rate throughout the whole volume of the mixed micelle suspension. This is achieved during dialysis by making sure that there is constant mixing inside as well as outside the dialysis chamber. Various mechanical systems have been devised to bring this about. In the 'Lipoprep' manufactured by Dianorm, a chamber holding 8 ml of fluid is mounted vertically between two membranes, and a magnetic follower is included in the chamber

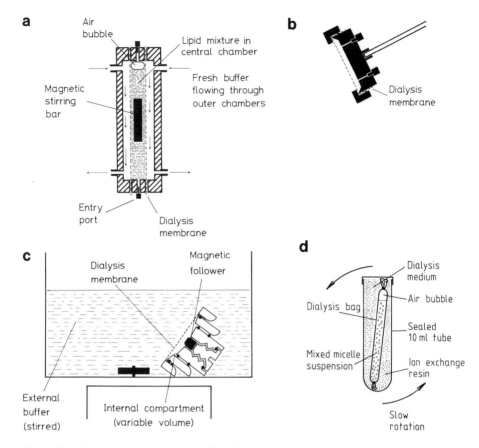

Figure 21. Different methods of dialysis. This figure shows four ways in which dialysis at controlled temperature can be achieved efficiently, with good mixing of aqueous solutions both outside and inside the dialysis compartment. (**a**) Triple cell dialyser. (**b**) Mini lipoprep. (**c**) Liverpool cell. (**d**) Bag in bottle.

which rotates about a horizontal axis collinear with the centre of the circular chamber (*Figure 21a*). Dialysis buffer is circulated around the outside of the membranes through spiral tracks moulded in solid Teflon. A simpler device for smaller volumes, the 'Mini Lipoprep', consists of a chamber formed by sliding a hollowed-out piston into a cup, the lower surface of which is cut away to reveal a membrane stretched over the bottom of the piston, with space for a small volume of fluid in between (*Figure 21b*). The chamber is thus a flat disc approximately 2 mm thick and 4 cm in diameter, containing a volume of 1 ml. The chamber is filled through inlet and outlet ports to leave a small bubble (~ 100 μl in volume), then sealed tightly, immersed in buffer at an incline, and rotated slowly about its axis. The movement of the bubble through the fluid during turning ensures complete mixing inside the chamber, at the same time as convection across the outer membrane surface ensures rapid dispersal of the detergent into the bulk fluid. In the author's laboratory, a cell has been designed in which a variable internal cell volume is mixed continually by a magnetic follower enclosed in the cell, the whole cell being immersed in dialysis medium (*Figure 21c*). Using any of these systems, rapid

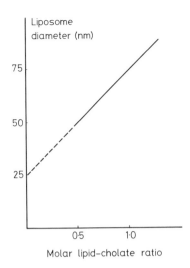

Figure 22. Graph showing the relationship between detergent:lipid ratio and liposome size for cholate-dialysed liposomes. Data derived from ref. 38.

and controlled dialysis can be achieved. The use of hollow fibre systems (e.g. Biofibres) has also been reported.

A modification of the dialysis technique has been described (33) which permits a smaller volume of dialysing buffer to be used, and makes use of the ability of materials such as XAD-2 resin, or Biobeads, to absorb out large quantities of detergent from solution (*Figure 21d*). In this technique, the mixed micelle solution ($\sim 1-2$ ml + air bubble) is contained in an ordinary dialysis tube, sealed at both ends, which is then put into a 15-ml screw-cap glass tube containing 10 ml of dialysis buffer and beads. The tube is capped tightly, and left to mix for 2 h on a rotary mixer which inverts the tube (including the air bubble) completely approximately five times per minute. The method is useful in cases where low molecular weight materials are to be entrapped, which would otherwise be lost in the dialysis medium, or which are too expensive to use in high concentration in a large volume. An extension of this method is to mix the beads directly with the micelle suspension. It has been demonstrated in certain cases that up to 40% of the lipid can be lost in this way, but the method has been used very successfully with other detergents. An illustration is given further on in Section 3.3.3.

The simplest method of reducing the detergent concentration rapidly is by direct addition of a large excess (e.g. 10- or 100-fold) of fresh medium. Clearly the disadvantage of this approach is that the final concentration of liposomes is very low, and will result in a very poor entrapment yield. Techniques such as ultracentrifugation might have to be used to concentrate the vesicles, and there would still be a significant quantity of residual detergent which would need to be removed by dialysis anyway.

Detergent may also be removed by column chromatography, using a Sephadex G-200 or G-50 column to give liposomes of a defined size (34,35). However, characterization of the other parameters controlling liposome dimensions has not been so extensive as for the dialysis method, and it is not clear whether the conditions of detergent removal are sufficiently well controlled to give reproducibility and homogeneity comparable

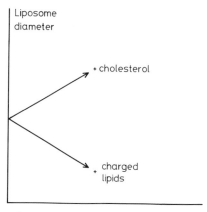

Figure 23. Effect of cholesterol and charge on liposome size.

with dialysis. In the following discussion, therefore, reference will be made to dialysis (36,37) as the method of choice.

In the cholate/phospholipid system the ratio of the two components determines very critically the size of the mixed micelles, which in turn would seem to determine the size of the liposomes finally formed after dialysis (38). For example, using egg yolk lecithin the size of the micelles increases from a diameter of 2 nm (no phospholipid) to a maximum of 20 nm at a molar ratio approaching 1.33 (lipid/cholate 4:3) when the transition between extended bilayer sheet and micelle takes place. Ratios of lipid to cholate higher than this value do not result in membrane solubilization. In parallel with these figures, mixed micelle preparations, ranging from 1.15 to 0.6 lipid/detergent (L/D) molar ratio give rise to liposomes of diameter 81 and 55 nm respectively (*Figure 22*). Interestingly, extrapolating this trend back to where L/D is zero (i.e. an infinite excess of detergent) gives a population of liposomes close to the minimum size limit of phospholipid vesicles. Thus, in theory one may be able to obtain liposomes of any diameter between 250–1000 Å (25–100 nm) by varying the ratio of egg PC to cholate from zero to 1.3 (the minimum and maximum values possible). The presence of excipients in the membrane (i.e. lipids or other materials not essential for formation of membrane structures) also influences the size of the resultant vesicles. The inclusion of cholesterol tends to steepen the curve [i.e. increase liposome size at higher lipid/detergent ratios (39)] while charged lipids have the opposite effect (*Figure 23*). Synthetic saturated lecithins cannot be solubilized easily by sodium cholate except in the presence of at least 10% charged lipid, in which case the vesicles formed are quite large (~1000 Å; 100 nm) even at low lipid/detergent ratio (i.e. large excess of detergent). Thus the factors determining the size of liposomes using this method are quite complex, and when using new combinations, a process of trial and error with different lipid/detergent ratios is advisable in order to obtain vesicles of the desired size. For best results it is desirable to dry down the detergent together with the lipids, and then dissolve the solids in saline at a temperature approximately 10°C above the phase transition of the highest melting phospholipid. The dialysis should also be carried out at this temperature.

Using a dialysis procedure such as described earlier (e.g. the Mini Lipoprep), the liposomes start to form in equilibrium with mixed micelles after about 45 min, when 30% of the detergent has been removed. Liposome formation is completed (giving 100% sealed vesicles and no mixed micelles) after 90 min, with 30% cholate still remaining in the reaction compartment. Presumably throughout this period of time the concentration of detergent in free solution remains approximately constant, in order that mixed micelles of constant size can coalesce to form a homogeneous population of vesicles.

After 90 min, however, the liposomes are still much larger in size than their final dimensions, as evidenced by their sedimentation coefficients (approx. 17 compared with 6.5 for egg lecithin vesicles). Dialysis for a further 24 h can reduce the concentration of detergent to less than 0.5% of its original value. A range of bile salts can be used for dialysis. Because of its high CMC, sodium cholate is favoured for its rapid rate of removal, while use of sodium deoxycholate may run less risk of denaturing proteins.

Preparation of cholate dialysis vesicles

(i)   In a 50-ml round-bottomed Quickfit flask introduce 10 mg of solid egg yolk phosphatidyl choline (dried down if necessary on a rotary evaporator) and 10 mg of sodium cholate, to give a phospholipid/detergent ratio of approximately 0.6.

(ii)   Dissolve the solids in 10 ml of ethanol/chloroform (1:1 v/v), and dry down on a rotary evaporator.

(iii) Redissolve the solids in 10 ml of absolute ethanol, or isopropanol, and dry down again. If other lipids are added, the temperature of drying should be above the highest transition temperature of the individual lipid components. At this stage, the solids should dry onto the wall of the vessel as a completely clear glassy film. To achieve this, take care not to overheat, or to dry down too fast.

(iv)   Gently introduce 2 ml of 10 mM phosphate buffer, pH 7.1, down the side of the flask, and dissolve the film carefully in the buffer by gently swirling around the wall of the vessel. (This must be carried out above the lipid transition temperature.) The solution obtained should be absolutely transparent from the start. If some cloudiness or opalescence is observed, resulting from the presence of contaminants, this may disappear after leaving to stand for an hour. For the preparation of entirely homogeneous unilamellar vesicles it is essential that the mixed micelle suspension be completely clear before starting the dialysis.

(v)   Prepare the dialysis cell, introduce the mixed micelle suspension and dialyse at an appropriate temperature for 24 h.

The method described here in which the detergent is dried down with the lipid is the one most easy to adopt for preparation of liposomes of different compositions. Cholesterol can be incorporated up to a mole ratio of 1:1 provided that the cholesterol is of the highest purity (preferably recrystallized twice from ethanol) and that the last traces of chloroform are removed by drying down several times from ethanol. (Alcohols are used in order to achieve co-dissolution of ionic bile salts and amphipathic lipids.) Materials to be entrapped inside the vesicles may be included in solution in the phosphate buffer. In the case of small molecules, to achieve the highest possible entrapment, the compound should be added to the dialysis medium in the outer compartment, for the first 90 min of dialysis until the liposomes are formed, whereupon ordinary medium can be used for the remaining 22 h.

Because an important application of the detergent preparation of liposomes is to incorporate proteins directly into membranes, a second example is given here based on the method described by Weder and co-workers (40) for incorporation of derivatized antibodies. The derivatization procedure, in which fatty acid chains are attached via amide linkages to various sites on the protein, serves to increase the hydrophobicity of the antibody molecule, so that it associates naturally with mixed micelles, and hence forms part of the liposome membrane itself, after dialysis. The starting reagent is an activated carboxylic acid, the *N*-hydroxysuccinimide ester of palmitic acid, which is synthesized according to the method of Lapidot *et al.* (41) and is described in Appendix I.

Conjugation of protein to fatty acids and incorporation into liposomes

(i)   Weigh out 20 mg of sodium deoxycholate and dissolve in 0.5 ml of 10 mM phosphate buffered saline to give a 4% solution. Readjust the pH to pH 7.1 if necessary.

(ii)   Measure out a volume of organic solution containing 20 μg of palmitic acid ester into a glass vial, and dry down under a stream of nitrogen.

(iii)   Transfer the 0.5 ml of bile salt solution to the vial, and shake gently to dissolve the ester.

(iv)   Add 0.5 ml of purified antibody containing 2.5 mg of protein (5 mg ml^{-1} in PBS).

(v)   Screw the cap on tightly, and incubate overnight at 37°C in a rotating mixer.

(vi)   Dissolve 18 mg of egg PC, 2 mg of egg PA and 5 mg of cholesterol in chloroform/methanol (2:1 v/v) and dry down in a 50 ml round-bottomed flask on a rotary evaporator.

(vii)   After drying, flush the flask with nitrogen, add 2 ml of phosphate buffered saline and swirl gently until all the lipid on the sides of the vessel wall has been suspended.

(viii) Add the 1 ml of bile salt solution of derivatized protein and shake gently until the solution has become clear.

(ix)   Dialyse at room temperature for 22 h as described previously.

Under these conditions, the antibody incorporates into the liposomes in virtually 100% yield, with a high retention of specificity.

3.3.2 *Alkyl glycoside dialysis*

In contrast to bile salts, the size of liposomes formed by dialysis of alkyl glycoside mixed micelles depends not on the lipid/detergent ratio, but rather on the chain length of the individual detergent molecules (42). This difference in behaviour may perhaps be explained by the different shapes and structures of the types of micelles formed by these two detergent classes, and by the fact that the micelle size, and detergent content varies much more widely with detergent concentration for bile salts than for other agents.

The most commonly used alkyl glycoside is *n*-octyl β-D-glucoside—an eight-carbon straight chain hydrocarbon attached to the five position of a glucose ring. This molecule has a high critical micelle concentration ($\sim$23 mM), permitting its rapid removal by dialysis. For effective solubilization of membranes, a detergent/lipid molar ratio of 5:1 is sufficient provided that removal of detergent is performed by carefully controlled dialysis. In circumstances where the conditions are not well controlled, it may be necessary to increase the ratio to 10:1, to avoid the formation of lipid aggregates and multilamellar vesicles. Preparation of the mixed micelle suspension is carried out either by co-deposition of lipid and detergent as a dry film prior to addition of buffer, or by addition of the detergent in solution to the dried lipid. Both synthetic and natural, neutral and charged lipids, as well as cholesterol, can be used. The concentration of total phospholipid may range from 10 to 50 mg/ml^{-1}. Liposomes formed at the higher lipid concentrations tend to be slightly larger in size. Using octyl glucoside and egg PC, unilamellar vesicles in the size range 180−240 nm can be obtained. After an hour 50% of the detergent is removed and less than 1% remains in the total suspension after dialysis for 10 h. As with cholate dialysed liposomes, the homogeneity of the final population is high (greater than 90% lying within a size window 100 nm wide).

Under reproducible conditions of dialysis flow rate, stirring, temperature etcetera, the size of liposomes can be varied by the use of glucosides with different alkyl chain lengths, and of mixtures of these glucosides. Thus substitution of octyl glucoside by heptyl glucoside reduces the diameter from 180 nm to 80 nm (measured by laser light scattering), with an intermediate value of 140 nm being obtained with a 1:1 mixture of the two detergents. Recently, decyl maltoside (containing a disaccharide head group) has been proposed as a useful alternative to octyl glucoside (43), since with its lower CMC ($\sim$2 mM), smaller quantities are required for effective solubilization. Liposome size is of the same order as with octyl glucoside, but unilamellar vesicles are not formed unless at least 10% charged lipid is incorporated into the bilayer. The methyl glucamide series of detergents can also give good results when used in the same way.

3.3.3 *Triton X-100 solublized Sendai virus particles* (contributed by A.Loyter)

In this last section, a method for preparation of vesicles is described, in which the HN and F glycoproteins of Sendai virus are incorporated into the membrane by removal of detergent (Triton X-100) from solubilized virus envelopes (*Figure 24*). The liposomes thus formed bind strongly to biological membranes containing the receptor for the HN protein, and fuse with these membranes in the same way that the envelope of the intact virus does (*Figure 25a*) (44,45). These vesicles can be prepared in such a way that non-viral material can be incorporated into the membrane (46), or entrapped inside, so that binding to many cell types can be achieved (*Figure 25b*), with subsequent introduction of these materials either into the cell membrane, or into the cytoplasm (see *Figure 26*).

Large quantities of fusogenic Sendai virions can be obtained by propagation in the allantoic cavity of 10-day-old fertilized chicken eggs (summarized in Appendix I). Incubation of intact Sendai virions with either ionic or non-ionic detergents results in the solubilization of the virus envelopes. The viral nucleocapsid remains detergent-insoluble as it is not part of the viral membrane. Active fusogenic reconstituted Sendai virus envelopes (RSVEs) can be obtained, however, only after solubilization of the virus particles with non-ionic detergents such as NP-40 or Triton X-100. Structurally intact viral envelopes can also be obtained after solubilization of intact virions with other detergents such as octyl glucoside, cholate, deoxycholate, or lubrol. However, treatment of virions with the latter detergents causes an irreversible inactivation of the

Preparation of liposomes

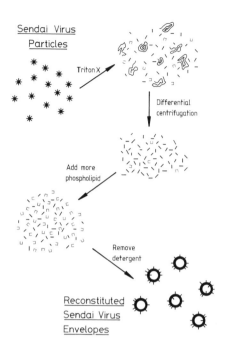

Figure 24. Preparation of RSVEs. Sendai virus particles are solubilized in Triton X-100, and the nuclear material is removed by centrifugation. After addition of extra phospholipid, the virus envelopes are reconstituted by removal of the detergent either by dialysis or by Biobead adsorption. The resultant vesicles have the virus HN and fusion proteins distributed throughout the membrane, oriented both inside and outside.

virus' fusogenic activity, and the RSVEs obtained after solubilization with these detergents, although able to bind and agglutinate cells, fail to fuse with cell membranes. The protocol for preparation of active fusogenic RSVEs is given below.

Preparation of fusogenic RSVEs

(i)   Prepare solution 'Na' containing 110 mM NaCl (0.644 g per 100 ml) and 50 mM Tris−HCl (0.6 g per 100 ml), pH 7.4.

(ii)   Prepare a solution of 10% Triton X-100 (w/v) in solution Na containing 0.1 mM phenyl methyl sulphonyl fluoride (PMSF, 1.7 mg per 100 ml) − dissolve 5 mg of PMSF in 600 μl of ethanol and add 200 μl to each 100 ml of Triton X-100 solution.

(iii)   Dissolve 5 mg of Sendai virus particles in 100 μl of 10% Triton-X solution. It is important to keep the ratio of Triton-X:viral protein (w/w) in the range 2:1, with a final concentration of Triton-X between 2−10%.

(iv)   Incubate the solution at 20°C for 30 min with shaking.

(v)   Centrifuge the solution for 1 h at 100 000 g. The pellet contains detergent-insoluble viral nucleocapsid and the viral M protein while the clear supernatant contains the viral F and NH glycoproteins (1−1.2 mg protein).

(vi)   Remove the detergent, either:

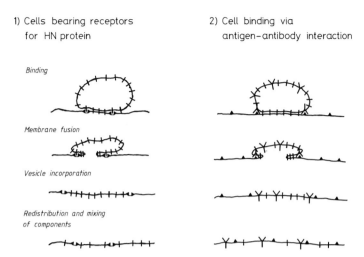

1) Cells bearing receptors
   for HN protein

2) Cell binding via
   antigen–antibody interaction

Binding

Membrane fusion

Vesicle incorporation

*Redistribution and mixing
of components*

Figure 25. Interaction of RSVEs with cell membrane. Cells bearing receptors for the HN glycoprotein can bind to RSVEs, which will then fuse with the cell membrane as a result of the action of the F protein. Target cells which do not have this receptor can be induced to bind to RSVEs by incorporation into the RSVE membrane of antibodies which have specificity for those particular cells. (**1**) Cells bearing receptors for HN protein. (**2**) Cell binding via antigen–antibody interaction.

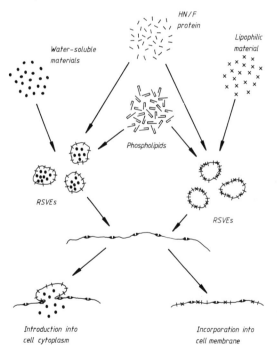

*HN/F
protein*

*Water-soluble
materials*

*Lipophilic
material*

Phospholipids

RSVEs

RSVEs

*Introduction into
cell cytoplasm*

*Incorporation into
cell membrane*

Figure 26. Incorporation of materials into cells using RSVEs. RSVEs can be used as vehicles for the delivery of material into cells, either into the cell cytoplasm, or into the cell membrane, as a result of fusion and mixing of RSVE membrane components with those of the cell.

(a)   by dialysis with Spectraphor-2 tubing. Dilute the supernatant with Triton X-100 solution to a volume of $0.5-1.0$ ml and then dialyse against 0.5 litres of buffer (pH 7.4) containing 10 mM Tris$-$HCl (12.1 g l^{-1}), 2 mM MgCl$_2$.6H$_2$O (0.4 g l^{-1}), 2 mM CaCl$_2$.2H$_2$O (0.3 g l^{-1}), 1 mM NaN$_3$ (65 mg l^{-1}) and 1.2 g l^{-1} of SM-2 Biobeads or 1 mg ml^{-1} of bovine serum albumin, for $48-72$ h. Centrifuge the turbid suspension at 100 000 g for 30 min, wash the pellet in solution Na, and resuspend in solution Na to give a protein concentration of 1 mg ml^{-1}. This suspension may be stored at $-70°$C.

Or

(b)   by removal of Triton X-100 by direct addition of SM-2 Biobeads. For each 10 mg of Triton X-100 in the detergent-solubilized virus preparation, add 70 mg of SM-2 Biobeads. After incubation for $2-3$ h at 20°C with vigorous shaking, add a second portion of 70 mg SM-2 Biobeads, and continue incubation for $12-14$ h at 20°C with vigorous shaking. After incubation, suck the soluble phase out with a 1 ml tuberculin syringe into which the SM-2 Biobeads cannot penetrate. Wash the SM-2 Biobeads twice with about 0.5 ml of solution Na. Centrifuge the turbid suspension for 30 min at 100 000 g, and resuspend the pellet in solution Na to give a protein concentration of 1 mg ml^{-1} and store at $-70°$C.

For each milligram of intact virus, about 300 μg of detergent-solubilized glycoproteins and $80-100$ μg of reconstituted envelopes are obtained. It will be noted that the detergent (Triton X-100) can be removed by two alternative methods: either by dialysis in Spectraphor-2 tubing for $48-72$ h or by direct addition of SM-2 Biobeads to the detergent-solubilized mixture of the viral phospholipids and glycoproteins. When the dialysis method is used, about $0.015-0.03\%$ Triton X-100 remains after dialysis for 60 h in the cold. When, alternatively, Triton X-100 is removed by the direct addition of SM-2 Biobeads, about 0.005% detergent remains in the RSVE suspension. Being hydrophobic, and possessing a very large surface area, the SM-2 Biobeads effectively adsorb macromolecules with hydrophobic characteristics such as detergents, phospholipids or polypeptides. However, as different macromolecules are adsorbed with different efficiencies, experimental conditions can usually be designed in which the Biobeads will mostly adsorb detergents and very little of the phospholipids and polypeptides present in the detergent solution, as in the case of the experimental system described above.

Electron microscopic observations, using the negative staining technique, reveal that the RSVE preparations obtained after removal of the detergent by the direct addition of Biobeads are somewhat more homogeneous and show vesicles with a larger diameter than those obtained after dialysis in Spectraphor tubing. However, both preparations are relatively homogeneous, and their size is very similar to that exhibited by a preparation of intact Sendai virions, that is $100-300$ nm. Only the two viral envelope glycoproteins, namely the HN and F polypeptides, are present in these RSVE preparations, as can be demonstrated by gel electrophoretic analysis.

For the efficient use of RSVEs as a biological carrier, it is important to establish an assay system which will allow the quantitative determination of the Sendai virus' fusogenic activity. This is especially important in experiments where loaded or 'hybrid'

RSVE are used or when the viral glycoproteins are subjected to various manipulations and modifications. The method commonly used for quantitative determination of the viral fusogenic ability has been to estimate its haemolytic activity. Indeed, it has been well established that the viral haemolytic activity reflects the viral fusogenic ability (47). However, the extent of virus-induced haemolysis depends also on the source and the age of the red blood cells, and, therefore, is not directly correlated to the viral fusogenic activity. Recently, it has been demonstrated that energy transfer and fluorescence dequenching methods can be used for a quantitative estimation of the RSVE or 'hybrid' vesicles' fusogenic activities (48,49). Fluorescent molecules such as octadecyl rhodamine B chloride (R18) or *N*-4-nitrobenzo-1,2-oxa-1,3-diazole phosphatidyl ethanolamine (N-NBD-PE) can be inserted at high surface densities into envelopes of Sendai virions or RSVEs, resulting in fluorescence reduction because of self quenching. Dilution of the fluorescent molecules due to fusion processes results in fluorescence dequenching. Increase in fluorescence correlates directly with the extent of virus−membrane fusion. Fluorescent-labelled RSVEs are prepared by adding the Triton X-100 supernatant from Step (v) (see method for preparation of fusogenic RSVEs) to a thin dry film of 85 μg NBD-PE (Avanti Biochemicals, Birmingham, Alabama), followed by shaking for 5 min, then continuing with Steps (vi) onwards as described previously.

In a population of actively fusogenic RSVEs, dilution of the quenching fluorophore can be observed by incubation with a large excess of red blood cell ghosts. Sheep, human, or chick red cells can all be used. Mixing of the lipids of the two membrane preparations after fusion leads to a redistribution of the fluorophore over a much larger surface area, resulting in dequenching, that is, a marked increase in fluorescence. This general technique will be discussed further in Chapter 5. Labelled RSVEs (approximately 1 μg) are incubated with ghosts ($\sim$200 μg of protein) in a volume of 150−200 μl for 30 min at 37°C. At the end of the incubation period, the degree of fluorescence is measured (excitation 460 nm, emission 543−546 nm) before and after addition of 0.1% Amonix-LO (Onyx Corp.). The degree of fluorescence in the presence of detergent is considered to represent an infinite dilution of the fluorophore, that is, 100% dequenching. Haemolysis is measured by incubating human (or other) erythrocytes (150 μl, 2−3% v/v) with 1 μg of RSVEs for 30 min at 37°C. The degree of cell lysis is estimated (before and after detergent addition) by measurement of the optical density of the supernatant at 540 nm.

Because it is impossible to remove the detergent completely from the RSVE preparations, their activity may be attributed—when a relatively high amount of detergent is left—to the residual amount of Triton X-100 remaining in the viral vesicle preparation, and not necessarily to their fusogenic activity.

In order to differentiate between lysis induced by vesicle-associated detergent and that induced by the viral glycoproteins, which in this case reflects its fusogenic activity, it is advisable to use various inhibitors that are known to inhibit specifically the viral fusogenic activity. This can be effected by pre-treatment (30 min, 37°C) of the virus particles with dithiothreitol (DTT), trypsin or PMSF (44,49,50). Treatment with trypsin (60 μg mg^{-1} viral protein) or 7 mM PMSF specifically inactivates the viral fusogenic activity without affecting its binding ability, while 3 mM DTT blocks both the viral binding and fusogenic activities. Under the conditions described above, a good preparation of fusogenic RSVEs can give about 60% fluorescence dequenching, or 90%

erythrocyte lysis. Both of these parameters are brought down to a few percent in the presence of inhibitors.

A determination of the lipid/protein (w/w) ratio reveals that a certain amount of the viral lipid is lost during the reconstitution process, especially when detergent is removed by dialysis in Spectraphor-2 tubing. However, this does not significantly affect the fusogenic properties of the RSVE, or their ability to fuse efficiently with cell membranes. Addition of lipid molecules such as phosphatidyl choline, with or without cholesterol, to the viral envelope reconstitution system affects mainly the size of the RSVEs obtained, which appear—through electron microscopic observation − to be much larger. Although experiments quantitatively analysing the effect of the addition of external lipids on the RSVEs' fusogenic activity are lacking, it seems that viral vesicles displaying a lipid/protein ratio higher than 70:1 mole/mole possess a very low fusogenic activity, although they are able to attach efficiently and specifically to cell plasma membranes.

DNA molecules can be enclosed in RSVEs (51,52) using the method described above, in which 100 μg of DNA is added to 100 μl of supernatant from the detergent-solubilized virus particles after spinning. The detergent is removed by direct addition of SM-2 Biobeads. Externally adsorbed DNA is digested by pancreatic DNase I (200 μg per 200 μg of viral glycoproteins, in a volume of 1 ml of solution Na with 5 mM $MgCl_2$ for 25 min at RT) prior to spinning at 100 000 g for 30 min at 4°C. Approximately 6−8% of the initial DNA is trapped inside the RSVEs. Essentially the same method can be employed for entrapment of RNA or polypeptides.

Preparation of 'hybrid' RSVEs can be performed, in which membrane proteins from other systems can be incorporated along with the viral proteins in the reconstituted vesicle. For example, incorporation of human erythrocyte Band 3 (53) (a transmembrane protein which can be isolated in pure form and functionally intact with non-ionic detergents) is carried out by mixing 420 μg of Band 3 protein (in 0.3 ml of 1.7% Triton X) with 1400 μg of virus envelope glycoproteins (0.7 ml of 2.0% Triton X-100 in solution Na containing 0.1 mM PMSF). Detergent removal can be carried out by any of the methods already outlined. It would seem that the maximum ratio (w/w) of non-viral/viral protein is about 1:2. Increasing the proportion of non-viral protein above this ratio can result in a significant decrease in fusogenic activity of the hybrid vesicles.

4. ACTIVE LOADING

In the methods outlined in the previous sections, it has been assumed up until now that water-soluble materials will be entrapped by introducing aqueous solutions of these materials before or at some stage during the manufacture of the liposomes. However, certain types of compound with ionizable groups, and those which display both lipid and water solubility, can also be introduced into liposomes after formation of the intact membranes (54,55). These compounds are often difficult to retain inside liposomes by normal means, since their lipophilicity leads to their being able to pass into and out of membranes readily, and thus equilibrating between the liposome interior and exterior. In the 'active loading' method, however, this behaviour is made use of by arranging that the conditions in the vesicle interior are such that the compound in question —usually a lipophilic amine—is always in the ionized form. Thus the solute, after having entered the liposome by diffusion through the membrane as the uncharged form, is converted to the fully charged species inside the liposome, and as such is unable to

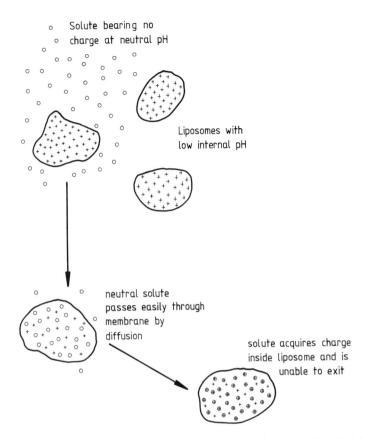

Figure 27. Principle of active loading. Active loading of liposomes is based on the principle that a low molecular weight solute, which under normal circumstances can diffuse easily across the membrane and into the liposome, can be made to undergo a change once inside the liposome so that it is unable to diffuse out again. Here, the principle is illustrated for species with a variable charge.

escape from the liposome, because its lipophilicity is very much reduced. Solute therefore accumulates inside the liposome as long as a pH difference is maintained between the inside and outside of the membrane (*Figure 27*). Such a difference can be brought about by entrapping a non-permeating buffer ion such as glutamate inside the liposomes, at a low pH, and replacing the extra-liposomal buffer by one which is iso-osmolar at pH 7.0. Alternatively, charged lipids may be incorporated into the membrane at low pH, followed by adjustment of the suspending medium to neutrality. A similar approach may be adopted using a potassium diffusion gradient, in which the membrane is made selectively permeable to potassium ions entrapped inside the liposome by incorporation of valinomycin into the lipid membrane (56,57). The feasibility of the approaches described above for bringing about uptake and retention of particular solutes has been demonstrated for anti-neoplastic agents (58).

5. PURIFICATION OF LIPOSOMES

Because liposomes are, of necessity, much larger than the materials they entrap, separation of unincorporated material can be achieved on the basis of size differences either by 'gel filtration' column chromatography, or by dialysis. In cases where proteins or DNA are being entrapped, or where there is concern that unentrapped material may form large aggregates, techniques such as centrifugation can be employed, making use of differences in the buoyant densities of liposomes and unentrapped solute.

5.1 **Column chromatographic separation**

Sephadex G-50 (Pharmacia LKB) is the material most widely used for this type of separation. The procedure is exactly identical to that used for desalting and protein separation as described in the literature for many decades. Two special points are worth noting with regard to the special use of Sephadex with liposomes.

(i)   There may be a small number of sites on the surface of the polydextran beads which can bind and interact with the liposome membrane. Although this interaction may not affect the flow characteristics of the liposome suspension through the column, there may be a small amount of lipid lost, and a degree of destabilization of the membrane leading to permeability changes and leakage of entrapped solute. This phenomenon is particularly noticeable if low lipid concentrations are applied to the column. The problem can be overcome either by making sure that the liposome sample size is not too small, or by pre-saturating the column material with 'empty' liposomes of the same lipid composition as the test sample either before or after packing the column. Usually 20 mg of lipid in the form of SUVs per 10 g of gel, followed by copious washing after the column is packed, is sufficient.

(ii)   Larger liposomes (>0.4 μm) may be sometimes retained in the column if the particle size of the gel beads is too small, or if the gel bed contains two many 'fines'. Medium or coarse grades of Sephadex (particle size $50-150$ μm) are preferable to fine for chromatography of MLVs, while all grades are suitable for SUVs.

A simple and convenient method is described in Chapter 3 for column separation of multiple samples using a centrifugation technique. The use of Sepharose to separate liposomes of different sizes is also discussed in Section 6.

5.2 **Dialysis**

Solute removal by dialysis can be performed by standard techniques, and their efficiency can be maximized by employing the modifications described in Section 3.3.1 for detergent removal. Several laboratories have reported the use of hollow fibre dialysis cartridges for large-scale liposome preparations (e.g. see ref. 59), and special attention will be given to this technique here, as the system is particularly valuable for the preparation of large batches of liposomes for clinical use.

Cartridges such as those manufactured by Travenol, Terumo, or Fresenius for hospital use as artificial kidneys can be employed for liposome dialysis. The experimental set-up is as shown in *Figure 28*, in which a peristaltic pump is used for recycling liposomes through the inside of the hollow fibres extending the length of the cartridge, while a second pump circulates fresh buffer around the outside of the fibres. The fibres are narrow capillaries, composed of either cuprophane or cellulose acetate membranes,

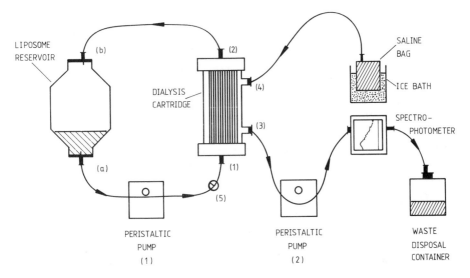

Figure 28. Recirculating dialysis sytem in operation for removing unentrapped solute. For liposomes prepared in batch sizes of the order of a litre, unentrapped solute can be removed efficiently by recirculation through a hollow-fibre dialysis cartridge of the type employed for human kidney dialysis. For low molecular weight compounds, greater than 90% of the unentrapped material can be cleared in 5 h.

and are capable of withstanding considerable pressure differences between the inside and the outside. Therefore numerous modes of operation are possible which aim to accelerate the flow of fluid through the membrane either by increased pressure inside the fibres, or by reduced pressure outside. However, whatever method is used, it is important to maintain a flow of fresh buffer around the outside of the fibres to remove extruded fluid, otherwise reverse dialysis will occur in the later stages as solute diffuses back from concentrated droplets adhering to the fibres.

The concentration facility can be used to good effect in the initial stages, particularly when working with volumes of liposomes larger than the capacity of the cartridge. A Travenol paediatric dialyser, for example, has an internal volume of 60−70 ml, but volumes of 100−150 ml liposomes can easily be accommodated in this cartridge by pumping the liposomes slowly in, while occluding the outlet, during which time the excess fluid will be 'squeezed out' through the membrane as in a diafiltration system. The membrane surface area of the dialysis cartridge described above is 0.8 m^2, with a membrane thickness of 11 μm. The inner diameter of each fibre is about 200 μm, and the maximum trans-membrane pressure exceeds 600 mm Hg. The molecular weight cut-off is several thousand daltons. The cartridges are available sterile, and larger sizes are obtainable.

In theory, once all of the liposome suspension is in the cartridge, there is no need to continue recirculation through the inside of the fibres, but just allow buffer to flow around the outside. In practice, however, a flow of fluid both through the inside and around the outside of the fibres is advisable in order to prevent polarization of the membrane, and to bring about continuous mixing of components inside the fibres for most efficient dialysis. The rate of dialysis naturally depends upon the molecular size of the solute molecule to be removed. Using a paediatric dialyser, for a polar solute of molecular weight 300, 5 h, with a flow rate of buffer of 5 ml min^{-1} (31 volume), is sufficient to remove 99.9% of the initial unentrapped material. For clinical (and other) uses, it is important to rinse the cartridge well with fresh buffer, to remove pyrogens; three litres of fluid each both inside and out has been found to be sufficient.

It is important to purge the buffer well with nitrogen before starting the dialysis. However, during the dialysis process, gas bubbles may form inside the fibres (possibly due to supersaturation with nitrogen) which can block the flow of individual capillaries if allowed to build up. A special trap should be incorporated into the liposome flow-line, therefore, in order to catch the air bubbles as they leave the cartridge, and prevent them from re-entering the capillaries. This trap can be used at the same time to introduce extra buffer into the system, so that one can operate the cartridge under excess pressure while still recycling the liposome suspension. Introduction of extra fluid at the rate of 5 − 10 ml min^{-1} is easily tolerated.

Care needs to be taken in removing liposomes from the column once the dialysis process has been completed. Because of the narrow diameter of the capillaries, the cartridge cannot be drained simply by admitting air into the top; attempting to blow the fluid out under excess gas pressure is also unsatisfactory, since the capillaries empty at different rates, and those which empty first pass all the remaining gas, preventing the pressure from building up sufficiently to flush out the remaining capillaries. The best way to remove the suspension from the column is by chasing it with fresh buffer. Because the flow rate through different capillaries varies, considerably more fluid than the cartridge capacity will have to be introduced before the last liposomes have left the column; usually most of the contents of a 60-ml dialyser can be collected in 200 ml.

The method is extremely suitable for the preparation of medium-size batches of liposomes for industrial and clinical use. Individual capillary dialysers can range in capacity from 60 ml to 150 ml, and can be used singly or multiply, in series or in parallel. For larger sizes, an alternative type of unit is available—namely a flatbed dialyser, in which many layers of membrane are mounted in a multiple sandwich close to each other, with liposomes and buffer flowing through alternate layers (see *Figure 29*). This dialyser (also manufactured for renal dialysis by Gambro) has a larger capacity, a higher flow rate, and a larger membrane area than the hollow fibre cartridges, resulting in faster dialysis under normal conditions, but it cannot be operated in the concentration mode. One other disadvantage is that the internal structure is not visible to the eye, making fault detection more difficult.

5.3 Centrifugation of liposomes

The density of pure egg phosphatidyl choline bilayers is reported by Johnson and Buttress (60,61) to be 1.0135 g ml^{-1}. This value increases slightly (to 1.0142 g ml^{-1}) with the inclusion of 50 mol% cholesterol, and the density can be expected to be even greater for saturated lipids, particularly below their phase transition temperature. The presence of proteins (with an average density of 1.35 g ml^{-1}) in the membrane will increase the density dramatically. Thus vesicles suspended in water or physiological saline, for example (density 1.0048 g ml^{-1}) may be expected to sediment under the influence of a high gravitational field. Small sonicated unilamellar vesicles of egg lecithin [radius 105.6 Å (21 nm diam)], have a calculated sedimentation coefficient (S) in water of 2.6 S

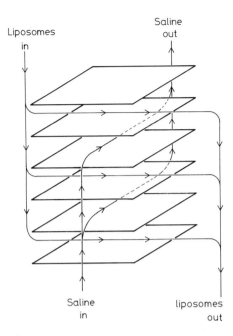

Liposomes
in

Saline
out

Saline
in

liposomes
out

Figure 29. Flatbed dialysis. An alternative to hollow fibre dialysis, in flat bed dialysis large sheets of membrane separate alternate layers of dialysate and dialysand. Flow rate and membrane area are greater than for hollow fibre systems.

increasing to 5.2 S when large amounts of cholesterol are incorporated (62). Thus, depending on the composition of the bilayer, SUVs in normal saline may be sedimented by spinning at 200 000 *g* for between 10 and 20 h in a preparative ultracentrifuge. Multi-lamellar vesicles may be spun down much more readily, in as little as an hour at 100 000 *g* or less depending on their size. This forms the basis of separation of SUVs from MLVs after probe sonication, and is described briefly in Section 3.1.7.

In situations where one is entrapping solute in liposomes at high concentration, or where solutions of high molecular weight compounds are used at concentrations iso-osmolar with physiological saline, the density of the suspending medium can easily approach, or exceed, that of the lipid, so that sedimentation is impossible. This problem may be overcome by diluting the liposomes in a medium of much lower density, so that the liposomes become heavier than the surrounding medium by virtue of their content of undiluted solute entrapped inside. The same principle has been applied, in reverse, to separate liposomes by flotation in a medium of increased density (see *Figure 30*). This method, employed by Fraley and co-workers (63) for purification of DNA-containing LUVs, is described in the following section.

5.3.1 *Ficoll flotation method for DNA*

(i)   Mix 0.5 ml of liposome suspension (prepared in buffered normal saline) with 1 ml of 30% (w/v) Ficoll (also in saline) to give a final concentration of 20% Ficoll in saline.

(ii) Transfer the liposome suspension to an ultracentrifuge tube.

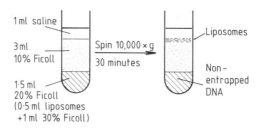

Figure 30. Ficoll floation. This figure illustrates a useful technique for separation of liposomes from unentrapped material of high molecular weight such as DNA or protein. An alternative method using metrizamide is also given in the text.

(iii)   Layer 3 ml of 10% w/v Ficoll gently on top of the liposome suspension.
(iv)   Cover the upper Ficoll layer with a layer of buffered saline.
(v)   Centrifuge in a swing-out rotor for 30 min at 100 000 *g* at room temperature.
(vi)   Collect liposomes at the interface between the saline and 10% Ficoll layers. Unentrapped DNA will remain in the lowest Ficoll layer.

A similar method to the above may be employed for separation from proteins, and is particularly useful for removal of free protein from liposomes after attachment of protein to the membrane surface by covalent linkage (see Chapter 4).

5.3.2 *Metrizamide flotation method for proteins* (contributed by F.J.Martin and T.D.Heath)

(a) *Reagent preparation*

Stock buffer

Prepare stock buffer containing 0.25 M each morpholino ethane sulphonic acid (Mes) and morpholino propane sulphonic acid (Mops) by weighing out 4.88 g of Mes and 5.2 g of Mops, and dissolve in 100 ml of distilled water. Adjust to pH 7.5 with 1 M NaOH.

5% Metrizamide solution

Dissolve 5 g of metrizamide in 50 ml of distilled water. Add 20 ml of stock buffer and check that the pH is still pH 7.5. Bring the volume to 100 ml, and measure the osmolarity with a vapour point osmometer or similar device. Adjust the osmolarity to 290 mOsm kg^{-1} by the addition of solid NaCl.

10% Metrizamide solution

Prepare 10% metrizamide solution in a similar manner, dissolving 10 g in 50 ml of water before making up to full volume.

20% Metrizamide solution

If required, prepare 20% metrizamide as above, except that only 8 ml stock buffer should be added, and the osmolarity should be adjusted to 265 mOsm kg^{-1}. The lowered osmotic activity for high concentration metrizamide solutions is essential to compensate for the dissociation of metrizamide dimers on dilution with the liposome suspension.

Sodium chloride solution (290 mOsm kg^{-1})

Dissolve approximately 0.5 g of NaCl in 50 ml of distilled water, add 20 ml of stock buffer, and make up to 100 ml. Recheck the pH and osmolarity.

(b) *Separation procedure*

(i)   Mix 0.5 ml of the liposome suspension with 1 ml of 10% (w/v) metrizamide and place in the bottom of a 5-ml ultracentrifuge tube.

(ii)   Carefully layer 3 ml of 5% metrizamide solution over the mixture.

(iii) Layer 0.5 ml of buffered saline over the 5% metrizamide (this requires more care than the first to achieve a sharp band). If only one gradient is to be run, prepare a balancing blank gradient.

(iv)   Spin at 200 000 *g* for 40 min in a swing-out rotor at room temperature, setting acceleration to minimum, with no braking at the end of the run.

(v)   After centrifugation, remove the liposomes from the interface of the upper and middle layers. (With care, liposomes may be aspirated in 0.5 ml.)

Metrizamide interferes with protein estimation by the Lowry method, so it should be removed by dialysis before such measurements are carried out. Alternatively, one can use one of the dye-binding assays for proteins which are not affected by metrizamide.

For larger volumes of liposomes, the method can be carried out starting off with 1 ml of liposomes mixed with 0.5 ml of 20% metrizamide. For separating small liposomes, the concentration of the middle layer should be increased to 10% (w/v), and the lower layer prepared by the addition of 0.5 ml of liposomes to 1 ml of 20% metrizamide. In this case, the osmotic strength is not so critical, since small liposomes are not osmotically sensitive. The gradients must be centrifuged at 200 000 *g* for 16 h and separated as described previously.

Those workers who do not have access to an osmometer on a routine basis can prepare dilute solutions which are nearly isotonic by assuming that a 30% solution (w/v) of metrizamide in distilled water is iso-osmotic with physiological fluids. This can then be diluted appropriately in isotonic buffers to obtain 5% and 10% solutions of metrizamide. An alternative to using metrizamide for these separations is Nycodenz (obtainable from Nycomed Pharma, Oslo, Norway), a non-ionic gradient medium that has similar properties to metrizamide. It has the advantage that it does not contain sugar residues, so it is less likely to interfere with certain types of ligand binding, and can be autoclaved rather than filtered if sterile solutions are required.

Other special methods involving centrifugation of liposomes are described elsewhere —a method for centrifugation of aggregated liposomes is reported in Chapter 3 on characterization of liposomes, for rapid determination of liposome entrapment.

6. FRACTIONATION OF LIPOSOMES

Liposomes of different sizes can be separated from each other by essentially the same three methods as used above for purificaton—that is chromatography, centrifugation and dialysis, although not all of them can be considered preparative techniques.

A method for purification of liposomes by ultracentrifugation is given in Section 3.1 on processing, in which a population of small unilamellar vesicles can be isolated free of contaminating multilamellar vesicles. Fractionation of vesicles over the whole size range can be achieved by a technique known as Sedimentation Field Flow Fractionation (SFFF) using a special rotor in which particles of different densities sediment into streams of fluid flowing at different rates perpendicular to the centrifugal force (64). Up to the present time use of the technique for liposomes has only been demonstrated for analytical studies.

Dialysis of liposomes to remove small vesicles from large at a predetermined cut-off value can be achieved by using the same polycarbonate defined-pore membranes as employed for extrusion (65). Thus, in theory, one can use these membranes to produce a population with both a defined upper and lower size limit. Unfortunately, in order to achieve complete removal of unwanted particles, dialysis needs to proceed for a very long time (several days), probably because diffusion of large particles through the pores is slow, and because the percentage area occupied by the pores in membranes currently available is relatively low.

Chromatography on Sepharose columns is often carried out to separate small unilamellar vesicles from MLVs (9), with the latter eluting in the void volume of Sepharose 2B and 4B columns. The 'CL' versions of these chromatography media have higher flow rates than the ordinary type. In Sepharose 6B columns, small MLVs are retained by the gel bed, and in theory MLVs of different sizes can be fractionated using this column. The column can be calibrated using latex beads of defined sizes. Liposomes which elute after the void volume usually come out as a single peak, but using electron microscopy, size differences can be detected between fractions taken from the leading and trailing edges of the peak. Unfortunately, there is considerable overlap in size profiles for different fractions, and clean separations are not possible. Thus, while SUVs and MLVs can be easily separated from each other, to date, there is no good method for preparative liposome fractionation according to size over the whole range.

7. PREPARATION METHODS ACCORDING TO LIPOSOME TYPE

7.1 MLVs

These are the simplest type of liposome to prepare. Their manufacture can be reproducibly scaled up to large volumes of production, and they are mechanically stable upon storage for long periods of time. In the laboratory they may be prepared by hand-shaking of lipids in aqueous media, and the ultimate size can be controlled by extrusion. On the industrial scale, even a small micro-fluidizer will easily cope with large quantities of high-melting lipids in high concentration.

Because the internal space of an MLV is occupied by lipid membrane, these vesicles are highly suited to acting as carriers for lipophilic compounds, which can be included in the phospholipid mixture at the time of drying down. On the other hand, the percentage capture of aqueous components is not always very high ($\sim 10\%$ for 0.2-μm MLVs 30 mg lipid ml^{-1}) when using conventional dispersion methods. Higher values can be achieved, however, if:

(i)   the MLVs are very large,

(ii)   the concentration of lipid in the aqueous medium is very high (~ 300 mg ml^{-1}),

or
(iii)   special methods are employed such as the 'DRV' method, or the more complicated two-phase systems such as MLV-REV, SPLV etc.

Under these circumstances, entrapments of 50% or more can be achieved.

7.2 SUVs

These liposomes, at the lower size limit for phospholipid vesicles, usually require a high energy input for their production, with the accompanying risks of lipid oxidation and hydrolysis. In the past they have been employed widely not so much for their convenience, but because they represent the only well-characterized vesicle population homogeneous in size and lamellarity. They are thermodynamically unstable, and are susceptible to aggregation and fusion, particularly below the phase transition temperature. The packing of phospholipids in SUVs appears to be different from that seen in membranes with a lower curvature. Their entrapped volume is small, and percentage entrapment of aqueous solutes is correspondingly low. With appropriate choice of lipid composition (see later) stable SUVs can be manufactured for use in applications where a high membrane surface area is needed, or where intermembrane transfer of lipophilic materials is required. Methods for the large-scale preparation of SUVs have not been developed. Bath sonication is not sufficiently powerful to deal rapidly with high lipid concentrations, and ethanol injection results in dilution of the lipids in large volumes of aqueous medium. Probe sonication, and the French pressure cell are the two methods recommended for small scale use.

7.3 IUVs and LUVs

Liposomes of this type have a high aqueous:lipid compartment ratio, so that larger volumes can be entrapped with a very economical use of membrane lipids. Because of the large size of the vesicles attainable, a high percentage capture can often be achieved, employing techniques such as the REV or double emulsion methods. Vesicles prepared in this manner are suitable for entrapment of aqueous materials, although the presence of only a single lipid membrane means that mechanical stability, and retention of solutes, may not be as high (particularly for low molecular weight compounds) as for MLVs or MLV-REVs. However, this type of vesicle lends itself very well to active loading of ionizable solutes (see Section 4).

LUVs and IUVs as a source of extended membrane monolayers may be prepared by all three types of dispersion method available.

Physical dispersion
    Careful swelling by the Reeves and Dowben method.
    Spontaneous vesiculation of acidic lipids.
    Fusion by freeze-thawing or calcium-mediated-interactions.
    High pressure extrusion.

Two-phase systems
    Ether injection.
    REVs, double emulsion systems.

Detergent dialysis

Bile salt or β-alkyl glycoside mixed micelles.

The use of IUVs as carriers for lipophilic compounds has nothing to recommend it in favour of MLVs, except in the case of incorporation of membrane proteins, especially where maximum exposure is required. Under these circumstances, detergent dialysis is the method of choice, often resulting in 100% incorporation of protein with full retention of its biological activity. Unfortunately, the nature of the dialysis technique is such that aqueous solutes are unlikely to be entrapped at high concentration inside the IUVs formed, unless some special mechanism is provided, such as active loading, or membrane association.

The size of detergent-dialysed liposomes can be controlled very precisely by adjustment of dialysis conditions, detergent:lipid ratio or chemical structure of the detergent, while LUVs made by other methods can be modified in size by membrane extrusion, in the same way as for MLVs.

8. STABILITY OF LIPOSOMES

Many different changes can take place in liposomes with the passage of time. The phospholipids can undergo chemical degradation—oxidation and hydrolysis—leading to a build-up of short-chain phospholipids and lyso-derivatives in the membrane. Either as a result of these changes, or otherwise, liposomes maintained in aqueous suspension may aggregate, fuse, or leak their contents.

Methods devised to overcome the problems of liposome instability fall into two categories—those designed to minimize the degradation processes which may take place, and secondly, those which contrive to help liposomes survive in the face of conditions which encourage these processes.

8.1 Prevention of chemical degradation

The level of oxidation can be kept to a minimum by taking the following precautions:

(i)   start with freshly purified lipids and freshly distilled solvents;
(ii)   avoid procedures which involve high temperature;
(iii)   carry out the manufacture process in the absence of oxygen;
(iv)   deoxygenate aqueous solutions with nitrogen;
(v)   store all liposome suspensions in an inert atmosphere;
(vi)   include an anti-oxidant as a component of the lipid membrane.

It may also be worthwhile including an iron chelator in the formulation, to prevent initiation of the free radical chain reaction (see Chapter 3 on characterization of liposomes). The anti-oxidant in most common use at the present time is α-tocopherol (vitamin E), a common non-toxic dietary lipid, although it has been suggested that β, γ, and δ-tocopherols may be more effective as long-term anti-oxidants (66) since they have a longer lifetime, and are not destroyed so easily in the process of radical neutralization. An alternative approach to the oxidation problem is to reduce the level of oxidizable lipids in the membrane, by using saturated lipids instead of unsaturated ones. Among phospholipids from natural sources, increasing unsaturation is in the order plant, egg yolk, mammalian origin. In the case of egg yolk lecithin, the degree of unsaturation of the phospholipid fatty acids can depend on the diet of the birds, and on the method of purification employed (see ref. 67 and Appendix I).

Preparations are possible which contain almost wholely 16:0, 18:0 and 18:1 with very little longer chain polyunsaturated material. The mono-unsaturated chains are much less susceptible to oxidation than polyunsaturated ones (see Chapter 3 on characterization of liposomes). Thus sphingomyelins, usually containing only a single double bond, can be expected to deteriorate more slowly than other membrane lipids of mammalian origin. Alternatively, unsaturated lecithins from natural sources may be converted to the saturated homologues by the process of catalytic hydrogenation (see Appendix I). Liposomes may also be prepared from entirely synthetic saturated compounds, such as DMPC, DPPC, and DSPC.

Hydrolysis of the ester linkages will proceed most slowly at pH values close to neutral. However, even at low pH, such as that required for active loading of drugs, hydrolysis can be kept to a minimum if scrupulous attention is paid to the removal of residual solvent from the dried lipids. The extent to which the presence of traces of lysolecithin (arising from hydrolysis) will jeopardize membrane integrity is not clear. All biological membranes contain small amounts of lyso-derivatives as a consequence of natural membrane turnover, recycling and cell triggering processes. Although exogenous lysolecithin incorporated into biological membranes from micelles is known to cause cell lysis very readily, it may be that lysolecithin co-existing in membranes with an equimolar quantity of free fatty acids will behave in a similar fashion to intact phospholipids. Indeed, liposomes can remain intact even when the entire outer leaflet has been converted to lyso-PC and fatty acid by enzymatic means, as long as there is no mechanism for rapid flip-flop of the lyso derivatives to the inner monolayer (68). Methods for avoiding hydrolysis altogether are the use of lipids which contain ether instead of ester linkages such as are found in membranes of halophilic bacteria (69). Hydrolysis *in vivo* as a result of enzymatic attack can be prevented by the use of sphingomyelin, or of phospholipid derivatives with the 2-ester linkage replaced by a carbamoyloxy function (70,71).

8.2 Prevention of physical degradation

Leakage and fusion of vesicles can occur as a result of lattice defects in the membrane introduced during their manufacture. Although reported particularly to occur in SUVs when prepared below the membrane phase transition temperature, there is evidence for packing defects being maintained in other types of vesicle (detergent dialysis vesicles, freeze-thawed vesicles) even above the phase transition temperature. These irregularities can be dispersed by a process termed 'annealing', which consists simply of incubating the liposomes at a temperature high enough above the phase transition temperature to allow differences in packing density between opposite sides of the bilayer to equalize by transmembrane flip-flop.

Even in annealed vesicles, aggregation, fusion etcetera can take place to significant extents over a long period of time. Aggregation (and sedimentation) of neutral liposomes is brought about by Van der Waals interactions, and tends to be more pronounced in large vesicles, where the increased planarity of the membranes allows greater areas of membrane to come into contact with each other. Although factors such as residual solvent and trace elements can enhance the process, for uncharged membranes it is a natural and unavoidable phenomenon, and the simplest way to overcome it is to include a small quantity of negative charge (e.g. 10% PA or PG) in the lipid mixture. It has also been reported that conversion of small vesicles to larger structures as a result of fusion processes can be markedly reduced by the presence of trace amounts of the phospholipid isomer 1,3-diacyl-2-phosphatidylcholine (72).

In large liposomes which are properly made, there is no reason why fusion should occur. Small unilamellar vesicles [<400 Å (40 nm) in diameter] are prone to fusion as a means of relieving stress arising from the high curvature of the membrane. Since this can occur particularly at the phase transition temperature, it is advisable to store liposome suspensions at a temperature away from the T_c, and it could be advantageous to include sufficient cholesterol in the membrane to reduce or completely remove the transition, particularly if it is in a temperature range close to that at which the liposomes will be stored or handled. For liposomes which have negative charge in the membranes, care must be taken to avoid high concentrations of metal ions. It may even by worthwhile including a metal ion chelator in the suspending buffer.

Permeability of liposome membranes depends very much on the membrane lipid composition, and on the solute which one has entrapped. Large polar or ionic molecules will be retained much more effectively than low molecular weight lipophilic compounds. In general, for both classes of compound, a rigid, more saturated membrane with a high molar ratio of cholesterol is the most stable with regard to leakage of solutes. As before, permeability can be increased at the phase transition temperature— particularly in the presence of certain proteins such as high density lipoprotein (HDL) apolipoproteins, tubulin or actin (73), and this phenomenon can be exploited to induce enhanced release from liposomes at a desired temperature (see Chapter 6 on biological systems).

Various methods have been employed which may increase the stability of liposomes by cross-linking membrane components covalently using methods such as glutaraldehyde fixation, osmification or polymerization of alkyne-containing phospholipids (74). Although these methods can increase the mechanical strength of the membrane, and render them less susceptible to disruption *in vivo* by serum components, the procedures themselves can act to introduce or preserve irregularities in membrane packing, which once present are not easily dissipated, so that permeability to small ions may well be increased. One 'polymerization' technique which does not restrict the relative mobility of adjacent phospholipids is the incorporation of long aliphatic branched-chain polymers (e.g. polyvinyl alcohols esterified with palmitic or stearic acid). These compounds, when incorporated up to about 10% by weight into PC membranes can substitute for cholesterol in reducing the leakage of medium-sized solutes (75), although it is not clear whether any further stabilization can be achieved by a combination of the two.

Another type of cross-linking, involving hydrogen bonding, is probably responsible for the increased stability of liposome membranes containing sphingomyelin, or 2-carbamoyl PC derivatives, in which the $-NH$ group in the interfacial region can stabilize packing by association with similar groups, or with the hydroxyl group of cholesterol.

The methods described previously for mechanical stabilization of the membrane may also be of value in protecting liposomes from damage during procedures such as freezing or lyophilization; these procedures consitute an alternative method of overcoming instability problems by transforming the liposomes into a solid or anhydrous form, where chemical degradation of lipid components or solutes is less likely to occur. At present, however, techniques which prevent membrane fracture upon freezing or drying have not yet been fully perfected. Methods for cryopreservation, analogous to those used for storage of cell lines, using high concentrations of sugar solution have achieved some success (76). The non-reducing disaccharide trehalose seems to be particularly effective. It appears to interact directly with the phospholipid membrane, probably via hydrogen bonding, and is able to alter the T_m of the dried phospholipids, thus preventing phase separations from occurring at the temperatures over which water is added back to the membrane preparations. Best results (up to 100% retention of solutes) are obtained when the sugar is in high concentration on both sides of the membrane, in order to prevent phase separations taking place independently (75).

9. PREPARATION OF STERILE LIPOSOMES

Liposomes cannot be sterilized by exposure to high temperatures, and are also sensitive to various types of irradiation, as well as chemical sterilizing agents, so the only method available for sterilizing after manufacture is filtration. Clearly, this procedure is only applicable to liposomes sufficiently small in size to pass easily through the pores of a 0.22-μm membrane. For larger liposomes, every stage must be carried out under aseptic conditions; the initial organic solution of lipids may be passed through membrane filters of regenerated cellulose (pore size 0.45 μm) and glass fibre (supplied by MFS − see Appendix II) before drying down, to remove microorganisms, spores, and pyrogenic materials.

10. REFERENCES

1. Bangham,A.D., Standish,M.M. and Watkins,J.C. (1965) *J. Mol. Biol.*, **13**, 238.
2. Mayer,L.D, Hope,M.J., Cullis,P.R .and Janoff,A.S. (1985) *Biochim. Biophys. Acta*, **817**, 193.
3. Reeves,J.P. and Dowben,R.M. (1969) *J. Cell Physiol.*, **73**, 49.
4. Lasic,D.D., Belič,A. and Valentinčič,T. (1988) *J. Am. Chem. Soc.*, **110**, 970.
5. Payne,N.I., Timmins,P., Ambrose,C.V., Ward,M.D. and Ridgeway,F. (1986) *J. Pharm. Sci.*, **75**, 325.
6. Imperial Chemical Industries Ltd (1978) Belgian Patent 866697.
7. Allergan Pharmaceuticals Inc. (1987) European Patent 86306014.1.
8. Mayhew,E., Lazo,R., Vail,W.J., King,J. and Green,A.M. (1984) *Biochim. Biophys. Acta*, **775**, 169.
9. Huang,C. (1969) *Biochemistry*, **8**, 344.
10. Barenholtz,Y., Gibbes,D., Litman,B.J., Goll,J., Thompson,T.E. and Carlson,F.D. (1977) *Biochemistry*, **16**, 2806.
11. Barenholtz,Y., Amselem,S. and Lichtenberg,D. (1979) *FEBS Lett.*, **99**, 210.
12. Hamilton,R.L., Goerke,J., Guo,L.S.S., Williams,M.C. and Havel,R.J. (1980) *J. Lipid Res.*, **21**, 981.
13. Friedman,J.E., Lelkes,P.I., Rosenheck,K. and Oplatka,A. (1986) *J. Biol. Chem.*, **261**, 5745.
14. Olson,F., Hunt,C.A., Szoka,F.C., Vail,W., Mayhew,E. and Paphadjopoulos,D. (1980) *Biochim. Biophys. Acta*, **601**, 559.
15. Mayer,L.D., Hope,M.J. and Cullis,P.R. (1986) *Biochim. Biophys. Acta*, **858**, 161.
16. Kirby,C. and Gregoriadis,G. (1984) *Biotechnology*, **2**, 979.
17. Deamer,D.W. and Barchfield,G.L. (1982) *J. Mol. Evol.*, **18**, 203.
18. Pick,U. (1981) *Arch. Biochem. Biophys.*, **212**, 186.
19. Kasahara,M. and Hinckle,P.C. (1977) *J. Biol. Chem.*, **252**, 7384.
20. Oku,N. and MacDonald,R.C. (1983) *Biochemistry*, **22**, 855.
21. Hauser,H. and Gains,N. (1982) *Proc. Natl. Acad. Sci. USA*, **79**, 1683.
22. Papahadjopoulos,D., Vail,W.J., Jacobson,K. and Poste,G. (1975) *Biochim. Biophys. Acta*, **394**, 483.
23. Batzri,S. and Korn,E.D. (1973) *Biochim. Biophys. Acta*, **298**, 1015.
24. Deamer,D.W. and Bangham,A.D. (1976) *Biochim. Biophys. Acta*, **443**, 629.
25. Deamer,D.W. (1978) *Ann. N.Y. Acad. Sci.*, **308**, 250.
26. Batelle Memorial Inst. (1979) British Patent Appl. No. 2001929A.
27. Kim,S. and Martin,G.M. (1981) *Biochim. Biophys. Acta*, **646**, 1.
28. Kim,S., Turker,M.S., Chi,E.Y., Sela,S. and Martin,G.M. (1983) *Biochim. Biophys. Acta*, **728**, 339.

29. Gao,K. and Huang,L. (1987) *Biochim. Biophys. Acta*, **897**, 377.
30. Szoka,F. and Papahadjopoulos,D. (1978) *Proc. Natl. Acad. Sci. USA*, **75**, 4194.
31. Pidgeon,C., Hung,A.H. and Dittrich,K. (1986) *Pharm. Res.*, **3**, 23.
32. Gruner,S.M., Lenk,R.P., Janoff,A.S. and Orton,M.J. (1985) *Biochemistry*, **24**, 2833.
33. Philippott,J., Mutaftschiev,S. and Liautard,J.P. (1983) *Biochim. Biophys. Acta*, **734**, 137.
34. Brunner,J., Skrabal,P. and Hauser,H. (1976) *Biochim. Biophys. Acta*, **455**, 322.
35. Allen,T.M., Romans,A.T., Kercret,H. and Segrest,J.P. (1980) *Biochim. Biophys. Acta*, **601**, 328.
36. Milsmann,M.H.W., Schwendener,R.A. and Weder,H.-G. (1978) *Biochim. Biophys. Acta*, **640**, 252.
37. Enoch,H.G. and Strittmatter,P. (1979) *Proc. Natl. Acad. Sci. USA*, **76**, 145.
38. Zumbühl,O. and Weder,H.-G. (1981) *Biochim. Biophys. Acta*, **640**, 252.
39. Rhoden,V. and Goldin,S. (1979) *Biochemistry*, **18**, 4173.
40. Harsch,M., Walther,P. and Weder,H.-G. (1981) *Biochem. Biophys. Res. Commun.*, **103**, 1069.
41. Lapidot,Y., Rappoport,S. and Wolman,Y. (1967) *J. Lipid Res.*, **8**, 142.
42. Weder,H.-G. and Zumbühl,O. (1984) In *Liposome Technology* Gregoriadis,G. (ed), CRC Press Inc., Boca Raton, Florida, Vol. 1, p. 79.
43. Alpes,H., Allmnan,K., Plattner,H., Reichert,J., Riek,R. and Schulz,S. (1986) *Biochim. Biophys. Acta*, **862**, 294.
44. Loyter,A. and Volsky,D.J. (1982) In *Membrane Reconstitution*. Poste,G. and Nicolson,G. (eds), Elsevier/North Holland Biomedical Press, Amsterdam, p. 215.
45. Vainstein,A., Hershkovitz,M., Israel,S., Rabin,S. and Loyter,A. (1984) *Biochim. Biophys. Acta*, **773**, 181.
46. Loyter,A. and Chejanovsky,N. and Citovsky,V. (1986) In *Methods in Enzymology*. In press, Academic Press Inc., London.
47 Maeda,Y., Kim,J., Koseki,I., Mekada,E., Shiokawa,Y. and Okada,Y. (1977) *Exp. Cell Res.*, **108**, 95.
48. Hoekstra,D., de Boer,T., Klappe,K. and Wilschut,J. (1984) *Biochemistry*, **23**, 5675.
49. Chejanovsky,N. and Loyter,A. (1985) *J. Biol. Chem.*, **260**, 7911.
50. Gitman,A.G. and Loyter,A. (1985) *Biochemistry*, **24**, 2762.
51. Loyter,A., Vainstein,A., Graessmann,M. and Graessmann,A. (1983) *Exp. Cell Res.*, **143**, 415.
52. Volsky,D.J., Gross,T., Sinajil-Kustynski,C., Bartzatt,R., Donahugh,T. and Kieff,E. (1984) *Proc. Natl. Acad. Sci. USA*, **81**, 5926.
53. Volsky,D.J., Cabantchik,Z.I., Beigel,M. and Loyter,A. (1981) *Proc. Natl. Acad. Sci. USA*, **77**, 7247.
54. Nichols,J.W. and Deamer,D.W. (1976) *Biochim. Biophys. Acta*, **455**, 269.
55. Mayer,L.D., Bally,M.R. and Cullis,P.R. (1986) *Biochim. Biophys. Acta*, **857**, 123.
56. Bally,M.B., Hope,M.J., Van Echteld,C.J.A. and Cullis,P.R. (1985) *Biochim. Biophys. Acta*, **812**, 66.
57. Mayer,L.D., Bally,M.B., Hope,M.J. and Cullis,P.R. (1985) *J. Biol. Chem.*, **260**, 802.
58. Mayer,L.D., Bally,M.B., Hope,M.J. and Cullis,P.R. (1985) *Biochim. Biophys. Acta*, **816**, 294.
59. Schwendener,R.A. (1986) *Cancer Drug Delivery*, **3**, 123.
60. Johnson,S.M. and Buttress,N. (1973) *Biochim. Biophys. Acta*, **307**, 20.
61. Johnson,S.M. (1973) *Biochim. Biophys. Acta*, **307**, 27.
62. Newman,G.C. and Huang,C. (1975) *Biochemistry*, **14**, 3363.
63. Fraley,R., Subramani,S., Berg,P. and Papahadjopoulos,D. (1980) *J. Biol. Chem.*, **255**, 10431.
64. Kirkland,J.J., Yau,W.W. and Szoka,F.C. (1982) *Science*, **215**, 296.
65. Bosworth,M.E., Hunt,C.A. and Pratt,C. (1982) *J. Pharm. Sci.*, **71**, 806.
66. Lambelet,P. and Löliger,J. (1984) *Chem. Phys. Lipids*, **35**, 185.
67. Klein,R.A. (1970) *Biochim. Biophys. Acta*, **219**, 496.
68. de Gier,J., Mandersloot,J.G., Schipper,N.P.M., van der Steen,A.T.M. and van Deenen,L.L.M. (1983) In *Liposome Letters*. Bangham,A.D. (ed.), Academic Press, London, p. 31.
69. Kates,M. and Kushwaha,S.C. (1976) In *Lipids*. Paoletti,R. *et al.* (eds), Raven Press, NY, Vol. 1: *Biochemistry*, p. 267.
70. Bali,A., Dawan,S. and Gupta,C.M .(1983) *FEBS Lett.*, **154**, 373.
71. Agarwal,K., Bali,A. and Gupta,C.M. (1986) *Biochim. Biophys. Acta*, **856**, 36.
72. Larrabee,A.L. (1979) *Biochemistry*, **18**, 3321.
73. Weinstein,J.N., Klausner,R.D., Innerarity,T., Ralston,E. and Blumenthal,R. (1981) *Biochim. Biophys. Acta*, **647**, 270.
74. Leaver,J., Alonso,A., Aziz,A.D. and Chapman,D. (1983) *Biochim. Biophys. Acta*, **732**, 210.
75. Ash,P.S. and Hider,R.C. (1980) UK Patent Application GB 2026340A.
76. Crowe,L.M., Womersley,C., Crowe,J.H., Reid,D., Appel,L. and Rudolph,A. (1986) *Biochim. Biophys. Acta*, **861**, 131.

CHAPTER 3

Characterization of liposomes

ROGER R.C. NEW

1. INTRODUCTION

The behaviour of liposomes in both physical and biological systems is determined to a large extent by factors such as physical size, chemical composition, membrane permeability, quantity of entrapped solutes, as well as the quality and purity of the starting materials, so it is important to have as much information as possible regarding these parameters before embarking on detailed investigations. The methods outlined in this chapter are divided into three sections—first, chemical analyses in which quantitative and qualitative tests are carried out on liposomal components both before and after manufacture; secondly, properties of intact liposomes are examined—material entrapped, permeability, lamellarity—and finally, methods for the estimation of liposome size are considered.

2. CHEMICAL ANALYSIS

2.1 Quantitative determination of phospholipid

Accurate measurements of phospholipid concentrations are difficult to make directly, since dried lipids can often contain considerable quantities of residual solvent, or if derived from liposomes, of other contaminating lipids. Consequently, the method most widely used for determination of phospholipid is an indirect one, in which the phosphate content of the sample is first measured (1).

2.1.1 Analysis of phosphorus using the Bartlett assay

In this method, phospholipid phosphorus in the sample is first acid-hydrolysed to inorganic phosphate. This is converted to phospho-molybdic acid by the addition of ammonium molybdate, and the phospho-molybdic acid is quantitatively reduced to a blue-coloured compound by amino-naphthyl-sulphonic acid. The intensity of the blue colour is measured spectrophotometrically, and is compared with calibration standards to give phosphorus and hence phospholipid content. The majority of phospholipid classes used for the preparation of liposomes contain exactly one mole of phosphorus per mole of phospholipid, hence the concentration of phospholipid can be derived directly from a measurement of the phosphorus content of the sample. However, an exception to this rule is cardiolipin which contains two moles of phosphorus per mole of phospholipid and so if the sample contains this lipid an appropriate allowance must be made in the calculations.

(a) *Preparation of reagents*

Sulphuric acid (5 Molar) reagent

Cautiously add 140 ml of concentrated sulphuric acid to approximately 300 ml of distilled water in a beaker in an ice bath. Mix carefully by stirring, and transfer to a 500-ml measuring cylinder. Make up to 500 ml with distilled water and mix. Store in a glass screw-capped bottle in an acid bin.

Phosphate standard solutions

Stock solution.

> Dry a sample of solid anhydrous potassium dihydrogen phosphate at 105°C for 4 h in a vacuum oven. Weigh out 43.55 mg of dried solid and transfer to a 100-ml volumetric flask. Dissolve in double-distilled water, and make up to 100 ml. The final concentration should be 3.2 μmol phosphorus ml^{-1}. For working standard solutions transfer 2.0, 3.0, 4.0, 5.0, 6.0, 7.0, and 8.0 ml of the stock phosphate solution into separate 100-ml volumetric flasks, and make each up to 100 ml with double-distilled water, to give solutions with concentrations of 0.064, 0.096, 0.128, 0.160, 0.192, 0.224, and 0.256 μmol phosphorus ml^{-1} respectively.

Ammonium molybdate − sulphuric acid solution.

> Add 5 ml of the sulphuric acid reagent to approximately 50 ml of distilled water and add 0.44 g of ammonium molybdate. Mix well and make up to 200 ml with distilled water.

1-Amino 2-naphthyl 4-sulphonic acid reagent.

> Use Fiske & Subbarrow reducer (Sigma Chemical Co.). Weigh out 0.8 g and dissolve in 5 ml of double-distilled water. Prepare fresh on day of use.

Hydrogen peroxide.

> Add 1 ml of 100% hydrogen peroxide (stored at 4°C) to 9 ml of distilled water, and mix well. Prepare fresh immediately before use.

(b) *Sample preparation*

Liposome suspension

Dilute the aqueous liposome suspension with double-distilled water to give a concentration of approximately 1 mg ml^{-1} of phospholipid.

Lipid solution

Dilute the lipid solution with chloroform to about 1 mg ml^{-1} of phospholipid. Use chloroform:methanol (2:1 v/v) if there is difficulty in dissolving it in chloroform alone.

Lipid solid

Prepare a 1 mg ml^{-1} solution of phospholipid in chloroform or 2:1 chloroform: methanol.

(c) *Assay method*

Caution Wear eye protection for this assay. Ideally the heating block should be housed in a fume hood.

(i)   Equilibrate heating block by pre-heating at 200°C for 30 min.
(ii)   Set up calibration curve by pipetting, into separate 16 × 150 mm disposable borosilicate tubes, 0.5 ml of each working standard solution, together with a blank (0.5 ml of double-distilled water).
(iii)   Prepare sample tubes in triplicate by adding 50 μl of the sample to each of three empty tubes. Dry down and resuspend in 0.5 ml of distilled water.
(iv)   Add 0.4 ml of sulphuric acid reagent to each tube, cover and incubate in a heating block in a fume hood at 180−200°C for an hour.
(v)   Allow the tubes to cool by standing them at room temperature.
(vi)   Prepare diluted hydrogen peroxide (10%) fresh.
(vii)   To each tube add 0.1 ml of diluted hydrogen peroxide and incubate on the heating block at 180−200°C for 30 min to achieve a clear solution. (If necessary, repeat addition and heating until solution is clear.)
(viii) Cool the tubes by standing them at room temperature.
(ix)   Add 4.6 ml of acid−molybdate solution to each tube and mix thoroughly by vortexing.
(x)   Add 0.2 ml of Fiske & Subbarrow reducer to each tube and mix thoroughly by vortexing after addition.
(xi)   Cover the tubes and place them in a boiling water bath.
(xii)   Leave the tubes in the bath for 7 min after boiling recommences.
(xiii) Cool the tubes to room temperature.
(xiv)   Measure absorbance of all tubes against distilled water, in 1-cm cuvettes at 800 nm. The concentration in the starting sample is ten times that read off the graph of the standard curve.

The Bartlett assay is very sensitive (measuring phosphate concentrations down to 30 nmol ml^{-1}) but is not terribly reproducible. One problem is that the test is easily upset by trace contamination with inorganic phosphate. A wise precaution is to carry out the assay using a set of borosilicate glass tubes which have been washed well, and which are not used for any other purpose. Always use double-distilled water for making up the solutions. Another source of error is the digestion step, where excessive bubbling and spitting can result in the dispersion or loss of material inside the tube, and unspecified quantities of aqueous solvent can evaporate away, changing the final volumes (and concentrations) when the final absorption measurements are made. To minimize errors introduced by these factors, the temperature of the heating block should be very closely controlled, so as not to exceed 200°C, but not to go below 180°C, to ensure proper digestion.

All samples should be estimated in triplicate, using three separate aliquots taken individually from each sample. The lipids (especially those in organic solution) should be drawn up in calibrated glass microcapillaries, not plastic-tipped 'Eppendorf'-type pipettes.

The sensitivity of the Bartlett assay to inorganic phosphate creates a problem with measurement of phospholipid liposomes suspended in physiological buffers, which usually contain phosphate ions. One way of overcoming this is to employ a more specific method which is unaffectd by inorganic phosphate.

2.1.2 *Stewart assay*

In the Stewart assay for phospholipids (2), the ability of phospholipids to form a complex with ammonium ferrothiocyanate in organic solution is utilized. The advantage of this method is that the presence of inorganic phosphate does not interfere with the assay. A simple conversion factor is used to translate absorbance values into milligrams of phospholipid. Since this factor differs for phospholipids of different headgroups, this method is not applicable to samples where mixtures of unknown phospholipids may be present. In particular, this method is especially unresponsive to phosphatidyl glycerol. In liposomes which contain only phosphatidyl choline and phosphatidyl glycerol, the Stewart assay could be used as a specific test for the former, and the proportion of phosphatidyl glycerol inferred after carrying out a Bartlett phosphate assay on the sample, which would measure total phospholipid.

(a) *Preparation of reagents*

Ammonium ferrothiocyanate solution (0.1 M).

Dissolve 27.03 g of ferric chloride hexahydrate and 30.4 g of ammonium thiocyanate in double-distilled water, and make up to 1 litre. The solution is stable at room temperature for several months.

(b) *Preparation of standard and sample solutions*

Make up 10 ml of solution of phospholipid in chloroform at a concentration of 0.1 mg ml^{-1}. Prepare samples at approximately the same concentration.

Assay method

(i) Pipette reagents and standard into 10-ml centrifuge tubes as follows:

Tube no.	Standard (ml)	Chloroform (ml)	Ferrothiocyanate (ml)
0	0.0	2.0	2.0
1	0.1	1.9	2.0
2	0.2	1.8	2.0
3	0.4	1.6	2.0
4	0.6	1.4	2.0
5	0.8	1.2	2.0
6	1.0	1.0	2.0

Prepare duplicates of each tube. Perform the same procedure with the test samples.
(ii) Vortex contents of each tube vigorously on a whirlimixer for 15 sec.
(iii) Spin each tube for 5 min at 1000 r.p.m. (approximately 300 g) in a bench centrifuge, remove the lower layer using a Pasteur pipette and retain it.
(iv) Read the optical density at 485 nm for samples and standards.
(v) Find the concentration in the test sample solutions by comparing with the standard curve.

Other methods for the measurement of phospholipids include a rhodamine complexation test (3), although in many people's hands this has been found to be very susceptible to interference. In cases where determination of different phospholipids in a mixture is required, a modification of the Bartlett assay can be adopted, in which the lipids are first chromatographed by TLC, and then stained with a molybdenum blue spray. The intensity of the blue colour developed can be measured using a plate-scanning densitometer. TLC systems are discussed in Section 2.2, and the technique as a whole is illustrated for the determination of lyso-derivatives (see Section 2.3).

2.2 Thin-layer chromatography (TLC) of lipids

Like all chromatographic methods, thin-layer chromatography of lipids employs two different, immiscible phases, and makes use of the fact that solutes in the liquid organic phase will have differing affinities for the hydrophilic solid phase. Thus, as the liquid phase runs through the solid gel, different lipids will be retained to different extents by the gel, and will spread out at different distances behind the solvent front. The composition of the mobile phase can be altered so as to be more or less hydrophilic, (e.g. by variation of the quantity of polar solvents, such as methanol in chloroform, or by the addition of water), in such a way as to alter the partition coefficient of solutes between the two phases. Acids or bases can also be added, which will alter the ionic charge on solute molecules, and alter the extent of their interaction with the stationary phase. For phospholipids, the most common stationary phase is silica gel, which is moderately hygroscopic, and consists of granules which under normal conditions are surface coated with a layer of tightly-bound water. The mobile phase is usually a mixture of solvents including chloroform. Under these circumstances, phospholipids are separated principally according to differences in the headgroup.

The lipids are visualized either by means of specific stains which are sprayed on to the plate, or non-specifically by methods such as charring, or iodine uptake. Using the phosphomolybdate method, quantities of phospholipid down to one microgram can be detected. In the case of lipids which absorb strongly in the UV, their presence may be inferred without the need for staining, by employing TLC plates which contain a fluorescent material (e.g. fluorescein) incorporated into the solid phase. Upon illumination with UV light, a dark spot will be seen on a light background, where the fluorescence of the fluorophore has been quenched by the lipid. Identification of the different lipids is based upon the speed at which they run (expressed as the R_f value), and is made by reference to standards which are run on the same plate as the test mixture.

TLC can give information both about the purity and the concentration of the lipids. If a compound is pure it should run as a single spot in all elution solvents. Phospholipids which have undergone extensive degradation can be observed as a long smear with a tail trailing to the origin, compared with the pure material which runs as one clearly defined spot. (A method has recently been reported for separating lipid peroxides and hydroperoxides by reverse phase TLC, see ref. 4.) Synthetic phosphatidyl cholines (PCs) usually give smaller, tighter spots than PCs from natural sources, which are composed of a mixture of components. Quantitation by scanning densitometry is described in the next section. Examples of elution patterns for different lipids on silica gel are given in *Figure 1*. Pre-coated TLC plates may be obtained from Sigma Chemical Co. Ltd, or Whatman.

(a) *Preparation of reagents.*

Solvent mixtures—make up fresh each day.

(i)   Chloroform:methanol:water (65:25:4 v/v).
        Measure out 65 ml of chloroform, 25 ml of methanol, 4 ml of double-distilled water, and mix thoroughly.

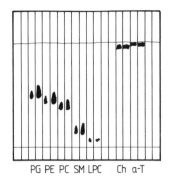

PG PE PC SM LPC   Ch α-T

Figure 1. Thin layer chromatography of lipids. This figure shows the relative positions of a few common lipids on a silica gel TLC plate after being run with a solvent such as chloroform:methanol:water 65:25:4 (v/v).

(ii)   Chloroform:methanol:water:ammonium hydroxide (65:35:2.5:2.5 v/v).
        Measure out 65 ml of chloroform, 35 ml of methanol, 2.5 ml of double-distilled water, 2.5 ml of ammonium hydroxide, and mix thoroughly.
(iii)   Chloroform:acetone: methanol:acetic acid:water (6:8:2:2:1 v/v).
        Measure out 30 ml of chloroform, 40 ml of acetone, 10 ml of methanol, 10 ml of glacial acetic acid, 5 ml of double-distilled water, and mix thoroughly.
(iv)   Ethyl acetate:cyclohexane (1:1 v/v).
        Measure out 50 ml of ethyl acetate, 50 ml of cyclohexane, and mix thoroughly.

Molybdenum blue spray reagent (from Sigma Chemical Company).
  Make up according to manufacturer's instructions.

Sulphuric acid spray (50% v/v).
  Carefully add 50 ml of a concentrated sulphuric acid to 50 ml of methanol. Mix in a 150-ml glass beaker surrounded by ice.

Potassium dichromate spray.
  Carefully add 70 ml of concentrated sulphuric acid to 30 ml of double-distilled water in a 150-ml glass beaker in an ice bath. Mix and bring to room temperature. Add potassium dichromate crystals and dissolve by stirring with a glass rod until saturation point is reached.

Iodine tank.
  Sprinkle approximately 10 g of solid iodine crystals on the bottom of a glass chromatography tank. Keep the tank closed with a well-fitting lid.

(b) *Sample preparation*

Solid lipids
Dissolve in chloroform to a concentration of 20 mg ml^{-1}. If the solution is cloudy, use chloroform:methanol (2:1 v/v) as the solvent.

Lipid solutions
Dilute the lipid solutions with an appropriate solvent to a concentration of 20 mg ml^{-1}.

110

Liposomes

(i)   Extract the lipid using acidified Bligh-Dyer procedure as outlined in Appendix I or,

(ii)   Dry down a small volume of the aqueous liposome suspension containing 2 mg of lipid under nitrogen and redissolve in 50 μl of methanol. Then add 50 μl of chloroform to give a concentration of 20 mg ml^{-1} of phospholipid.

(c) *Separation method*

Sample application

(i)   Place the TLC plate on a pre-heated hot-plate (50°C).

(ii)   Spot the samples exactly 2.5 cm from the lower edge of the plate—this is the spotting origin.

(iii) Spot 10 μl of each standard in separate lanes using a 10-μl microdispenser with disposable tips.

(iv) Spot 10 μl and 5 μl of the sample in separate lanes on each plate with 10-μl and 5-μl microdispensers.

(v)   Prepare three more plates in an identical manner, to have one for each solvent system.

Development

(vi) Line the inside of four tanks with Whatman No. 1 filter paper sheet. Pour 100 ml of each solvent mixture into separate tanks and let the tanks equilibrate for 1−2 h, or until the filter paper is completely soaked with solvent.

(vii) Carefully place one plate into each of the equilibrated tanks and cover.

(viii) Allow the solvent to ascend to within 3−4 cm of the top of the plate—about 50 min for polar systems, 30 min for neutral solvents.

(ix) Remove the plates from the tanks and mark the solvent front with the point of a pencil.

(x)   Air-dry the plates for 15 min in a fume hood before spraying to visualize the lipids.

(d) *Visualization of lipids*

Phospholipids

Spray the plate evenly with molybdenum blue spray, using back-forth and up-down motions across the entire plate, but take care to avoid saturating the plates. Air-dry the plates in a fume hood for 10 min. Phosphate-containing compounds will stain blue within 15 min. Record the results obtained.

Cholesterol and α-tocopherol

Spray the plate (can be the same plate as above) evenly with 50% sulphuric acid. Place the plate in an oven for 10 min. Cholesterol will stain red-brown, α-tocopherol will stain gold-brown. Egg yolk, PC, and PG will stain brown, while LPC will remain blue.

α-Tocopherol

(i)   As soon as the plate is dry, spray it with Emmerie-Engel reagent. Prepare fresh reagent before use by mixing equal volumes of 0.2% ethanolic ferric chloride hexahydrate with 0.5% 2,2'-dipyridyl. α-Tocopherol will stain red.

or (ii) As soon as the plate is dry, spray it with phosphomolybdate spray, made up by dissolving 3 g of dodeca molybdophosphoric acid in 10 ml of ethanol. α-Tocopherol (and phospholipids) will stain blue-black.

All lipids

Place the dried plates in an iodine tank for 5 – 15 min until dark brown spots, marking the position of the lipids, are fully developed.

Densitometry

The positions of spots which have been visualized by any of the above methods may be recorded by running under a TLC plate scanning densitometer. If the recording apparatus does not have its own integrator, quantitation may be achieved by obtaining a photocopy of the chart recorder trace, cutting out relevant peaks with scissors, and weighing each accurately, in order to evaluate the area under each peak corresponding to the concentration of the material. In cases where a single component is to be estimated, comparison of peak heights with a standard curve obtained from standards of a range of concentrations spotted on to the same plate is reliable with certain concentration limits.

2.3 Quantitation of lysolecithin: densitometry

Lysolecithin is the major product of hydrolysis of lecithin, in which one fatty acid chain is lost (either from the 1- or 2- position) by de-esterification (see Chapter 1, *Figure 1c*). Ideally, estimation of phospholipid hydrolysis by quantitation of lysolecithin could be carried out by HPLC, where the column outflow can be monitored continuously by UV absorbance to obtain a quantitative record of the eluted components. Unfortunately, because many natural phospholipids have fatty acids which are unsaturated (and therefore absorb) to different extents in the 1- and 2-position, it is difficult to relate peak height accurately to absolute quantities of lyso-phosphatidyl choline (LPC), since one does not know the absorbance of the fatty acid species that have been retained on the glycerol bridge. Consequently, methods are preferred which permit detection of LPC via the phosphate group, after first separating the hydrolysis product (LPC) from the parent PC by TLC. The spots can either be stained with iodine, then scraped off and the phosphate measured directly, or they can be measured by scanning densitometry as described here. Hydrolysis products of other phospholipids can be estimated in the same way.

2.3.1 *Preparation of reagents*

(a) *TLC elution solvents*

Prepare elution solvents as described for systems (ii) and (iii) referred to in Section 2.2(a).

System (ii): 65 ml of chloroform, 35 ml of methanol, 2.5 ml of double-distilled water and 2.5 ml of ammonium hydroxide.

System (iii): 30 ml of chloroform, 40 ml of acetone, 10 ml of methanol, 10 ml of glacial acetic acid and 5 ml of double-distilled water.

(b) *Molybdenum blue spray reagent*

(i)   Prepare 4.2 M sulphuric acid by carefully adding 4.5 ml of concentrated sulphuric acid to 15 ml of double-distilled water in a 25-ml measuring cylinder. Mix well, then make up to 20 ml.

(ii) Add 10 ml of 4.2 M sulphuric acid to 10 ml of the molybdenum spray reagent (Sigma Chemical Co.) just prior to spraying the TLC plates.

2.3.2 *Sample preparation*

(a) *Lyso-phosphatidyl choline standards*

Prepare a chloroform/methanol (2:1) solution of LPC with a final concentration of 3.2 mg ml^{-1}. Prepare five 2-fold dilutions in small glass screw-capped vials.

(b) *Liposomes*

In a small glass vial, dry down 50 μl of liposomes ($\sim 20-50$ mg ml^{-1}) and resuspend in 100 μl of chloroform/methanol solvent.

2.3.3 *Method of analysis*

(i) Spot 10 μl of each standard and sample (duplicates) on to the origin of channels of two TLC plates. The quantities of lysolecithin in the standards will be 64, 32, 16, 8, 4, and 2 μmol respectively.

(ii) Run the plates in separate tanks containing elution solvents for systems (ii) and (iii) as in Section 2.3.1 (a) for 45−60 min, until the solvent front has reached the top of the plates.

(iii) Remove the plates from the tanks and dry them well in a stream of air.

(iv) Just before scanning the plates in a scanning densitometer, spray the plates with molybdenum blue spray reagent in a well-ventilated fume hood. Wipe and rinse out the spray container well with distilled water immediately after use.

(v) Set up the TLC plate scanning densitometer to measure the absorbance at 800 nm. Scan each plate, channel by channel, to give peaks corresponding to each lysolecithin standard and sample.

(vi) Measure peak heights and plot against LPC concentration. At the lower concentrations, the curve obtained approximates to a straight line, and the quantity of LPC in the test samples can be read off accurately.

2.4 Estimation of phospholipid oxidation

Oxidation of the fatty acids of phospholipids, in the absence of specific oxidants, occurs via a free radical chain mechanism. The initiation step—abstraction of a hydrogen atom from the lipid chain—can occur most commonly as a result of exposure to electromagnetic radiation or trace amounts of contamination with transition metal ions. Although any lipid can form radicals by this mechanism, the most susceptible to degradation are lipids containing double bonds, since the unsaturation permits delocalization of the remaining unpaired electron along the lipid chain, and lowers the energy of this state (and hence increases the likelihood of its being formed). Polyunsaturated lipids are thus particularly prone to oxidative degradation.

In the presence of oxygen, a stable diradical, the process can develop further by the formation of hydroperoxides, which are relatively long lived, but break down spontaneously to aldehydes, with concomitant fission of the fatty acid chain on either side of the carbon−oxygen bond. A few of the processes which can take place are illustrated for a hydrocarbon containing a single double bond in *Figure 2a*.

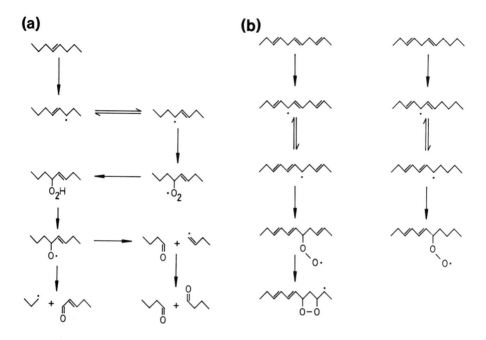

Figure 2. Lipid oxidation. (**a**) Radical abstraction in fatty acids with an isolated double bond, showing some of the chain-breaking reactions which can occur upon exposure to oxygen. (**b**) Formation of peroxides in polyunsaturated fatty acid chains. Di-unsaturated acids are less likely to form cyclic peroxides than tri-unsaturates, since rearrangements giving the di-oxygen radical adjacent to a—CH_2—CH=CH—group in diunsaturates can only take place if the opportunity to form conjugated dienes by abstraction is sacrificed.

In hydrocarbons which contain multiple unsaturations, additional processes can occur, such as the formation of endoperoxides, and rearrangement of double bonds in the molecule (see *Figure 2b*). Natural phospholipids, even polyunsaturated ones, contain only double bonds which are non-conjugated (i.e. non-adjacent), and thus have a UV absorbance peak at a very short wavelength (200−205nm). Abstraction of a hydrogen atom from a methylene group between two double bonds spreads the unsaturation over five carbon atoms and results in the formation of a conjugated diene which is energetically more favourable than the two double bonds in isolation. The radical intermediate is stabilized either by the addition of oxygen, or by abstraction of a hydrogen atom from a nearby hydrocarbon molecule. In lipids with high numbers of unsaturations, this process can take place many times in the same molecule, giving rise to conjugation of three or more double bonds (conjugated trienes).

In mixtures of lipids containing fatty acids with either single or multiple unsaturations, oxidation of the polyunsaturated fatty acids will take place preferentially, since the intermediates can be stabilized much more readily. For the purpose of measurement of phospholipid oxidation, the process can be considered to develop in three stages:

(i)   conjugation of isolated double bonds;
(ii)   formation of peroxides;

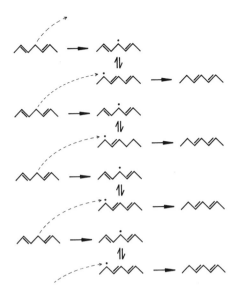

Figure 3. Rearrangement of double bonds by radical abstraction. After a hydrogen atom has been removed from a methylene group in between two double bonds, the radical species thus formed is stabilized by delocalization of the free electron over the five carbons encompassing those two unsaturations. This species can itself abstract a hydrogen atom from adjacent lipid molecules, to continue the chain reaction, but the hydrogen will return to a different carbon in the chain from the one that left, so that the double bonds can end up adjacent to each other (that is, conjugated—a configuration energetically more favourable than being isolated from each other by a methylene group). Note that this process can take place without the participation of oxygen.

(iii) aldehyde production and chain scission.

It is worth noting that, in the absence of oxygen, a free radical chain reaction can take place which will result in extensive formation of conjugated dienes and trienes (stage i) without the appearance of any peroxides whatsoever (*Figure 3*). On the other hand, the later stages of the process take place with the consumption of dienes, so that with the passage of time, it is possible to see an increase in end-products at the same time as a reduction in the components of the earlier stages. Consequently, no single test is sufficient on its own to determine the extent of oxidation of phospholipids, and even performance of several tests will still only give an approximate idea of the relative rate of the oxidation process, rather than absolute quantities.

The simplest method to perform monitors the formation of conjugated dienes during the free radical chain initiation phase. These are easily detected, since they absorb in the UV at 230 nm (5). Typical absorbance spectra are shown in *Figure 4*. Extensive phospholipid degradation may be signalled by a third peak at 270 nm, due to the formation of conjugated trienes.

Separate tests measure the levels of the two types of peroxides which can be formed—hydroperoxides and cyclic peroxides (or 'endoperoxides'). The latter are detected by reaction of their breakdown product at elevated temperatures (malondialdehyde) with thiobarbituric acid (TBA) giving a red chromophore which absorbs at 532 nm (6) (see

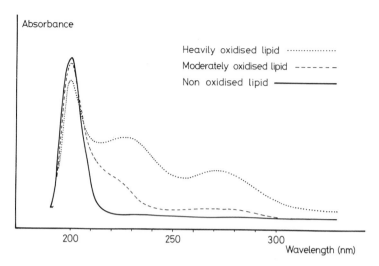

Figure 4. UV spectra of oxidized phosphatidyl cholines. Changes in the UV absorbance of lipids are the first sign of the occurrence of radical chain reactions which can lead to oxidation. The presence of conjugated dienes is indicated by the appearance of a peak around 230 nm, while the formation of conjugated trienes (as a result of abstraction of two hydrogen atoms per molecule) gives rise to a new peak at 270–280 nm.

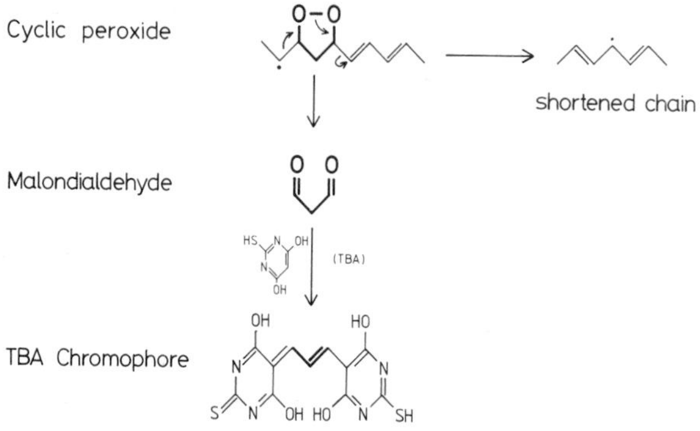

Figure 5. Detection of cyclic peroxide via malondialdehyde. Elimination of malondialdehyde from a chain containing a cyclic peroxide can result in reformation of the chain—i.e. chain shortening rather than chain breaking. The presence of malondialdehyde itself is detected colorimetrically by reaction with thiobarbituric acid (TBA) to give the coloured adduct shown.

Figure 5). The detection of hydroperoxides is based on their susceptibility to reduction by iodide as follows:

$$2H^+ + ROOH + 3I^- \rightarrow H_2O + ROH + I^-_3$$

Under the conditions used in this method, only lipid hydroperoxides react with iodide, thus excluding the endoperoxides that form malondialdehyde from the assay. Since in each case, only one type of peroxide is being measured, the results of a single test

116

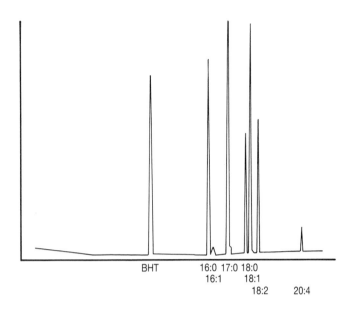

Figure 6. Gas−liquid chromatography of fatty acids of egg phospholipid. Separation of derivatized fatty acid chains from egg lecithin on a Carbowax 20M GLC column. Ideally, the homologues of the same chain length should all have eluted off before the saturated homologue of the next highest chain appears. Note that the anti-oxidant butylated hydroxytoluene (BHT, included to minimize oxidation during the derivatization procedure) comes off in approximately the C14 position. Natural phospholipids, however, contain no fatty acids shorter than C16. Normal heptadecanoic acid is added to the mixture to aid in the interpretation, since this odd-numbered chain fatty acid is absent from phospholipid mixtures from natural sources.

can only give an indication of the presence or absence of oxidative processes.

The final stages of lipid oxidation involve breakage and shortening of polyunsaturated fatty acid chains. The decrease in length of these acyl chains may be detected by gas-liquid chromatography, where fatty acids are drawn through the column at different rates according to their chain number and degree of unsaturation. The phospholipids are first hydrolysed, to liberate the free fatty acids, which are then derivatized with boron trifluoride to aid their passage down the column. In natural egg lecithin for example, the longest and most unsaturated fatty acids are 20 and 22 carbons long (see *Figure 6*). Reduction in quantity of chains of this length is indicative of lipid degradation. Appearance of shorter chains as a result of scission may be less easy to detect, since these compounds are much more volatile, and will appear with the solvent peak. Different methods of derivatizing may overcome this problem.

2.4.1 *UV absorbance method (5)*

Because the method involves spectroscopic measurements at short wavelengths, care must be taken to avoid interference due to the presence of other chromophores. This is a particular problem with materials which have a strong absorbance between 200−300 nm. Such compounds may be separated from phospholipids by acid Bligh-Dyer extraction as outlined in Appendix I. In the event that α-tocopherol or α-tocopherol hemisuccinate (which also absorb strongly in the UV) are present in sufficient quantities to interfere with this test, separation may be effected by passage through a Waters Sep-Pak silica gel column (see Appendix I).

(a) *Preparation of samples*

Lipid solids

Weigh out approximately 2 μmol of phospholipid (around 1.5 mg). Suspend in 0.1 ml of distilled water.

Lipid solutions

Measure out enough sample to contain 5 mg of phospholipid. Dry down using a stream of nitrogen. Suspend in 0.3 ml of ethanol, and take off two aliquots of 0.1 ml. Store in screw-capped vials.

Aqueous liposome suspensions

Measure out enough sample to contain 6 μmol (4.5 mg) of phospholipid but not more than 0.3 ml. Bring the volume to 0.3 ml with ethanol, and take off two aliquots of 0.1 ml each. Store in screw-capped vials.

(b) *Assay method*

(i)   To each 0.1-ml sample add 3 ml of absolute ethanol.
(ii)   Scan the UV absorbance spectrum of each sample (including those from Bligh-Dyer or Sep-Pak extractions) over the range 200−350 nm on a UV absorption spectrophotometer using 1-cm quartz cuvettes. Use ethanol as the blank.
(iii) Take the absorbance with ethanol in both cuvettes as the baseline value, and measure the absorbance at 233 nm.
(iv)   Using a molar extinction coefficient of 30 000 for dienes, calculate the value for μmol ml^{-1} diene in the sample from the OD reading at 233 nm. OD/ϵ = concentration in mol litre^{-1}.
(v)   Calculate the oxidation index according to the formula:

$$\% \text{ oxidation} = \frac{[\mu\text{mol ml}^{-1} \text{ diene}]}{[\mu\text{mol ml}^{-1} \text{ phosphate}]} \times 100.$$

2.4.2 *TBA method (for endoperoxides)*

(a) Preparation of reagent

1,1,3,3, Tetraethoxypropane (TEP) stock (mol. wt 220)
    Weigh out 44 mg of TEP (50 μl) and dissolve it in 100 ml of double-distilled water to give a concentration of 2 mM. Prepare weekly, and store refrigerated at 4°C.

TEP working solution (0.2 mM)
    Take 1 ml of stock TEP solution and add to 9 ml of double-distilled water. Mix well. Prepare fresh each time.

TBA reagent
    Weigh out 0.5 g of thiobarbituric acid and 0.3 g of sodium dodecyl sulphate. Dissolve

118

and make up to 100 ml with double-distilled water. Prepare fresh each time.

Ferric chloride solution
  Weigh out 0.27 g of ferric chloride and dissolve in 100 ml of distilled water. Store in a screw-capped bottle at 4°C.

Butylated hydroxytoluene solution (BHT)
  Dissolve 0.22 g in 100 ml of absolute ethanol, and store at 4°C.

Glycine buffer
  Dissolve 2.23 g of glycine hydrochloride in distilled water, and make up to 100 ml volume. Adjust the pH to pH 3.6 with 1 M sodium hydroxide.

(b) *Preparation of samples*

Lipid solids

Weigh out 5 mg of phospholipid and suspend in 0.5 ml distilled water.

Lipid solutions

Measure out enough sample to contain 1 mg of phospholipid. Dry down using a stream of nitrogen. Suspend in 1 ml of double-distilled water.

Aqueous liposome suspensions

Measure out enough sample to contain 5 mg of phospholipid, but not more than 0.5 ml. Bring the volume to 0.5 ml with double-distilled water.

(c) *Assay method*

(i)   Prepare tubes as follows:

Tube no.	TEP (ml)	Water (ml)	TEP (nM)
Blank	0.00	0.10	0
1	0.01	0.09	2
2	0.02	0.08	4
3	0.03	0.07	6
4	0.04	0.06	8
5	0.05	0.05	10
6	0.06	0.04	12

and three samples containing 1 mg of phospholipid in 0.1 ml of distilled water.

(ii)   Vortex all tubes.

(iii)   To each tube add the following in order, mixing well after each addition:

ferric chloride solution   0.1 ml
BHT solution   0.1 ml
0.2 M glycine buffer   1.5 ml
TBA reagent   1.5 ml

(iv)   Loosely cover all the tubes, and place them in a boiling water bath in a fume hood for approximately 15 min.

(v)   Cool the tubes containing samples to room temperature in a cold water bath.

(vi)   To each tube add 1 ml of glacial acetic acid, followed by 2 ml of chloroform.

Mix well and centrifuge for 20 min at 4°C to separate the two phases. Allow the sample to come to room temperature.

(vii) Take off the upper layer and measure the optical density at 532 nm against a water blank.

(viii) Subtract the average blank absorbance from standards and samples and calculate the TBA reactivity of the sample as μmol of TEP equivalent per μmol of phospholipid by comparison with TEP standard curve.

2.4.3 *Iodometric method (for hydroperoxides)*

(a) *Preparation of reagents*

Potassium iodide solution (7.2 M)

Dissolve 6 g of potassium iodide crystals in 5 ml of double-distilled water in a 25-ml glass conical flask, and stand it in an ice bath. Purge with a stream of nitrogen for 15 min.

Acetic acid:chloroform (3:2 v/v)

Mix together 12 ml of glacial acetic acid and 8 ml of chloroform. Degas in a bath sonicator for 10 min, and purge with a stream of nitrogen for 5 min.

Cadmium acetate solution

Dissolve 0.5 g of cadmium acetate in 100 ml of distilled water. Take care—cadmium acetate is a suspected carcinogen. Purge with nitrogen for 10 min, then store under nitrogen at 4°C in a screw-capped bottle.

Cumene hydroperoxide solution

Dissolve 15 ml of 80% cumene hydroperoxide in 100 ml of ethanol (degassed by sonication) to give a concentration of 0.8 μmol ml^{-1}. Store the stock solution under nitrogen at 4°C in a screw-capped bottle. To prepare a standard curve, measure out volumes of 250, 200, 150, 100, and 50 μl into 10-ml test tubes.

(b) *Preparation of samples*

Dry down approximately 5 mg of phospholipid in a stream of nitrogen in a 10-ml glass centrifuge tube.

(c) *Assay method*

(i)   To each of the samples, and to the cumene hydroperoxide standards, add 1 ml of acetic acid:chloroform mixture.

(ii)   Add 50 μl of potassium iodide solution to each tube, flush with nitrogen, cap tubes, vortex well, and incubate in the dark for 5 min at room temperature.

(iii) Add 3 ml of cadmium acetate to each tube, flush with nitrogen, cap tubes, mix well, and spin at 1000 g for 10 min.

(iv)   Read the upper phase at 353 nm.

2.4.4 *GLC method*

Columns for GLC work come in two different forms—packed columns, in which the liquid phase is coated on an inert granular support which is packed into a coiled tube of glass or stainless steel, and 'capillary' columns, which are much narrower in bore, and longer, made out of a glass or fused silica capillary which contains no packing, but where the liquid phase is coated directly onto the inner surface of the capillary wall itself. The capillary columns, also known as 'wall-coated open tubular' (WCOT), are a relatively recent innovation, and are becoming increasingly popular, as they are more sensitive than packed columns, since the background noise is markedly reduced. Essentially the same liquid phases which are suitable for separating fatty acids on packed columns can be used on capillary columns, with slight differences in operating conditions (oven temperature, etc.).

Cyanosilicone liquid phases (e.g. Silar 10C) are often used for analytical studies of fatty acids, since they can separate the *cis*- and *trans*-isomers of unsaturated fatty acids. For the purpose of investigating changes in composition due to oxidation, however, such resolution is unnecessary, and even counterproductive, since it can make identification of the smaller peaks more difficult, especially if there is overlap between homologues of different chain length. In this case adoption of a phase such as the polyglycols (e.g. Carbowax 20M) is preferable. Using this coating, homologous fatty acids of the same chain length but different number of unsaturations are separated, with the parent saturated fatty acid eluting first followed by the mono-, di-, tri- etc. unsaturated fatty acids eluting later. The important point is that these homologues should run close enough together that they are all off the column before the saturated homologue of the next higher chain length arrives. In the case of mixtures of fatty acids where only even chain lengths are found, this can easily be achieved (see *Figure 6*), so that interpretation of the GLC spectrum can be carried out without recourse to additional analytical techniques. Other liquid phases can be employed (e.g. methyl silicone—SP2100, OV1) in which case the members of the homologous series elute in reverse order, e.g. C18:2, C18:1 then C18:0. For precise information of operating conditions for individual columns, refer to the manufacturer's instructions. A more detailed discussion of GLC (and other techniques) for structural analysis of lipids is given in reference 7. A protocol for obtaining derivatized fatty acids from phospholipids is given below.

(a) *Reagent preparation*

(i)   Weigh out 20 mg of BHT (butylated hydroxytoluene—2,6 di-tert-butyl *p*-cresol) and dissolve it in 1 ml of methanol to give a 2% solution. Store at −20°C.

(ii)   Aliquot 12% BF_3 in methanol (obtained commercially from Sigma) into 2-ml portions in sealed screw-capped vials under nitrogen, and store at −20°C.

(iii) Weigh out 20 mg of pentadecanoic acid and dissolve it in 1 ml of methanol to give a 2% solution. Store in a screw-capped vial at −20°C.

(b) *Sample preparation*

Lipid solids

Weigh out 5 mg of dry phospholipid into a 15-ml glass screw-capped test-tube.

Lipid solutions

Measure out sufficient volume to contain 5 mg of lipid, and dry down under a stream of nitrogen in a 15-ml glass screw-capped test-tube.

Liposomes: Bligh-Dyer extraction

(i)   To 0.4 ml of liposome suspension add 0.5 ml of chloroform and 1 ml of methanol in a glass test-tube. Vortex for 30 sec.

(ii)   Add 0.5 ml of chloroform, and 0.6 ml of 1 M HCl and vortex again.

(iii)   Transfer to a separating funnel (volume 5 ml) and stand to allow the layers to separate fully.

(iv)   Draw off the lower organic layer, and measure out a volume containing 5 mg of lipid (assuming complete extraction) into a 15-ml glass screw-capped test-tube.

(v)   Dry down under a stream of nitrogen.

(c) *Assay method*

(i)   To a glass test-tube containing 5 mg of dried lipid add 2 ml of 12% BF_3 in methanol, 200 μl of 2% BHT solution in methanol, 50 μl of 2% pentadecanoic acid solution (standard—1 mg) in methanol.

(ii)   Flush the tube with nitrogen and tightly seal it with the screw-cap.

(iii)   Heat it in a water bath at 80°C for 5—10 min.

(iv)   Allow the tube to cool, then open it carefully, add 1 ml of distilled water, then 2 ml of hexane, and vortex.

(v)   Spin down on a bench centrifuge for 10 min.

(vi)   Remove the supernatant organic layer, and dry it down under a stream of nitrogen.

(vii)   Redissolve the residue in 1 ml of hexane, to give a solution of approximate concentration 5 mg ml^{-1}. Store at −20°C until ready for use.

2.5 Analysis of cholesterol

2.5.1 *Qualitative examination of cholesterol purity: GLC*

In an analogous manner to phospholipids, cholesterol and its autoxidation products can be examined using capillary columns of flexible fused silica (8). A typical coating would be bonded poly(dimethyl siloxane). Both parent compound and derivatized sterols may be passed down the column, although the life of the column is extended when using TMS derivatives, since these materials are less likely than free sterols to stick to the column in trace amounts. TMS ethers can be prepared using a similar experimental set-up to that described for fatty acid derivatization, in which the sterols are treated with a large excess of a mixture of bis-(trimethylsilyl) acetamide, hexamethyl disilazane and chlorotrimethyl silane in a ratio 40:40:15 v/v. For this the reactants ($\sim$2 ml) need to be heated to 80°C in a tightly-sealed 15-ml screw-capped test-tube for 15 min. Then run the sterols on a 12 mm × 0.2 mm column (carrier gas flow rate 1.5ml min^{-1}) with the oven temperature starting at 240°C for 2 min, rising at a rate of 5°C/min to 290°C. Cholesterol comes off first followed by any oxygenated degradation products.

2.5.2 *Quantitation of cholesterol: ferric perchlorate method*

In this method (9), cholesterol in either free or esterified form gives a purple complex with iron upon reaction with a combined reagent containing ferric perchlorate, ethyl acetate, and sulphuric acid. Absorbance at 610 nm is directly proportional to the quantity of cholesterol present, in the range 0—80 μg.

(a) Reagent preparation

Cholesterol reagent

Dissolve 260 mg of ferric perchlorate in 300 ml of ethyl acetate in a one-litre conical flask. Cool the mixture to 4°C in an ice bath, then slowly add 200 ml of concentrated sulphuric acid, in small portions with constant mixing in the cold, making sure that the temperature does not rise above 45°C at any time. Store at 4°C in the dark in glass screw-capped bottles. The solution is stable for one to two years.

Stock cholesterol solution

Dissolve 100 mg of cholesterol in 50 ml of glacial acetic acid. Cholesterol dissolves only very slowly. Dissolution may be aided by sonicating for approximately 20 min. Store below 0°C in sealed bottles.

(b) Sample preparation

Cholesterol standards

Measure out the following volume of stock cholesterol solution into test tubes: 0.8 ml, 0.6 ml, 0.4 ml, 0.2 ml, and 0.1 ml. Make each up to 1 ml with glacial acetic acid. For the blank use glacial acetic acid. Transfer 50 µl of each into separate clean glass 10-ml screw-capped test-tubes.

Liposome sample

Prepare the lipid extract from 0.1 ml of liposomes using a Waters C18 Sep-Pak minicolumn as described in Appendix I and collect in a total volume of 10 ml of cholesterol. Take three separate 1-ml samples of this extract and dry them down in separate 10-ml glass screw-capped test-tubes. Redissolve the lipids in 50 µl of glacial acetic acid.

Lipid solution

Take 100 µl of organic solution up into a 0.1-ml glass pipette and introduce the solution into a small glass vial. Dilute 10-fold by adding 0.9 ml of cholesterol (from a 1-ml glass pipette) and mix well. Using a glass microcapillary, transfer three 50-µl samples into three fresh 10-ml glass screw-capped test-tubes and dry down. Redissolve the lipids in 50 µl of glacial acetic acid.

(c) Assay method

(i) Add 5 ml of cholesterol reagent to each tube containing standards and samples. Vortex well and loosely screw the caps on to the tubes.

(ii) Set the water bath to boil. When the temperature has reached 100°C, immerse the tubes in the bath and leave them in for exactly 90 sec.

(iii) Remove all the tubes quickly, and allow them to cool rapidly by standing them in a beaker of cold water. Take care to treat all the tubes identically, as the intensity of colour is critically dependent upon the length of time spent at an elevated temperature. Spin them, if necessary, in a bench centrifuge at 1000 g for 15 min to remove cloudiness.

(iv) Read the absorbance at 610 nm on a spectrophotometer, using zero standard as

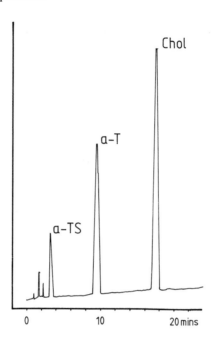

Figure 7. HPLC of neutral lipids. High performance liquid chromatographic separation of cholesterol, α-tocopherol and its hemisuccinate on a reverse phase (C18) HPLC column with methanol as eluting solvent. Phospholipids may be removed from the sample mixture by Sep-Pak separation prior to application. Individual components can be accurately quantitated by comparison of peak heights with those of a range of standards of known concentration.

a blank to obtain zero reading. Quantities of cholesterol in each of the standards are 0.08, 0.06, 0.04, 0.02, and 0.01 mg. Draw a standard curve to read off the quantity of cholesterol in the sample tubes.

2.6 Quantitation of α-tocopherol: HPLC

α-Tocopherol (α-T) itself can be quantitated using specific tests for reducing power, involving the formation of iron complexes as in the Emmerie-Engle reagent (see Section 2.2 on TLC). However, the use of alternative anti-oxidants such as the more stable hemisuccinate derivative of tocopherol precludes the use of this type of test, since the hemisuccinate derivative displays little reducing activity until after it is hydrolysed to the free tocopherol, a process which takes place only slowly. Quantitation using HPLC is the simplest alternative general purpose method for looking at both of these compounds, as well as their breakdown products. The system described here provides a rapid separation of these components (important for quantitative determinations), and has the advantage that cholesterol can be measured simultaneously if desired (*Figure 7*).

2.6.1 *Sample preparation*

(a) *Standards*

Weigh out 5 mg of α-tocopherol (or hemisuccinate derivative α-TS), and dissolve in 5 ml of methanol (HPLC grade). Measure out aliquots of 0.5 ml, and dilute each appropriately to give final concentrations of 1, 0.8, 0.6, 0.4, 0.2, and 0.1 mg ml^{-1} in methanol; prepare these fresh just before use, and keep in screw-capped vials to prevent evaporation.

(b) *Test material*

(i)   Mixture of lipids in chloroform solution. Introduce exactly 0.2 ml of chloroform solution of lipids onto a silica gel Sep-Pak minicolumn, and flush through with 20 ml of chloroform. Evaporate to dryness under a stream of nitrogen, and dissolve the residue in 1 ml of methanol. Store in capped vials.

(ii)   Liposome suspension. Measure out 0.1 ml of aqueous suspension, and evaporate it to dryness under a stream of nitrogen. Dissolve the residue in 0.2 ml of chloroform:methanol (2:1) and process this solution as for lipid mixtures, in order to remove phospholipid and entrapped material.

2.6.2 *Assay method*

(i)   Install a Spherisorb ODS (C18) reverse phase HPLC column in an appropriate HPLC apparatus and set the detector to measure at 214 nm.

(ii)   Run methanol (degassed by bath sonication for 15 min) through the column for 30 min, at a flow rate of 1 ml min^{-1}.

(iii)   Introduce the first four standards on to the column (from lowest to highest concentration) at intervals of 4 min. Use an injection volume of 20 μl. The first sample will be detected at 4−8 min (cholesterol comes off later).

(iv)   After the standards have been eluted, continue to apply the remainder to the column in the same manner as described in Step (iii).

(v)   Apply three replicates of 20 μl each of the test samples, allowing 15 min for each run.

(vi)   Plot the peak height against concentration to check the linearity of the standard curve.

(vii)   Calculate the concentrations of α-T(S) in test sample from the peak height by reference to the standard curve.

3. PROPERTIES OF INTACT LIPOSOMES

3.1 Determination of percentage capture

It is clearly essential to measure the quantity of material entrapped inside liposomes before going on to study the behaviour of this entrapped material in physical or biological systems, since the effects observed experimentally will usually be dose-related. After removal of unincorporated material by the separation techniques described in the last chapter, one may assume that the quantity of material remaining is 100% entrapped, but the proportion may change upon storage, or after subsequent manipulations are performed. For long term stability tests, therefore, and also for the initial stages of development of new liposome formulations, or new methods of preparation, a technique is needed for separating free from entrapped material which is rapid, which requires small quantities of sample, and which can be performed on many different samples simultaneously. The two methods described below fulfil all these criteria.

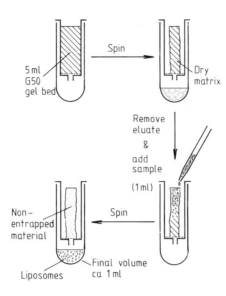

Figure 8. Removal of unentrapped material by minicolumn centrifugation. This method can be used both as a means of purification of liposomes on a small scale, and analysis of a liposomal suspension to determine percentage entrapment. The method is satisfactory for solutes less than 7000 molecular weight.

3.1.1 *Minicolumn centrifugation method*

This is based on the method of Fry *et al.* (10).

(a) *Column preparation*

(i)   Allow 10 g of Sephadex G-50 (medium) to swell in 120 ml of 0.9% NaCl in a glass screw-capped bottle for at least 5 h at room temperature. Store at 4°C until required for use.

(ii)   Remove the plungers from 1-ml disposable plastic syringes (one for each sample) and plug each barrel with a Whatman GF/B filter pad.

(iii) Rest each barrel in a 13 × 10 mm centrifuge tube.

(iv)   Fill the barrels to the top with hydrated gel, using a Pasteur pipette with the tip removed.

(v)   Place the tubes containing the columns in a bench centrifuge, and spin at 2000 r.p.m. for 3 min to remove excess saline solution. After spinning, the gel column should be dry and have come away from the sides of the barrel. The height of the bed should be level with the 0.9 ml mark (see *Figure 8*).

(vi)   Empty the eluted saline from each collection tube.

(b) *Processing of samples*

(i)   Apply exactly 0.2 ml of liposome suspension (undiluted) dropwise to the top of the gel bed. Take care not to let the sample trickle down the sides of the column bed.

(ii)   Spin the columns at 2000 r.p.m. for 3 min in a bench centrifuge to expel the void volume containing the liposomes into the centrifuge tube.

126

(iii)   Remove eluates from each tube and set aside for assaying.

(iv)   Apply 0.25 ml of saline to each column, spin as previously, and remove eluates from each tube. (This sample may or may not contain liposomes, depending upon size, lipid composition, etc. Unentrapped solute should not come out at this stage.)

(v)   Load 0.1 ml of saline onto each column. Centrifuge to recover all eluate, containing the unentrapped material. (Repeat this step if necessary until all free material has come off the column.)

(vi)   Measure the concentration of free or entrapped material in eluates by standard methods (see below).

(vii) Measure the phosphate concentrations in the manner outlined in Section 2.1.

Another advantage of this technique is that the liposomes can be recovered with practically no dilution, since the excess void fluid which would supplement the liposome volume has already been drained off in the previous spin. Larger sizes of column can be used (5–10 ml syringe barrels) capable of handling 1 ml of liposome suspension at a time, so that for small-scale experiments, this technique can be used as a preparative method for the separation of liposomes from unentrapped solute (see Chapter 2, Section 5.1).

3.1.2 *Protamine aggregation*

This method (11,12, and illustrated in *Figure 9*) may be used for liposomes of any composition (both neutral or negatively charged). However, a preliminary test should be carried out beforehand to check that the solute material entrapped does not itself precipitate in the presence of protamine after release from liposomes.

Assay method

(i)   Place 0.1 ml of liposome suspension [20 mg ml^{-1} lipids in normal (0.9% NaCl) saline] in a 10-ml conical glass centrifuge tube.

(ii)   To the liposome suspension add 0.1 ml of protamine solution (10 mg ml^{-1}). Mix it using a micropipette and allow it to stand for 3 min.

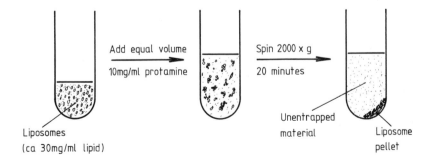

Figure 9. Protamine aggregation method for determination of the percentage entrapment. Liposomes may be precipitated in the presence of protamine, leaving unentrapped material remaining in the supernatant after centrifugation.

(iii) To the mixture add 3 ml of saline. Spin in a bench centrifuge (using a swing-out rotor) for 20 min at 2000 *g* at room temperature.

(iv) Take off the top 2 ml of supernatant, and assay the concentration of free, unentrapped compound by standard methods.

(v) Remove the remainder of the supernatant from the tube and discard. Resuspend the pellet in 0.6 ml of 10% Triton X-100. When liposomal material is properly dissolved (after warming if necessary), make the solution up to a final volume of 3.2 ml with normal saline and assay the concentration of entrapped material by standard methods.

3.1.3 *Measurement of liposomal contents*

Measurement of coloured contents can be carried out by the addition of 10 μl of liposomes to 2 ml of ethanol, or, if the agent is not soluble in ethanol (e.g. potassium chromate) by dissolution of 10 μl of liposomes with 10 μl of 10% Triton X-100 (with warming for 30 min if necessary, for example, for cholesterol-containing liposomes), followed by dilution to 2 ml. Solutions obtained by either method are sufficiently clear for direct spectrophotometric determinations to be made. Radioactive markers are easily measured by addition of liposomes to a scintillation cocktail—see Appendix I.

Having prepared a suspension of liposomes containing a known quantity of entrapped solute, with minimal solute remaining in the extra-liposomal buffer, many applications require a knowledge of the rate at which the liposomal contents leak out into the medium. Under normal circumstances, with a well-formulated preparation, leakage will be very slow and the techniques described above will suffice. Where the projected use of a liposome preparation is a pharmaceutical application for drug delivery or targeting, the marker for measuring release in long term stability tests will naturally be the incorporated pharmacological agent itself. For other applications, however, particularly those in which release is induced deliberately as a result of a biochemical or biophysical change, it may be desirable to use special 'purpose-designed' markers which simplify measurement and interpretation in such experiments. These are discussed in the next section.

3.2 Determination of percentage release

3.2.1 *Choice of marker*

Liposomes are often used as model membrane systems to investigate the effects of biological modulators directly on membrane properties such as fluidity, phase transition, or permeability, and changes in these parameters can give quantitative and qualitative information about membrane interactions. Relatively rapid changes in permeability can be brought about by, for example, calcium-mediated fusion, interaction with polyanions or anaesthetics, passage through the phase transition temperature, ligand-mediated induction of a phase change, or antibody/complement mediated lysis of antigen-bearing liposomes. The ways in which these phenomena are employed are discussed in following chapters.

In the study of the release of water-soluble markers from liposomes, it is advisable to choose the marker to be used with some care. Ideally, one would like to start by using a molecule which does not pass through intact membranes, which is highly

Table 1. Water-soluble markers for liposomal entrapment studies.

Detection method	Marker	Molecular weight
Optical density	Sodium chromate	162
	Ponceau red	760
	*Arsenazo III	776
	Cytochrome C	13000
	Haemoglobin	64500
Fluorescence	Fluorescein	319
	*Carboxyfluorescein	362
	*Calcein	620
	Fluorescein dextran	4000−2 million
Enyzmatic	*Glucose	181
	*Isocitrate	258
	*Soybean trypsin inhibitor	22000
	*Superoxide dismutases	16 500, 30 000
	*Horseradish peroxidase	40 000
Radiolabel	[^{14}C]glucose	183
	[^{99m}Tc]DTPA	492
	[^{111}In]bleomycin	1153
	[^{14}C]inulin	5000
	[^{125}I]PVP	10000−360000
	[^{3}H]DNA	millions

*Use of these markers avoids the need for liposome separation procedures.

water-soluble, with a very low solubility in organic media, which does not associate with membranes in any way so as to destabilize or aggregate them, and which can be easily separated from liposomes by conventional methods. *Table 1* presents a list of compounds, classified according to method of detection, which are widely used, and are generally considered to meet the specifications outlined above. Most of these materials can be measured easily after separation from liposomes by the methods already described. However, there are some agents (marked with an asterisk) whose extra-liposomal concentration can be determined *in situ* without the need for a separation step. These are discussed in the following sections.

3.2.2 *Arsenazo III*

This method, developed by Weissman and colleagues (13) for the investigation of liposome integrity *in vitro* and *in vivo*, makes use of the fact that the spectral characteristics of the dye, arsenazo III (2,7-bis[2-arsonophenylazo-1,8-dihydroxy-naphthalene-3,6-disulphonic acid) change markedly in the presence and absence of calcium ions. Calcium binds strongly to the dye and produces a spectral shift from 560 nm to 606/660 nm, that is, a change from red to blue which is clearly distinguishable by eye (see *Figure 10*). This property of the dye can be used in two ways. In the first, liposomes can be made incorporating the purified calcium-free dye. Under these circumstances, both the entrapped and non-entrapped dye will be coloured red, and will have low absorbance at 660 nm. The liposome suspension can be diluted out so

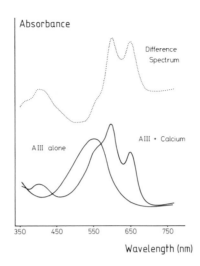

Absorbance

Difference
Spectrum

AIII · Calcium

AIII alone

350   450   550   650   750

Wavelength (nm)

Figure 10. Spectral characteristics of arsenazo III. The addition of calcium ions to the strongly coloured arsenazo III dye induces a spectral shift which permits one to distinguish between dye outside liposomes and that entrapped inside liposomes, without recourse to physical methods of separation, since calcium does not enter into the liposome within the time course of the determination.

that interference due to scattering is minimized. Upon addition of calcium, the unentrapped dye absorbs while the encapsulated dye remains unchanged. Upon addition of Triton X-100, the liposome membrane is disrupted sufficiently to allow calcium ions access to the internal compartment, and the whole of the dye present in the sample changes its absorbance. The difference in readings before and after Triton X-100 gives the proportion of dye entrapped. The determination can be carried out in a more sophisticated way using a double-beam spectrophotometer (or a single beam with the blanks measured sequentially), in which samples treated in various ways are compared as shown in *Table 2a*. In this way, allowance is made for any differences in scattering due to lysis of liposomes by detergent.

The problem with this method is the requirement for free calcium ions. A number of liposome compositions are sensitive, to a greater or lesser degree, to the presence of calcium which may make interpretation of the results rather difficult. These objections can be overcome in part by carrying out a test in reverse, that is by preparing liposomes containing the dye plus calcium bound to each other in stoichiometric proportions, and measuring the portion released by addition of EGTA, which competes with arsenazo III for calcium ions and converts the complex to the free dye. The scheme of measurements is shown in *Table 2b*. Methods for purifying arsenazo III (to remove any free calcium before use) and for quantitation are given in ref. 13.

This method is a very useful and versatile one. It can be used to monitor processes where normal separation procedures are not feasible, for example fusion between liposomes, or interaction with living cells. Its only drawback is that, because the indicator system is calcium-dependent, membrane processes which themselves rely on interactions with calcium ions are not amenable to rigorous study.

Table 2. Arsenazo III method for measurement of percentage release.

(a)
Readings:

(1) Blank: 2 ml AIII liposomes (1:100)
    Sample: 2 ml AIII liposomes (1:100)

(2) Blank: 2 ml AIII liposomes (1:100) + 200 μl saline
    Sample: 2 ml AIII liposomes (1:100) + 200 μl calcium

(3) Blank: 2 ml AIII liposomes (1:100) + 200 μl saline   + 100 μl Triton X-100
    Sample: 2 ml AIII liposomes (1:100) + 200 μl calcium + 100 μl Triton X-100

Measure OD at 660 nm
Set reading (1) to zero

$$\frac{(3)-(2)}{(3)} \times 100 = \% \text{ dye entrapped}$$

(b)
Readings:

(1) Sample: 2 ml AIII/Ca liposomes (1:100)
    Blank: 2 ml AIII/Ca liposomes (1:100)

(2) Sample: 2 ml AIII/Ca liposomes (1:100) + 200 μl EGTA
    Blank: 2 ml AIII/Ca liposomes (1:100) + 200 μl saline

(3) Sample: 2 ml AIII/Ca liposomes (1:100) + 200 μl EGTA + 100 μl Triton X-100
    Blank: 2 ml AIII/Ca liposomes (1:100) + 200 μl saline + 100 μl Triton X-100

Measure OD at 660 nm
Set reading (1) to zero

$$\frac{(3)-(2)}{(3)} \times 100 = \% \text{ dye entrapped}$$

Prepare liposomes in 3 mM AIII (2.3 mg ml^{-1})
            or 3 mM AIII/Ca (CaCl$_2$ concentration 0.45 mg ml^{-1})

Dilute liposomes 1 in 100 in calcium-free buffer before adding to cuvette

Concentration of added calcium: 100 mM (15 mg ml^{-1} CaCl$_2$) to give a final concentration of 10 mM
Concentration of added EGTA: 100 mm (38 mg ml^{-1}) to give final concentration of 10 mM
Concentration of added Triton X-100: 5% v/v (0.26% final concentration)

3.2.3 *Carboxyfluorescein and calcein*

Carboxyfluorescein (CF) is very much less lipophilic than its parent compound fluorescein, and has a much lower tendency to associate with phospholipid membranes. Consequently its use in place of fluorescein as a marker for the aqueous compartment of liposomes is to be recommended. However, the property of carboxyfluorescein which lends itself to *in situ* determinations of liposome contents is its ability to self-quench (14), that is, at high concentrations, the fluorescence of carboxyfluorescein is very much reduced compared with diluted samples, probably because of intermolecular interactions.

Thus the basic approach when using this compound is to prepare liposomes containing a high quenching concentration of CF (usually about 100 mM in 10 mM Tris or Hepes,

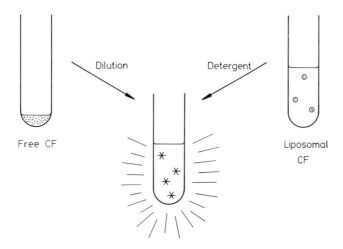

Figure 11. Unquenching of carboxyfluorescein fluorescence. At high concentrations, carboxyfluorescein (CF) fluoresces only weakly because of self-quenching (**top left**). Reducing the concentration of CF by diluting CF solutions results in increased fluorescence (**centre**) as this quenching is reduced. CF which is entrapped inside liposomes at high concentration will give no increase in fluorescence, however, even when the liposome suspension is diluted, since the concentration of the entrapped liposomal contents remains unchanged (see **right-hand side**). On the other hand, CF which has leaked out of the liposomes into the extraliposomal medium will display enhancement of fluorescence upon dilution of the suspension. This may also be accomplished by lysis of diluted liposomes with detergent. This technique can be used to investigate liposomal leakage and permeability processes over time ranges too short to permit separation of components by physical means.

which is iso-osmolar with physiological saline), and then dilute the suspension approximately 10 000-fold, whereupon the non-entrapped material will fluoresce intensely, while the entrapped CF gives no signal (*Figure 11*). Upon addition of Triton X-100, the liposomal contents will be released, and, being diluted in the suspending medium, it will also fluoresce, so that the concentration of CF initially inside, as well as outside the liposomes can be deduced. As long as the final concentration of CF in the measuring cuvette is $3-30\ \mu M$ (when fluorescence is completely unquenched— see *Figure 12*), the fluorescence readings will be directly proportional to the concentration.

Preparation and utilization of carboxyfluorescein liposomes
(i)   Prepare Tris-buffered saline containing 10 mM Tris ($1.21\ \mathrm{g\ l^{-1}}$) and 140 mM NaCl ($8.2\ \mathrm{g\ l^{-1}}$).
(ii)   Prepare a solution of CF (purified first as described in Appendix I) at a concentration of 100 mM ($36.2\ \mathrm{mg\ ml^{-1}}$) in 10 mM Tris ($1.21\ \mathrm{mg\ ml^{-1}}$).
(iii)   Use this solution to prepare liposomes by any of the standard methods.
(iv)   Remove unincorporated dye by passage through a Sephadex G-50 column equilibrated in Tris-buffered saline and pre-saturated with empty liposomal lipids. (It has been reported that CF liposomes are susceptible to leakage in high gravitational fields, so use of the minicolumn centrifugation method for separation may be inadvisable.)

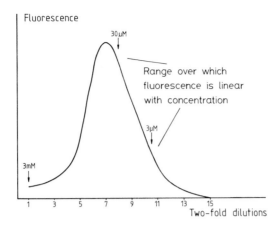

Figure 12. Change in carboxyfluorescein fluorescence with increasing dilution. Self-quenching of carboxyfluorescein is not completely relaxed until the solution has been diluted to 30 μmol concentration. For accurate quantitative work, it is important to carry out determinations at a concentration range below 30 μM. (Data provided by R.R.C.New and R.E.Stringer.)

(v)   Determine the concentration of CF in the purified liposomes by dissolving 100 μl in 100 μl of 10% Triton X-100 (v/v), making the volume up to 2 ml and measuring the optical density at 470 nm. Compare the reading obtained with a standard curve prepared over the range 0.1−1 mM of CF.

(vi)   On the basis of the measured concentration, dilute the liposomes in Tris-buffered saline to give a final concentration of 2 mM CF.

(vii)   Place 10 μl of the liposome suspension in a glass fluorescence cuvette.

(viii) Add 2.0 ml of Tris-buffered saline and mix well.

(ix)   Insert the fluorescence cuvette (clear on all four sides) into the sample holder of a spectrofluorimeter (excitation wavelength set at 470 nm, emission at 520 nm). Ensure that the sample is equilibrated to the temperature of the instrument.

(x)   Measure fluorescence emission to obtain a zero baseline reading. (At this stage, agents under study which may have a bearing on the efflux rate of CF from liposomes can be added.)

(xi)   Measure the fluorescence (F) at regular time intervals after the start.

(xii)   At the end of the experiment, add 100 μl of Triton X-100 (20% v/v) to give a final concentration of 2%. Mix well by pipetting, and take a final fluorescence reading (F_t). Correct for the increase in volume of 5% upon addition of Triton X-100.

Carboxyfluorescein has been used very widely for studying release of entrapped solutes from liposomes in the presence of biological fluids (15), and during the course of these investigations, a number of features have come to light where some caution needs to be exercised in the use of this method. These are as follows.

(a)   The fluorescence intensity of CF (in its unquenched state) is pH-dependent, increasing by a factor of 2 over the range pH 6−8, as the proportion of the ionized species of the molecule increases.

133

(b)   Leakage of CF from liposomes is enhanced at low pH as the molecule becomes more highly protonated.

(c)   Serum components bind CF, and some quenching can occur; in the presence of Triton X-100 this quenching is enhanced, particularly with human serum (16).

With the use of strongly buffered solutions at neutral pH, and the inclusion of appropriate controls, allowances can be made for these potential inaccuracies. In cases where difficulty is still encountered, the use of calcein (also a self-quenching fluorophore) as an alternative may be investigated, since the fluorescence of this, although similar to CF (both are fluorescein derivatives) is less pH-dependent (17). Its interaction with serum components has not been widely documented, however, and at high dilutions, calcein seems to be more prone to artefacts such as adherence to glassware, etc. Purification procedures for these compounds are given in Appendix I.

3.2.4 *Glucose*

Extraliposomal concentrations of unlabelled glucose can be determined by enzymatic means, making use of the coupled ATP-driven conversion of glucose to 6-phospho-gluconate (18). The reaction is catalysed by hexokinase and glucose-6-phosphate dehydrogenase (G-6-PDH) with the reduction of NADP to NADPH, the formation of which can be monitored by an increase in optical density in the UV (at 340 nm). The scheme is shown in *Figure 13*. One mole of NADPH is formed for every mole of glucose available, so the concentration of extraliposomal glucose is linearly related to the UV absorbance of NADPH. The procedure to follow for carrying out measurements is as follows.

(a) *Glucose release assay*

Preparation of reagents

Tris-buffered saline (TBS) (100 mM Tris, 64 mM NaCl, 3.5 mM $MgCl_2$, 0.15 M $CaCl_2$).

Measure out Tris   1.21 g
            NaCl   0.37 g
            $MgCl_2.6H_2O$ 70 mg
            $CaCl_2.2H_2O$ 22 mg
Dissolve in 100 ml of double-distilled water, mix well and adjust to pH 7.5 with 1 M HCl.

Stock ATP solution (10 mM)
    Dissolve 100 mg of ATP in 20 ml of Tris-buffered saline. Subdivide into 1-ml aliquots and store frozen at $-20°C$ until required for use.

Stock NADP solution (5 mM)
    Dissolve 75 mg of NADP in 20 ml of TBS. Subdivide into 1-ml aliquots and store frozen at $-20°C$ until required for use.

Hexokinase (yeast) solution (Boehringer-Mannheim)
    Dissolve 2 mg in 1 ml of distilled water. Remove ammonium sulphate by dialysis overnight in the cold against 2 litres of distilled water, then make up to 15 ml with TBS and subdivide into 1-ml aliquots and store at 4°C (do not freeze).

Glucose

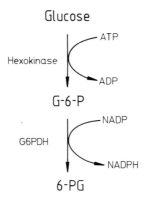

Figure 13. Enzymic detection of glucose. This is another way of monitoring the leakage of components from liposomes without need for separation of the individual components. Glucose which has been released from liposomes is converted, via glucose-6-phosphate to 6-phosphogluconate, with concomitant conversion of NADP to NADPH which may be monitored spectrophotometrically in the UV.

Glucose-6-phosphate dehydrogenase (yeast) solution (Boehringer-Mannheim)
Dissolve 1 mg in 1 ml of distilled water. Remove ammonium sulphate by dialysis overnight in the cold against 2 litres of distilled water, then make up to 15 ml with TBS and subdivide into 1-ml aliquots and store at 4°C (do not freeze).

Glucose assay reagent (prepare fresh on the day of use)
Add together 1 ml each of stock ATP, NADP, hexokinase, and G-6-PDH solutions and mix well. (Sufficient for 8 assays.) In addition, prepare 1 ml of reagent as above, with one of the components (e.g. ATP or NADP) replaced by TBS, to act as a control blank.

(b) *Preparation of liposomes*

(i)   Make up 300 mM glucose solution by dissolving 0.54 g of glucose in 10 ml of distilled water.
(ii)   Using the glucose solution from Step (i), prepare liposomes by any of the standard methods at a concentration of approximately 10 mg lipid ml^{-1}.
(iii)   Remove unentrapped glucose by the minicolumn method (see Section 3.1.1).
(iv)   Dilute the liposomes approximately 20-fold by addition of 500 μl of the liposome suspension to 10 ml of TBS.

(c) *Assay method*

(i)   For each sample to be studied, introduce 500 μl of glucose assay reagent into a 1.5-ml quartz semi-micro cuvette (clear on two sides).
(ii)   To an additional cuvette, introduce 500 μl of control buffer (either TBS or assay reagent minus one component).
(iii)   Make the volume of cuvettes up to 1 ml with 500 μl of buffer. At this stage, additional agents may be added to each mixture to study their effect on the release process.
(iv)   Place both cuvettes in a double-beam spectrophotometer, with the control cuvette in the reference beam.

(v)   To each cuvette add 100 μl of liposome suspension, and mix well by gently pipetting, then immediately take a reading of the difference in optical density between blank and sample at 340 nm. (Set instrument reading to zero at this stage if desired.)

(vi)   Incubate the reaction mixture in the cuvettes at room temperature for 30 min, then measure the difference in optical density again at 340 nm as before.

(vii)   Determine the quantity of total glucose separately by adding 25 μl of 50% (v/v) Triton X-100 to 100 μl of liposomes and mixing well (with warming if necessary). Then add 500 μl of TBS, mix well, and add 500 μl of assay reagent. Incubate for 10 min at room temperature, then measure the optical density (relative to control) at 340 nm. Make a correction for the increase in volume due to the added Triton X-100.

The main disadvantage of the enzymatic determination of glucose is that it is not instantaneous, so that it is not suitable for use in kinetic studies where release of glucose is fairly rapid. The principal application for this method to date has been the quantitation of immune lysis of antigen-bearing liposomes in the presence of antibody and complement (19). Even then, difficulties may be encountered as a result of interference by the presence of high levels of endogenous glucose in animal sera. Most workers routinely dialyse these sera to remove glucose before addition to the assay mixture. It is possible that the use the glucose-6-phosphate in place of glucose as the entrapped agent could overcome this problem. In this case, hexokinase and ATP would no longer be required. Another marker which has been used in a similar way is isocitrate [detected by isocitric dehydrogenase (20)].

The advantage of the method is that it is inexpensive, requires simple instrumentation, will handle many samples at one time, and is relatively insensitive to small variations in experimental conditions.

3.2.5 *Other markers for entrapment and release*

Presumably, satisfactory methods analogous to the above could be devised, in which substrates of other enzyme systems are entrapped inside liposomes—horseradish peroxidase, alkaline phosphatase, esterase—or in which chromogenic substances are coupled with NADPH production (PMS and MTT are well-known candidates). Where one wished to investigate entrapment of larger molecules such as proteins, the enzymes themselves could be used as markers, using the enzymic activity of the mixture to differentiate between contents and protein outside the liposomes. One would have to bear in mind that the process of entrapment itself could have a deleterious effect on the activity of the enzyme, although this can be allowed for if comparison is made between activity before and after solubilization of liposomes. It is conceivable, however, that this approach could give a misleading picture if the liposome membrane were selectively leaky to small molecules such as the substrates, but not the enzymes themselves, in which case liposomally-entrapped enzymes would contribute to the 'extra-liposomal' reaction.

Such a reservation does not arise in the case of soybean trypsin inhibitor, a protein with a molecular weight of 22 000, which is not an enzyme itself, but is able to inhibit the enzymic activity of trypsin. Protein concentration is assayed by observing inhibition in the generation of a chromophore resulting from the action of trypsin (21), which itself has no access to the internal contents of liposomes.

3.3 **Determination of entrapped volume**

The entrapped volume of a population of liposomes (in μl mg^{-1} phospholipid) can often be deduced from measurements of the total quantity of solute entrapped inside liposomes assuming that the concentration of solute in the aqueous medium inside liposomes is the same as that in the solution used to start with, and assuming that no solute has leaked out of the liposomes after separation from unentrapped material. In many cases, however, such assumptions may be invalid. For example, in two-phase methods of preparation, water can be lost from the internal compartment during the drying down step to remove organic solvent. On other occasions, water may enter, or be expelled from the liposome as a result of unanticipated osmotic differences. The best way to measure internal volume, therefore, is to measure the quantity of water directly, and this may be done very conveniently by replacing the external medium with a spectroscopically inert fluid, and then measuring the water signal, for example by NMR (22). Briefly, the liposomes, prepared in aqueous solution consisting of ordinary water, are spun at high centrifugal force (200 000 g for 6 h) to give a tight pellet, from which the supernatant is decanted off to remove every drop of excess fluid (including some liposomes if necessary). The pellet is then resuspended in deuterium oxide (D_2O). The permeability of the membrane to water is such that H_2O and D_2O equilibrate very rapidly throughout the whole volume of the medium. A small aliquot is removed for quantification of phospholipid, and the remainder is used to obtain an NMR scan of H_2O, the peak height of which can be related to concentration by comparison with standards containing known amounts of H_2O in D_2O.

3.4 **Lamellarity**

For intermediate or large unilamellar vesicles (where the membrane thickness is small compared with the diameter), the entrapped volume E, expressed in terms of volume/mole lipid is dependent on the radius of the vesicles according to the relation:

$$E = \tfrac{1}{3}Ar$$

where A is the area of the membrane occupied by one mole of lipid. Thus, an estimate can be made of degree of unilamellarity simply by measuring the average particle diameter (see next section) and comparing the calculated value for E with that obtained by experimental determination as described in the preceding section. If the experimental value is less than that expected from theory, then either the suspension contains a significant proportion of small vesicles (for which the above simple relation does not hold), or some of the vesicles are multilamellar.

An alternative way of reaching the same conclusion is to measure the proportion of phospholipid which is exposed on the outside surface of the vesicles. For large unilamellar vesicles this will be exactly 50% (i.e. the outer leaflet, but not the inner leaflet of the single bilayer membrane). For SUVs the proportion will be even higher, since the lipid distribution is asymmetric, while for MLVs the value will be lower than 50%.

The proportion of lipid at the surface can be measured either chemically, or by direct spectroscopic means. In the first method, a small quantity of PE is incorporated into the liposome membrane during preparation, which is then derivatized in the intact vesicle by reaction with trinitrobenzene sulphonic acid (TNBS)—only the outer exposed PEs react. In parallel, aliquots of the same liposomes are disrupted in Triton X-100 before derivatization, in which all the PE molecules present are available for reaction. Both samples are then made up to the same concentration in acidified Triton X-100 to stop further reaction, and the optical density measured (the derivatized product absorbs more strongly than the reactants). A suitable method, modified from Barenholtz *et al.* (23), is given below.

3.4.1 *TNBS method for measurement of liposome surface area*

(a) *Reagent preparation*

(i) Prepare 0.8 M NaHCO$_3$ buffer by dissolving 6.7 g of NaHCO$_3$ in 100 ml of distilled water. The pH of the solution should be pH 8.5.

(ii) Dissolve 15 mg of TNBS in 1 ml of distilled water.

(iii) Prepare Triton X-100 solutions as follows:
   (i) 10 ml of 1.6% Triton X-100 in 0.8 M bicarbonate buffer
   (ii) 10 ml of 1.2% Triton X-100 in 1.5 M HCl
   (iii) 10 ml of 0.4% Triton X-100 in 1.5 M HCl

(b) *Standards*

Dissolve 5 mg of PE in 7.5 ml of chloroform/methanol and dispense into stoppered tubes in volumes of 20, 40, 80, 120, 160, and 200 μl. Dry each down under a stream of nitrogen, and resuspend in 0.6 ml of normal saline.

(c) *Assay method*

(i) Measure out four aliquots of liposomes containing not more than 0.2 mg (0.25 μmol) of PE, and make up to 0.6 ml in normal saline.

(ii) Add 0.2 ml of bicarbonate buffer to each of two samples (call these (A) and (B)).

(iii) To the remaining two samples (C) and (D), and to each of the standards, add 0.2 ml of 1.6% Triton X-100 in bicarbonate buffer.

(iv) Add 20 μl of TNBS to all tubes, mix, stopper and incubate in the dark at room temperature for 30 min.

(v) To sample tubes (A) and (B) add 0.4 ml of 1.2% Triton X-100 in 1.5 M HCl to stop the reaction. Mix well.

(vi) To the remaining tubes add 0.4 ml of 0.4% Triton X-100 in 1.5 M HCl and mix well.

(vii) Keep in the dark, and measure the optical density at 410 nm within an hour of acidification. The standard curve should be linear over the range chosen.

This technique assumes that the PE is evenly distributed throughout both leaflets of the bilayer, and can give misleading results when working with populations containing small liposomes, since in membranes of high curvature, the PE may be located in higher concentration on the inner leaflet than on the outer, because of the difference in shape of the PE molecules, and its different packing characteristics.

The direct spectroscopic method involves measuring the phosphorus NMR signal of the PC headgroups of a population of liposomes before and after addition of manganese ions to the external medium (24). Manganese ions interact with the phosphorus on the outer surface of the bilayer membrane such that the resonance signal is broadened beyond detection; thus the remainder of the peak height is due to phospholipid headgroups inside the vesicle, which do not come into contact with managanese ions. Direct comparison of the size of the two signals readily reveals the proportion of phospholipid in the outer leaflet.

On the basis of the simple considerations outlined above, it is not possible to determine the exact number of lamellae in a liposome population, since one cannot tell what is the precise configuration that has been adopted by the internal lamellae. A liposome with all its lamellae concentrated around the outside will have far fewer bilayers than one in which the lamellae are located in the centre, even though both liposomes have the same weight of internal lipid, since the quantity of lipid required to form each lamella depends upon its diameter. The reader is referred elsewhere (22) for a more detailed discussion of the subject.

4. SIZE DETERMINATION OF LIPOSOMES

Methods for determining the size of liposomes vary in complexity and degree of sophistication. Two techniques are discussed here—one based on quasi-elastic laser light scattering, and one involving 'direct' observation of the liposomes by electron microscopy. Undoubtedly the most precise method is that of electron microscopic examination, since it permits one to view each individual liposome, and given time and patience, and the skill to avoid numerous artefacts, one can obtain exact information about the profile of a liposome population over the whole range of sizes. Unfortunately, it can be very time-consuming, and requires equipment that may not always be immediately to hand. In contrast, laser light scattering is very simple and rapid to perform, but suffers from the disadvantage of measuring an average property of the bulk of the liposomes, and even with the most advanced refinements it may not pick up or describe in any detail small deviations from a mean value or the nature of residual peaks at extremes of the size range. For large liposomes, of course, information on size distribution is accessible using a Coulter counter. This technique is not discussed here since it does not measure the whole range of liposome sizes, and uses a fairly standard piece of apparatus for which information is available elsewhere.

All the methods mentioned above require items of very costly equipment. If only an approximate idea of size range is required, then gel exclusion chromatographic techniques are to be recommended, since the only expense incurred is that of buffers and gel materials. The reader is referred elsewhere (25) for further details. If one wishes to make comparisons between liposome populations of identical composition and concentration, and only relative rather than absolute values are required, then the even simpler method of optical density measurement (i.e. turbidity due to scattering) can be employed. This may be useful if one requires a rapid check for whether liposomes are being reduced in size during sonication, extrusion or micro-fluidization processes, etc.

4.1 **Negative stain electron microscopy** (Section contributed by P.J.Bugelski, J.M.Sowinski and R.L.Kirsh)

Negative stain electron microscopy is a useful method for addressing questions concerning size distribution of liposomes, and although obtaining quantitative data is laborious, negative staining is a reliable technique, which is simple to perform and requires only limited specialized equipment which should be available in any electron microscopy laboratory. In addition, negative staining can also provide information on whether liposomes produced in a particular manner are multi- or unilamellar.

4.1.1 *Principles of negative staining*

Negative staining is one of the most popular methods for visualizing particulates by electron microscopy. Although staining procedures may vary, all techniques are based on embedding the particulates in a thin film of an electron dense 'glass'. When the films are examined by electron microscopy, the relatively electron-transparent particulates will appear as bright areas against a dark background, hence the term negative stain. First described for visualizing viruses (26), negative staining has been used for a wide variety of micro-organisms (27), cells (28), macromolecules (29), and liposomes as well. Two methods are commonly employed in negative staining—the spray method (30) and the drop method (31). This section will be limited to describing a single variation of the drop method. This variation is the technique most commonly used with liposomes and is the easiest to perform. The reader is directed to the original references, or to the chapter on negative staining by Haschemeyer and Myers in the excellent series on electron microscopy edited by Hayat (32) for descriptions of the other methods. The spray method is not recommended because, in the opinion of the authors, it is much less reliable in terms of quality of preparation and because the shear forces which the specimen must undergo during atomization may alter the size distribution of a preparation of liposomes.

In simple terms, to perform the drop method an electron microscope grid is covered with an electron-transparent support film, and a drop of liposomes suspended in buffer is placed on the grid. Time is allowed for some of the liposomes to attach to the film and then the excess buffer and liposomes are drawn off and quickly replaced with a drop of solution of a heavy metal salt. The bulk of the heavy metal salt solution is also then drawn off, leaving behind a thin film of stain. This film dries rapidly, forming an amorphous electron dense layer in which the liposomes are embedded. This is shown schematically in *Figure 14*. Four liposomes, two of which are aggregated, are shown attached to the support film and embedded in the layer of negative stain. When the grid is thoroughly dried, and photographed in an electron microscope, micrographs like *Figure 15* are produced where multilamellar liposomes are shown against a dark background. The contrast in the micrographs is provided by the heavy metal ion of the negative stain which strongly scatters electrons. The objective aperture of the

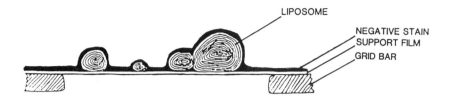

Figure 14. Embedding of liposomes in a thin film of electron-dense heavy metal stain.

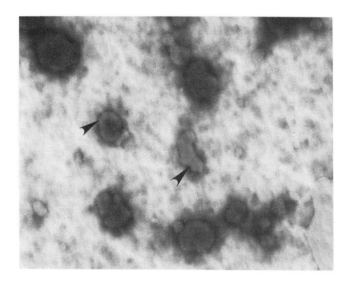

Figure 15. Negative stain electron micrographs of multilamellar vesicles.

microscope is used to remove selectively some of the elastically scattered electrons. Because the probability of elastic scattering is directly related to atomic number, the phospholipid bilayers of the liposomes, which are composed principally of carbon and hydrogen, will elastically scatter relatively few electrons. Thus, when the image is formed, the liposomes appear in negative contrast. Although procedural details differ for the other methods of negative staining, all result in the particulates being embedded in a thin layer of negative stain and the resulting electron micrographs are essentially the same. The following sections will describe the various materials which are used in negative staining and provide step by step instructions on how to produce and evaluate negative stain electron micrographs of liposomes.

4.1.2 *Support films*

The first step in performing negative staining is to prepare the coated grids which will support the stain and liposomes. Two plastics are commonly used in preparing coated grids—formvar and collodion (33). The authors have had very good success with collodion (Parloidin, nitrocellulose) films and will limit discussions to this method. It is a bit more difficult to cast formvar films and they are less stable in the electron beam. The procedures for using formvar are, however, essentially the same.

(a) *Casting collodion films on glass*

Collodion is commercially available as a 1% stock solution (Ernest F.Fullam, Inc., Schenectady, NY, USA). When nitrocellulose films are cast on glass and subsequently floated off on to a water surface, they produce a stable, easily manipulated film (see *Figure 16*).

(i)   Clean glass slides with a detergent solution. Household dishwashing detergent is best. Rinse well with glass distilled water.

(ii)   Air-dry the slides.

(iii)   Fill a Coplin jar with the 1% nitrocellulose solution in amyl acetate that has been filtered twice with Whatman No. 4 filter paper.

(iv)   Dip cleaned slides into nitrocellulose solution. Pull the slide from the jar at a steady rate and hold at a 30° angle above the jar for 60 sec.

(v)   Rest the slides at an angle against a support with the bottom edges resting on absorbent paper.

(vi)   Allow the slides to dry—at least 30 min.

(vii)   Lay the slides flat and with a clean razor blade score the plastic film approximately 5 mm from the bottom edge.

(viii) Fill a square staining dish to overflowing with distilled water. Just before use run a piece of lens paper across the surface of the water to remove any surface contamination.

(ix)   Hold the scored slide several centimetres above and parallel to the water surface.

(x)   With a narrow-tipped plastic pipette, drop a 10% hydrofluoric acid solution along the scored edge of the film. Slowly, the film will strip way from the slide floating on the dilute hydrofluoric acid.

(xi)   When approximately 5 mm of the film has been lifted off, gently breathe on the film and immediately, but slowly, lower it into the water-filled dish at a 30° angle.

(xii)   A visible film will float on the water surface. The film should appear silver grey. If it is golden, violet, or any other colour, the film is too thick and should be discarded.

(xiii) Place the grids, one at a time, dull side down on the floating film. When as many grids as can comfortably fit on the floating film have been laid down, touch each grid with the end of a clean wooden stick to ensure contact with the film surface. Be careful not to touch the stick area on to any area of film not covered by a grid as the film will adhere to the wooden stick.

(xiv)   Cut a piece of Parafilm, slightly larger than the floating film and lay the clean side over the film upon which the grids are resting. Holding the Parafilm at one end with a pair of clean forceps, use a clean wooden stick to push the Parafilm, grids and film below the surface of the water. Then flip it completely over while under water and pull it out so that the film exits the water surface at a 30° angle.

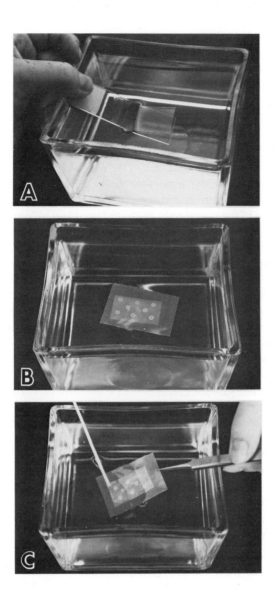

Figure 16. Casting of collodion films on glass.

Be sure your fingers do not touch the water at any time during the procedure to avoid contaminating the film.

(xv) Place the Parafilm with attached grids in a clean Petri dish and allow to dry for several hours before use.

(b) *Coating collodion films with carbon*

Plastic films are slightly unstable and tend to weaken and shrink under the electron beam. To minimize this problem, collodion-coated grids can be stabilized by evaporating a thin layer of carbon or silicon monoxide on to the collodion coating. Once a suitable plastic film, pale grey in colour (30−50 nm thick) has been cast on to the grids and allowed to dry adequately, they can be coated in the following manner using a standard vacuum evaporator. Because of the variety of makes and models of vacuum evaporator, it is not practical to describe the procedures in detail. The reader is referred to the instruction manual of the evaporator available in the laboratory for details of its operation.

(i)   For evaporating carbon, two carbon rods are required. The tips of both rods may be sharpened or only one sharpened while the other remains flat. The rods are aligned so that their ends abut. One rod is clamped in a stationary holder while the second is held in a movable holder. Light tension is placed on the rods through the use of a spring. The complete carbon evaporation assembly including the placement of the carbon rods, grids, and porcelain indicator is shown in *Figure 17a*, while the approximate vertical distance from carbon rods to grids is illustrated in *Figure 17b*.

(ii)   The collodion-coated grids are placed on filter paper at a distance of approximately 15 cm from the carbon rods. To determine when an appropriate amount of carbon has been deposited, place a drop of high vacuum oil on a piece of white porcelain and lay on the filter paper next to the grids. This porcelain piece will serve two purposes—first, it will act as a visual indicator of the amount of carbon deposited; secondly, it will help to hold the filter paper in place during the venting process. When a thin grey film becomes visible on the porcelain during the evaporation, an appropriate amount of carbon has been deposited.

(iii)   Pump down the vacuum evaporator to a pressure of $<10 \ \mu$Hg.

(iv)   Rapidly bring the current through the rods to around 30 A. An arc should jump between the rods and carbon will be evaporated. Evaporation for 30 sec should be sufficient. Never look directly at the arc. High levels of UV light are produced by carbon arcs which can permanently damage the retina.

(v)   Isolate the evaporation chamber from the pumps.

(vi)   Very slowly ventilate the chamber so that the grids on the filter paper are not disturbed by a sudden rush of air. Grids are now ready for use.

(iii) *Coating collodion films with silicon monoxide*

Silicon monoxide can also be used to stabilize the collodion film. It is evaporated onto the grids in a manner similar to carbon, except that a tungsten basket is used. The basket is clamped between the electrodes in the evaporator and a small (1 mg) piece of silicon monoxide is placed in the basket (*Figure 18*). When the evaporator is at vacuum a current is passed through the basket until it glows white hot. Silicon monoxide is evaporated off and will coat the grids. Evaporation for 30 sec is usually sufficient and the thickness of the film is determined in a manner identical to that for carbon. Silicon monoxide is thought to give results superior to carbon when staining liposomes (34).

(iv) *Treating coated grids to make them hydrophilic*

Carbon and silicon monoxide coated grids are relatively hydrophobic. To allow liposomes to stick to their surface and to allow negative stain to form a thin film, they should be rendered hydrophilic. There are a variety of ways in which this can be

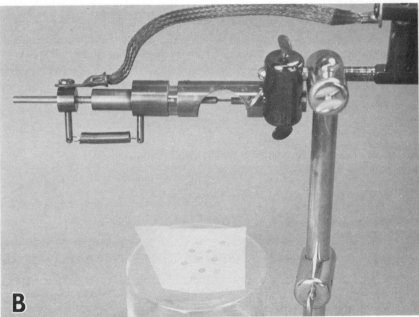

Figure 17. Coating of collodion films with carbon. (**A**) Complete carbon evaporation assembly. (**B**) Distance of carbon rods to grids.

145

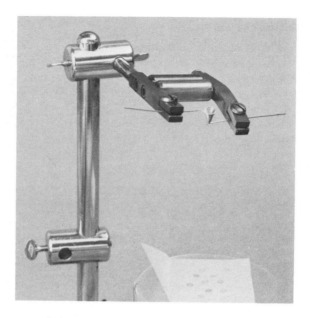

Figure 18. Coating of collodion films with silicon monoxide.

accomplished, both physical and chemical. Of the available methods, for example glow discharge, UV light, albumin, polylysine, etc., the authors have found treatment of grids with bacitracin (35), a polypeptide antibiotic, to provide quick, reliable results. The recommended procedure is described in the next section. Any of the other methods will also give good results but none of them surpasses bacitracin in terms of ease, speed and reliability.

4.1.3 *Negative stains*

Materials used as negative stains should fulfil the following criteria:

(i)   non-reactivity with the specimen and its surrounding buffer;
(ii)   high solubility;
(iii)   high electron density, and
(iv)   stability under the electron beam (32).

Two heavy metal salts which fulfil these criteria and have been used extensively with liposomes are phosphotungstic acid (PTA) and ammonium molybdate (AM). AM is reported to be significantly superior to other stains for a variety of membranous structures (36). Other heavy metals are used as negative stains (e.g. uranium salts, ref. 32), but PTA and AM give good results with most liposomes and commonly used buffers.

In both PTA, $H_3PO_4.12WO_3.24H_2O$, and AM, $(NH_4)_6Mo_7O_{24}$, the staining species are anions which do not react wth phospholipids or phosphate ions. Both agents are used at concentrations between 0.5% and 2% and are usually titrated to a pH ranging from pH 6 to pH 8. Negative stains are usually not buffered but are simply dissolved in distilled water and titrated up from their original low pH to the working range with dilute NaOH of KOH. When staining liposomes, strict control of pH or concentration is usually not necessary, good results being obtained over a wide range of values.

As mentioned previously, PTA and AM are both anionic and thus do not bind to liposomes composed of either neutral (e.g. PC) or negatively-charged phospholipids (e.g. PA, PG, or PS). Caution must be exercised if these stains are used with liposomes containing positively-charged lipids (e.g. stearylamine) as the anionic stains can cause aggregation and precipitation of the liposomes. In the case of positively-charged liposomes, cationic uranyl salts, for example uranyl acetate, $UO_2(CH_3COO)_2.2H_2O$, $0.2-0.5\%$ a pH $4-5.5$, are useful. It should be noted, however, that uranium salts are precipitated by phosphate ions, and liposomes prepared in phosphate buffers should be 'washed' before being stained.

4.1.4 *The two-step drop method*

(i)   Clamp a coated grid in a pair of anticapillary forceps (Dumont, No. N4 Inox) coated side up.

(ii)   Place a drop of bacitracin (0.1 mg ml^{-1} in water, Sigma Chemical Co.) on the grid.

(iii)   After a minute, draw off the drop by touching the edge of the grid with the torn edge of a piece of filter paper (Whatman No. 1) held at $90°$ to the plane of the grid (*Figure 19*). The grid can be left to dry at this point or you can proceed immediately.

(iv)   Cover the grid with a drop of the liposome preparation and allow time for some of the liposomes to stick to the grid. The time required is largely dependent on the concentration of lipid in the liposome preparation. With vortex-shaken multilamellar liposomes at a lipid concentration between 1 and 5 mg ml^{-1}, $1-2$ min is usually sufficient. With lower lipid concentrations, longer times may be necessary and with higher concentrations it may be necessary to dilute the preparation with a buffer prior to applying the drop to the grid.

(v)   After some of the liposomes have adhered, draw off the bulk of the drop of liposomes with filter paper, in an identical manner as was used to draw off the bacitracin.

(vi)   *Without allowing the film remaining on the grid to dry*, replace the drop of liposomes with a drop of negative stain (1% AM, pH 7.0 in distilled water).

(vii)   After a minute, draw off the drop of stain and allow the grid to dry thoroughly.

If the liposomes are suspended in a buffer which is incompatible with the desired negative stain, prior to applying the drop of stain the grid can be washed by applying and drawing off several drops of a compatible buffer. Caution should be exercised in using this technique if quantitative data on liposome size distribution is desired, because liposomes can be osmotically active and can shrink or swell as water diffuses in or out of the liposome (37); the washing medium should be iso-osmotic with the interior of the liposomes.

A few authors have used osmium tetroxide as a fixative prior to negative staining of liposomes. Osmium tetroxide reacts with unsaturated lipids and reduced osmium precipitates out. In membranes, although this reaction presumably takes place in the hydrophobic regions of the bilayer, reduced osmium precipitates in the hydrophilic regions, giving the membranes their characteristic 'rail-road track' appearance. There

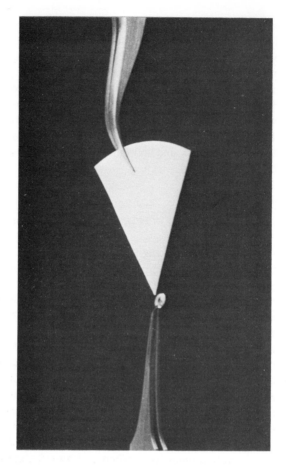

Figure 19. Removal of drop of liquid from grid with filter paper.

have been no systematic studies to determine whether osmium fixation offers any advantages in negative staining of liposomes. In the opinion of the authors, any possible benefit of osmium fixation (e.g. increased contrast) is outweighed by the extra handling required, the potential for the generation of additional artefacts and the minimal, but real, health risks associated with osmium tetroxide. (Osmium tetroxide is toxic and is volatile at room temperature both in crystalline form and from aqueous solutions.)

The sticking of liposomes to the grid is probably electrostatic in nature. Because liposomes are generally less than 1 μm in diameter, Brownian motion will keep them in suspension and they will not settle down on the grid. The number of liposomes which stick will thus depend on the net charge on the grid, the net charge on the liposomes, the time allowed for attachment and probably most importantly, on the thickness of the film left behind when the drop is drawn off with filter paper. If there is a mixture of two or more populations of liposomes with different net surface charges, it is possible that sticking will not be random. Caution should be exercised in interpreting results in cases such as this.

4.1.5 *Estimation of lamellicity*

One of the least understood phenomena concerning negative staining of liposomes is the visualization of multilamellar structures. As closed phospholipid vesicles, liposomes are well known for their limited permeability. Although small, uncharged solutes and water can quickly equilibrate across lipid bilayers, larger molecules, especially charged, complex ions are excluded or equilibrate very slowly (38). Since the ions used for negative staining are large and charged, they would not be expected to permeate liposomes. It has long been observed, however, that the negative staining of multilamellar liposomes often results in alternating light and dark concentric lamellae being revealed (39). The mechanism which allows visualization of concentric lamellae is unknown. Encapsulation of negative stains inside MLVs by including the stain in the aqueous phase as liposomes are formed does not give results which are different from simply staining 'empty' MLVs. It has been suggested that the apparent multilamellicity is an artefact of negative staining (40). This may well be so, but we feel that even if it is true, this artefact can be exploited to reveal qualitative information on the interior structure of liposomes. A possible mechanism which could explain the apparent multilamellicity seen by negative staining for known unilamellar vesicles (proven by freeze etch and thin section studies) (40) involves piling up of liposomes, and the effects of surface tension during drying out of the negative stain. We have found that multilamellar structures are most commonly seen in clumps of liposomes (*Figure 20*). Similarly, allowing a film of liposomes to dry completely prior to application of the layer of negative stain or staining with a very thick layer of dilute negative stain (i.e. placing 5 μl of 0.1% AM in water on the grid and allowing it to dry down without removing any of the stain with filter paper) will result in many more liposomes appearing multilamellar than will normal negative staining of the same preparation of liposomes. The authors believe that when the liposomes dry one on top of the other, when they dry slowly or when they dry prior to negative staining they flatten out. A 'rough' surface results. The negative stain follows this rough surface, resulting in the alternating light and dark 'lamellations' when the grid is viewed. Multilamellations that are seen when liposomes are partially overlapping thus do not reflect the actual internal structure. When a liposome is isolated from its neighbours, however, the multilamellicity seen reflects, albeit probably not very accurately, the internal structure of the liposomes. For example, although the 4 nm periodicity seen in *Figure 21* is roughly equal to the 5 nm periodicity of liposome membranes demonstrated by X-ray diffraction, the light lamellae, which should correspond to the lipid bilayers, are much more than the 4.3 nm determined by X-ray diffraction.

Whether this theory is accurate or not, interpretation remains problematical. Because of the variegate nature of negatively-stained specimens, one can never be sure that multilamellar structures will be revealed. If, however, an isolated multilamellar structure is seen, it is highly suggestive that the sample is multilamellar.

4.1.6 *Size analysis of negatively-stained liposomes*

(a) *Electron microscopy*

When the grid is throughly dry (15 − 30 min) it is ready for examination in the electron

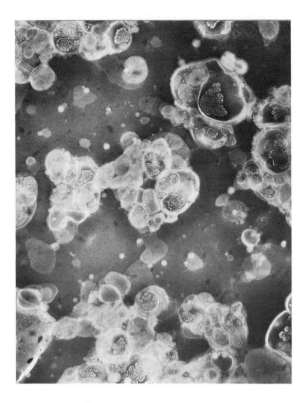

Figure 20. Appearance of 'multilamellar' structures in liposome clumps.

microscope. Grids should be viewed and photographed within 1 or 2 days of preparation because recrystallization of the negative stain can occur on storage. Storing grids in a desiccator may provide some increase in storage time. It is beyond the scope of this chapter to review the use of electron microscopes, and the reader is directed to the instruction manual for the particular microscope available, and that most important resource—the resident electron microscopist!

When negative staining is performed properly, a monolayer of liposomes embedded in negative stain will be spread across the grid. This can result in the formation of artefactual aggregates of liposomes (*Figures 20–22*). These aggregates are indistinguishable from those formed prior to staining and will make assessment of the particle size distribution impossible. It is a good idea to examine the liposomes by phase contrast or dark field light microscopy to check for aggregates. To ensure adequate spreading and good staining it is advisable to prepare three or four grids each from several dilutions of liposomes and with different adherence times. In this way you will avoid having to go back to restain the liposome preparation. One of the idiosyncrasies of negative stain is the variability of the resulting preparations. Two grids prepared side by side, in exactly the same manner, can give vastly different results. This same variation can even occur between adjacent holes on the same grid (*Figure 22*)! When the grid is first examined in the electron microscope it is best to scan around at low

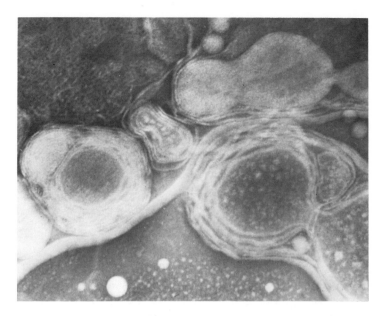

Figure 21. Periodicity of multilamellar structures.

magnification to find areas that show good staining characteristics. Once these areas are found, you can go up to high magnification for photography.

When photographing the grids, it is important what use you expect for the electron micrographs. If you wish only to document that you indeed have a liposome preparation, then only photograph a few fields. You should, however, make multiple exposures of each field, stepping through from slightly 'underfocus' to slightly 'overfocus'. In this way you will ensure that you have micrographs which will best illustrate the preparation. For example, the contrast between lipid bilayers in multilamellar vesicles is higher at slightly 'overfocus' than when exactly focused.

If, on the other hand, particle size distributions are desired, you should photograph as many fields as practicable. As long as you are close to focus, you will be able to use the micrographs for measurement even though they would be unsuitable for illustration purposes. Because the magnification of an electron microscope is only approximate and can vary slightly each time a current through the lenses is turned on, it is best to make all exposures for a size distribution measurement at a single photographic session, at the same magnification setting and at the same acceleration voltage. A micrograph of a calibration grating replica should be made at the same setting and at the same session. Grating replicas come in several scales; 2160 lines mm^{-1} are available from most electron microscopy supply houses. You should choose one which will give several lines in the micrograph when made at the setting used to photograph your liposome preparation. This micrograph should be used to calibrate all measurements made on the liposomes.

The use of negative stain electron microscopy greatly facilitates estimation of the liposome size range at the lower end of the frequency distribution. For small unilamellar liposomes, sizing by negative staining compares favourably with measurements determined by physical techniques (41) as well as other electron microscopic techniques such as freeze fracture (42). However, for larger size, or heterogeneous populations of liposomes, negative staining electron microscopy offers real advantages over freeze-fracture techniques. In the latter technique, the fracture plane passes through liposomes that are oriented randomly in the frozen specimen, hence the potential for non-midline sections. Thus, the observed diameter measurements may differ from the real liposome diameter. Negative staining electron microscopy obviates this problem in that the complete structure is observed, although the liposome morphology may be deformed from spherical. Irregular or ellipsoid shapes can be treated mathematically to correct for perimeter irregularities and estimate the size of the original spherical liposome (see below).

(b) *Computer-assisted measurement*

In recent years, significant advances have been made in image analysis. Not only has the hardware and software improved, but the computer has also assisted in making image analysis more versatile. Additionally, the cost has decreased dramatically, making this technology available to many more investigators. In analysing liposome size distributions, image analysis can greatly speed up data collection and significantly improve the accuracy of the measurements. Although fully automated systems (which gather data by converting video images to digital) have the potential to accelerate data collection, they require uniform specimens—for example, the liposomes must not be in clumps and the staining must be of even density. Although it is theoretically possible to generate such specimens, or to make up for variation in the specimen using erosion/dilution routines, the authors have found it much more practical to use semi-automatic techniques. In semi-automatic image analysis, a digitizing table is used to delineate the structures to be measured. The computer then manipulates the data to generate a variety of parameters for the structure. The authors have found that only a few minutes are required to take measurements from hundreds of liposomes. Differentiation between individual liposomes from aggregates is possible, and dirt and stray crystals can also be excluded. Further advantages of computer-assisted analysis are accuracy and calibration. Because the actual surface contour of the stained liposomes is followed when entering data, the measurements are more accurate than those obtained by simple major axis – minor axis measurements. Calibration is often automatic, the micrographs of the grating replica are used to enter a known length into the computer and this length is then used as a reference for all measurements and calculations.

The stepwise procedures for generating liposome size distribution using a Zeiss Videoplan image analysis system are as follows.

(i)   Set up the computer to be calibrated from the grating replica.
(ii) Set up the data analysis program to provide at least the following parameters: area, diameter of an equivalent circle of the same area, and the major and minor axis of the structure. Other parameters can be included if necessary.
(iii) Calibrate the computer from a 20 × 25 cm photographic print of the micrograph of the grating replica.
(iv) Digitize each liposome by tracing around its outline with the stylus on 20 × 25 cm

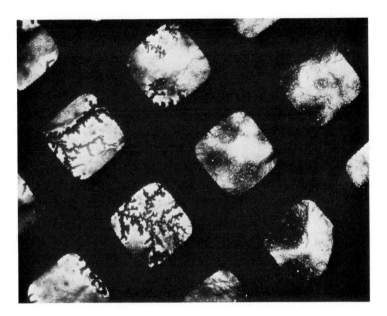

Figure 22. Artefactual aggregates of liposomes.

photographic prints made at the same magnification as the print of the grating replica. Trace individual liposomes as well as those included in any aggregates.

(v)   If aggregates are present and believed to be real and not an artefact of staining (as checked by phase contrast or darkfield light microscopy of wet mounts of liposomes performed prior to staining) their outline should be traced.

After data collection, the results can then be displayed and analysed by the same computer or transferred to another. *Figure 23* is a histogram showing the size distribution of a preparation of multilamellar liposomes obtained by image analysis of negatively-stained grids. The distribution is unimodal and the mean equivalent diameter is 0.1 μm. In this case the data have not been corrected for the effects of flattening. For a more accurate diameter, the values should be multiplied by a factor of 0.71 as described by Olsen *et al.* (43). (This correction factor was determined for multilamellar vesicles with a mean diameter of 0.27 μm. It may not be valid for other types of liposomes or size ranges.) If all that is required is the shape of the distribution and an estimation of size, the data can be used directly as gathered.

(c) *Manual measurement of liposome sizes*

For hand measurement of liposome sizes the method of Olsen *et al.* is recommended (43). Electron micrographs are prepared as described, and are printed as enlargements. The micrographs of the grating replica are printed to the same enlarger setting for calibration purposes. Measurements of the diameter of the liposomes are made with a vernier calliper and the general procedure used with computer-assisted measurement is followed.

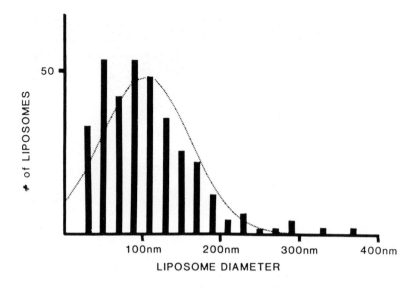

Figure 23. Size distribution of an MLV preparation obtained by image analysis of negative-stained grids.

(i)   Print 20 × 25 cm enlargements from the electron micrograph negatives.

(ii)   Measure the diameters of individual liposomes. A minimum of 500 liposomes should be measured for each preparation.

(iii)   For ellipsoid or irregular-shaped liposomes the diameter can be calculated as the mean of the diameters of the major and minor axis.

(iv)   Twice the surface area of the measured disc-shaped structure is assumed to be equivalent to that of the spherical liposome. Back extrapolating, the diameter of the original liposome is 0.71 times the diameter of the measured disc.

(v)   In order to determine the size distribution, measured liposome diameters are arbitrarily assigned into predetermined size ranges. Summing the number of measured liposomes in each range gives a frequency distribution for statistical evaluation.

4.2 Sizing by photon correlation spectroscopy (laser light scattering) (Section contributed by N.I.Payne and R.R.C.New)

4.2.1 Principle of the technique

Photon correlation spectroscopy (PCS) is the analysis of the time dependence of intensity fluctuations in scattered laser light due to the Brownian motion of particles in solution/ suspension. Since small particles diffuse more rapidly than large particles, the rate of fluctuation of scattered light intensity varies accordingly (*Figure 24*). Thus, the translational diffusion coefficient (D) can be measured, which in turn can be used to determine the mean hydrodynamic radius ($\bar{R}_h$) of the particles using the Stokes−Einstein equation.

Depending upon the power of the laser, it is possible to measure particles in the range of about 3 nm up to about 3 μm. Data can be presented either as mean particle size

Small particle

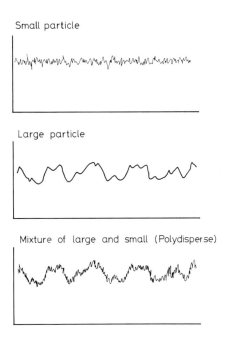

Large particle

Mixture of large and small (Polydisperse)

Figure 24. Fluctuations in light intensity in relation to particle size. Fluctuations in light intensity of a beam of laser light passing through a suspension of particles occur because of movement of the particles in the beam as a result of Brownian motion. The movement of large particles in a fluid medium is slower than for small particles, and the fluctuations in light intensity are therefore correspondingly slower.

(assuming spherical equivalence), or alternatively a more rigorous analysis can yield details of two or more subpopulations of particles in solution/suspension. A number of instruments are available with which to perform PCS. Typical instrumentation is manufactured by Malvern Instruments Ltd, Malvern, UK (44). A laser light source (helium−neon or argon) is focused on the contents of a highly polished, fine quality glass cuvette (either cylindrical or, preferably, square cross-section). The cuvette is housed within a thermostatically-controlled goniometer cell. The temperature of the goniometer cell should preferably be within $\pm 0.1°C$ of that required. This will minimize any potential errors due to variation in fluid viscosity, but more importantly, will minimize random convection currents superimposed on the Brownian movement which could lead to substantial errors in particle size measurement.

The sample under examination can be suspended in a range of dispersion media, the only information required being viscosity and refractive index of the medium. These can be determined using a suitable viscometer (e.g. U-tube) and refractometer respectively, ideally at the temperature at which PCS is being performed (typically $25°C$). To eliminate potential light flare at the surfaces through which the laser beam passes, it is recommended that the sample cell is immersed in a liquid which matches the refractive index of the dispersion fluid. Scattered laser light from the sample is detected by a photomultiplier assembly usually situated at an angle of 90° relative to the laser beam.

Once the signal has been recorded in terms of a series of photomultiplier bursts over

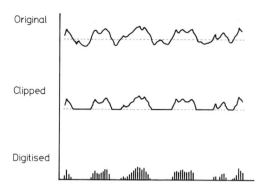

Figure 25. Processing of signal fluctuations. The signal is processed so that intensity variations with time can be stored and manipulated easily by the computer in the form of a series of digits.

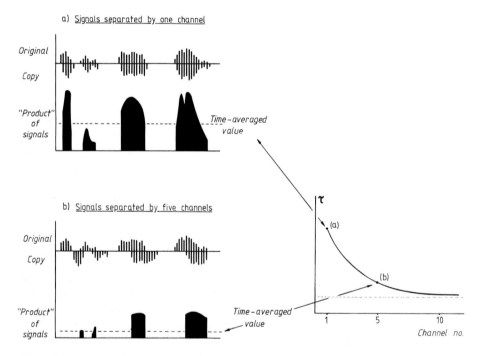

Figure 26. Derivation of correlation curve. The correlation curve (from which the diffusion coefficient is deduced) is a measure of how quickly a signal becomes completely out of phase with itself. This is determined by multiplying the signal with a copy of itself separated from the original by a given period of time (i.e. a set number of channels). The product, expressed as the cumulative area under the curve obtained, is plotted for each time separation against the channel number corresponding to that separation. The result is a curve which decays exponentially to a mean value at a rate which depends on the rapidity of the signal fluctuations (and hence the particle size). The value to which it decays will depend on the intensity of the signal fluctuations.

a period of time (*Figure 25*), a mathematical process called 'correlation' is carried out, in which the similarity between the signal and itself separated from the original by a time delay is measured. In essence this is performed by multiplying the amplitudes

156

of the signal and its time-delayed copy together at different time points to give a correlation function. As the signals become more and more out of phase with each other (i.e. the time separation is increased), their randomness with respect to each other results in a decay of the correlation function to a constant value (*Figure 26*). The correlation function at any given time separation is described mathematically as

$$G_{(t)} = <N>^2(1-Be^{-\Gamma t})$$

in which N is the intensity of the signal averaged over many sample times, B is a constant determined by mechanical constraints of the apparatus and the sampling procedure, and Γ, the decay constant, is $2DK^2$, where K is the scattering vector (dependent on the detection angle, etc.) and D is the diffusion coefficient of the particles causing the fluctuation. Having obtained a value for the diffusion coefficient, particle radius can then be determined by inserting D in the Stokes−Einstein equation thus:

$$D = kT/6\pi\eta\bar{R}_\text{h}$$

where k = Boltzmann's constant, T = absolute temperature, η = solvent viscosity, $\bar{R}_\text{h}$ = mean hydrodynamic radius.

The lower limit of particle size detection using PCS is dictated by a number of factors, but principally the power of the laser. Use of a helium−neon laser, rated nominally at about 35 mW, will generally restrict particle size measurements down to about 10 nm (although the Malvern will generate 'data' below this limit). To achieve greater sensitivity, a more powerful laser is required, for example, typically a 2 W argon ion laser is used. This will enable the analysis of, for example, micellar systems (45).

4.2.2 Instrument and sample handling

The technique of performing PCS analysis requires stringent attention to detail. Ideally, the photon correlation spectrometer should be housed in its own room and mounted on a vibration-free surface. The environment should be dust-free and strong light sources falling in the vicinity of the photomultiplier assembly should be avoided. Glass cuvettes should be soaked in chromic acid for 24 h before use and then rinsed thoroughly with 0.22-μm-filtered distilled water prior to drying—ideally in a laminar flow cabinet.

Samples for measurement should, in general, be constituted in media which have been passed through a suitable 0.22-μm filter. This should normally eliminate problems usually attributable to 'dust'. However, it is generally worthwhile monitoring the 'clarity' of the constitution medium by checking the photon count at the photomultiplier assembly. The ratemeter facility on the Malvern 4600M enables a rational choice of sample concentration and can give a rapid indication of poor sample preparation. For the Malvern system a ratemeter count of 30 000−150 000 is generally regarded as usable and this can be achieved by suitable dilution of the sample and/or adjustment of the photomultiplier aperture. When observed from the side, the laser beam should pass cleanly through the sample with no evidence of 'flare' adjacent to the beam within the

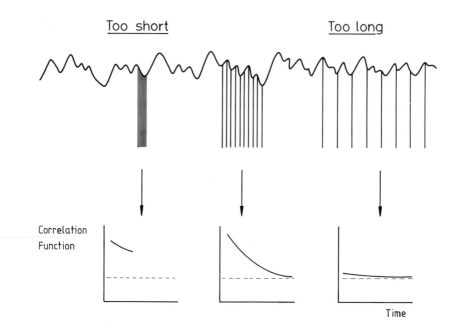

Figure 27. Choice of channel sample time. The choice of channel sample time is important in ensuring that sufficient information is acquired to enable a full plot of the correlation curve to be obtained. Ideally, the channel sample time should correspond roughly to the frequency of signal fluctuations. Although correlation curves can be drawn when the sample time is long or short, values derived for the diffusion coefficient will not be so accurate.

confines of the cuvette. Large variations in the ratemeter count usually indicate poor sample preparation with the presence of very coarse material in suspension. To avoid interference by very large aggregates which are unrepresentative of the preparation as a whole, one option is to pass all liposome suspensions through a 5-μm pore size (or smaller) membrane filter before measurement (providing this does not adversely affect the sub-5-μm liposome population).

4.2.3 *Channel sampling time*

In the Malvern photon correlator, the increments of time separating the two copies of the signal to be processed are termed 'channels'; increasing the time separation between the signals is accomplished by increasing the number of channels, or 'channel number', up to a maximum of 60. Although the total number of channels is pre-set, the overall separation can be varied by altering the channel width, or channel 'sampling time'. The choice and configuration of sample time is quite critical in achieving meaningful results. Choosing excessively long sample times will result in only a few channels responding to scattered light. Conversely, excessively short sample times will 'select' only a proportion of the total signal (*Figure 27*). For example, in linear mode (see below), a sample of 100 nm in diameter might need a sample time of 20 μsec. In log mode, however (see below), the majority of samples (unless very polydisperse) would be measurable using sample times in the range 1 − 20 μsec. Comparison of 'calculated' and 'measured' background values generally gives a reasonable indication of the correct choice of sample time. In an ideal system, these values should be identical although in practice the calculated background is invariably lower.

The percentage difference between calculated and measured background values gives a good indication of data accuracy. More recent instrumentation from Malvern (Malvern 4700) can generate data using four separate sample times simultaneously.

4.2.4 *Data analysis*

Various methods of data analysis are available depending upon the types of samples being studied. For relatively monodisperse populations of particles the method of cumulants analysis is available (for references, see 44) using the Malvern PCS24 software package. Using this procedure, both linear and quadratic curves are fitted to the semi-log plot of the correlation function to obtain a direct estimate of the variance of the particle size distribution. For this type of analysis, correlator channels should generally be spaced at intervals separated by a linear progression of sample time (60 channels in total).

However, many situations arise where the sample is either very polydisperse, skewed, bi- or multi-modal and therefore not described adequately by simple mean and polydispersity values. In these instances, the Malvern 4600M allows correlator channels to be spaced at intervals separated by a logarithmic progression of sample times (26 channels in total; $\sqrt{2}$ progression of delays as a multiple of fundamental sample time).

This type of correlator configuration has been proposed as optimal for the processing of signals from polydisperse systems where the correlation function is the sum of many exponentials of different decay times (see ref. 44).

Using the Malvern PCS14 polydispersity analysis program (44), the information contained within an experiment can be expressed in terms of a set of 'delta functions'; these are of infinitely narrow width. The data fitting process comprises setting upper and lower limits to the particle size distribution within which between two and nine delta functions can be placed. The relative weighting of exponentials about a given delta function is then allocated accordingly. In practice, data analysis consists of starting at a 'low resolution' situation (a small number of delta functions) and increasing the number of delta functions within the specified upper and lower limits until an unstable 'fit' is obtained. This is evidenced by large deviations between positive and negative values in the exponential weight distribution profile and should be avoided in favour of a profile showing predominantly positive values.

This so-called exponential sampling method (44), whilst giving additional information about the particle size distribution, should not be regarded as a rigorous particle size analysis *per se*. Rather, the method should be regarded only as a means of probing the data until a 'best fit' is obtained. Thus, an exponential weight distribution rather than a true particle size distribution is derived since the method does not take into account such effects as, for example, scattering efficiency as a function of particle size.

The validity of the exponential sampling technique is questionable if poor quality raw data are obtained in the first instance. High quality data can be obtained on the Malvern PCS by setting a stringent 'rejection factor'. This effectively eliminates all data which does not approximate closely to the calculated correlogram. The operator can set the rejection factor manually and hence effectively select the quality of the final data. However, by setting a very high rejection factor, a large number of experimental results may be rejected. In this instance it would be necessary to execute a large number of individual experiments (probably in excess of 20) to ensure that adequate numbers are accepted. It is generally regarded that a small number of experiments accepted using a high rejection factor are preferable to a large number of experiments accepted at a low rejection factor.

4.2.5 *PCS in laboratory usage*

We have used PCS (in conjunction with other sizing techniques) as a routine tool for monitoring liposome preparations and particularly to detect potential changes in particle size during stability studies of liposome suspensions (46). It was in this context that certain limitations of the technique were revealed during an analysis procedure. As highlighted earlier, upper and lower particle size limits are set, usually automatically, and as a result of this, delta functions are fixed at related intervals between these two figures. In practice, slight changes in the sample, for example, the occasional aggregate, can affect markedly the placing of the upper and lower size limits so that the relative positioning of delta functions is also changed. Thus comparative evaluation of the differential weightings is difficult since the positioning of the delta functions with time may vary slightly.

The approach the authors took ultimately was to wait until the conclusion of the stability study, isolate the sample which had the broadest exponential weight distribution profile, and then use the upper and lower size limits from this sample for all the other samples tested. Thus all the weightings at each specified delta function could be compared.

5. REFERENCES

1. Bartlett,G.R.J. (1959) *J. Biol. Chem.*, **234**, 466.
2. Stewart,J.C.M. (1959) *Anal. Biochem.*, **104**, 10.
3. Schiefer,N-G. and Neuhoff,V. (1971) *Hoppe-Seyler's Z. Physiol. Chem.*, **352**, 913.
4. Terao,J., Asano,I. and Matsushito,S. (1985) *Lipids*, **20**, 312.
5. Klein,R.A. (1980) *Biochim. Biophys. Acta*, **210**, 486.
6. Gutteridge,J.M.C. (1975) *Anal. Biochem.*, **69**, 518.
7. Ackman,R.G. (1986) In *Analysis of Oils and Fats*. Hamilton,R.J. and Rossel,J.B. (eds), Elsevier Applied Science Publishers, London, p. 137.
8. Brooks,C.J.W.,MacLachlan,J., Cole,W.J. and Lawrie,T.D.V. (1984) In *Proceedings of Symposium on Analysis of Steroids*. Szeged, Hungary, p. 349.
9. Wybenga,D.R., Pileggi,V.J., Dirstine,P.H. and Di Giorgio,J. (1970) *J. Clin. Chem.*, **16**, 980.
10. Fry,D.W., White,C. and Goldman,D.J. (1978) *Anal. Biochem.*, **90**, 809.
11. Rosier,R.N., Gunter,T.E., Tucker,D.A. and Gunter,K.K. (1979) *Anal. Biochem.*, **96**, 384.
12. Gunter,K.K., Gunter,T.E., Jarkowski,A. and Rosier,R.N. (1982) *Anal. Biochem.*, **120**, 113.
13. Weissman,G., Collins,T., Evers,A. and Dunham,P. (1976) *Proc. Natl. Acad. Sci. USA*, **73**, 510.
14. Weinstein,J.N., Yoshikami,S., Henkart,P., Blumenthal,R. and Hagins,W.A. (1977) *Science*, **195**, 489.
15. Weinstein,J.N., Klausner,R.D., Innerarity,T., Ralston,E. and Blumenthal,R. (1981) *Biochim. Biophys. Acta*, **647**, 270.
16. Weinstein,J.N., Ralston,E., Leserman,L.D., Klausner,R.D., Dragsten,P., Henkart,P. and Blumenthal,R. (1984) In *Liposome Technology*. Gregoriadis,G. (ed.), CRC Press Inc., Boca Raton, Florida, Vol. 3, p. 183.
17. Allen,T. (1984) In *Liposome Technology*. Gregoriadis,G. (ed.), CRC Press Inc., Boca Raton, Florida, Vol. 3, p. 177.
18. Kinsky,S.C. (1974) In *Methods in Enzymology*. Fleischer,S. and Packer,L. (eds), Academic Press, Inc., London, Vol. 32, p. 501.
19. Alving,C.R., Shichijo,S. and Mattsby-Baltzer,I. (1984) In *Liposome Technology*. Gregoriadis,G. (ed.), CRC Press Inc., Boca Raton, Florida, Vol. 2, p. 157.

20. Crowe,L.M., Womersley,C., Crowe,J.H., Reid,D., Appel,L. and Rudolph,A. (1986) *Biochim. Biophys. Acta,* **861**, 132.
21. Schlieren,H., Rudolph,S., Finkelstein,M., Coleman,P. and Weissman,G. (1987) *Biochim. Biophys. Acta,* **542**, 137.
22. Pidgeon,C., Hunt,A.H. and Dittrich,K. (1986) *Pharmaceut. Res.,* **3**, 23.
23. Barenholtz,Y., Gibbes,D., Litman,B.J., Goll,J., Thompson,T.E. and Carlson,F.D. (1977) *Biochemistry,* **16**, 2806.
24. Hope,M.J., Bally,M.B., Webb,G. and Cullis,P.R. (1985) *Biochim. Biophys. Acta,* **812**, 55.
25. Van Renswoude,A.J.B.M., Blumenthal,R. and Weinstein,J.N. (1980) *Biochim. Biophys. Acta,* **595**, 151.
26. Hall,C.E. (1955) *J. Biophys. Biochem. Cytol.,* **1**, 1.
27. Verschueren,H., Dekegel,D., Gilquin,C., DeMaeyer,S. and Lafontaine,A. (1981) *Ann. Microbiol. (Paris),* **1**, 307.
28. Katsumoto,T., Takayama,H. and Takagi,A. (1987) *J. Electron Microsc.,* **27**, 1.
29. Sleytr,U.B. and Plohberger,R. (1980) In *Electron Microscopy at Molecular Dimensions. State of the Art and Strategies for the Future.* Baumeister,W. and Vogell,W. (eds), Springer-Verlag, Berlin, p. 36.
30. Horne,R.Q., (1965) In *Techniques for Electron Microscopy.* 2nd edn, Kay,D.H. (ed.), Davis, Philadelphia, p. 311.
31. Huxley,H.E. and Zubay,G. (1960) *J. Mol. Biol.,* **2**, 10.
32. Haschmeyer,R.H. and Myers,R.S. (1972) In *Principles and Techniques of Electron Microscopy Biological Applications.* Hayat,M.A. (ed.), Van Nostrand Reinhold Co., New York, Vol. 2, p. 101.
33. Hayat,M.A. (1970) In *Principles and Techniques of Electron Microscopy Biological Applications.* Hayat,M.A. (ed.), Van Nostrand Reinhold Co., New York, Chapter 5, p. 323.
34. Larrabee,A.L., Babiarz,J., Laughlin,R.G. and Geddar,A.D. (1978) *J. Microsc.,* **114**, 319.
35. Gregory,D.W. and Pirie,B.J.S. (1973) *J. Microsc.,* **99**, 251.
36. Muscatello,V. and Horne,R.W. (1968) *J. Ultrastruct. Res.,* **25**, 73.
37. Chapman,D. (1984) In *Liposome Technology.* Gregoriadis,G. (ed.), CRC Press Inc., Boca Raton, Florida, Vol. 1, p. 1.
38. Szoka,F. and Papahadjopoulos,D. (1980) *Annu. Rev. Biophys. Bioeng.,* **9**, 467.
39. Bangham,A.D., Hill,M.W. and Miller,N.G. (1974) *Methods Membr. Biol.,* **1**, 1.
40. Miyamoto,V.K. and Stoeckenius,W. (1971) *J. Membr. Biol.,* **4**, 252.
41. Huang,C.H. (1969) *Biochemistry,* **8**, 344.
42. Watts,A., March,D. and Knowles,P.F. (1978) *Biochemistry,* **17**, 1792.
43. Olsen,F., Hunt,C.A., Szoka,F.C., Vail,W.J. and Papahadjopoulos,D. (1979) *Biochim. Biophys. Acta,* **557**, 9.
44. Malvern Loglin Correlator K7027 Reference Manual—Release 2 (1983).
45. Roe,J.M. and Barry,B.W. (1983) *Int. J. Pharm.,* **14**, 159.
46. Payne,N.I., Browning,I. and Hynes,C.A. (1986) *J. Pharm. Sci.,* **75**, 330.

CHAPTER 4

Covalent attachment of proteins to liposomes

FRANCIS J.MARTIN, TIMOTHY D.HEATH, and ROGER R.C.NEW

1. INTRODUCTION

Protein-conjugated liposomes have recently attracted a great deal of interest, principally because of their potential use as targeted drug delivery systems (1,2) and in diagnostic applications (3−5). Antibodies are the most commonly conjugated proteins, although the techniques can readily be applied to other proteins such as *Staphylococcus aureus* protein A (2), plant lectins (6) and enzymes (7). Some early methods for conjugation used amino-reactive homobifunctional reagents such as glutaraldehyde and diethyl suberimidate (8). Water-soluble carbodiimides have also been used to catalyse the formation of an amide linkage between amino groups (from phosphatidyl ethanolamine present in the liposome membrane) with carboxyl groups of proteins (9). However, as reviewed in detail elsewhere (10), these procedures are generally inefficient and likely to bring about protein polymerization. The preferred methods are designed in such a way that the protein can react only with the liposome surface. The attachment of horseradish peroxidase to liposomes is a good example of this approach (11). The carbohydrate on the enzyme is oxidized with sodium periodate to create reactive aldehydes after blocking the free amino groups with fluorodinitrobenzene. The enzyme is then coupled to liposomes by reaction of the aldehyde groups with the primary amino group of phosphatidyl ethanolamine present in low concentration in the liposome membrane. The bond is formed by reductive amination with sodium borohydride. The product, characterized biochemically and electron microscopically, consists of intact liposomes with horseradish peroxidase attached to their outer surface. A similar approach is applicable to the coupling of proteins to liposomes containing glycolipids (e.g. PI or cerebrosides). In this case, hydroxyl groups of glycolipids incorporated into the liposome during preparation are oxidized to aldehydes under mild conditions using using sodium periodate, and subsequently reacted with primary amino groups in the unmodified protein. Schiff base formation between the amino groups on the liposome and aldehyde groups of the protein proceeds satisfactorily at neutral pH and the linkage can be stabilized by the addition of cyanoborohydride, which converts the imine linkage to that of a secondary amine.

Protein conjugation methodology has been revolutionized by the development of three efficient and selective reactions that can be used in aqueous media. In the first of these reactions, carboxyl groups, activated with *N*-hydroxysuccinimide, react with amino groups to produce an amide bond. In the second reaction, pyridyldithiols react with thiols to produce disulphide bonds. In the third reaction, maleimide derivatives react

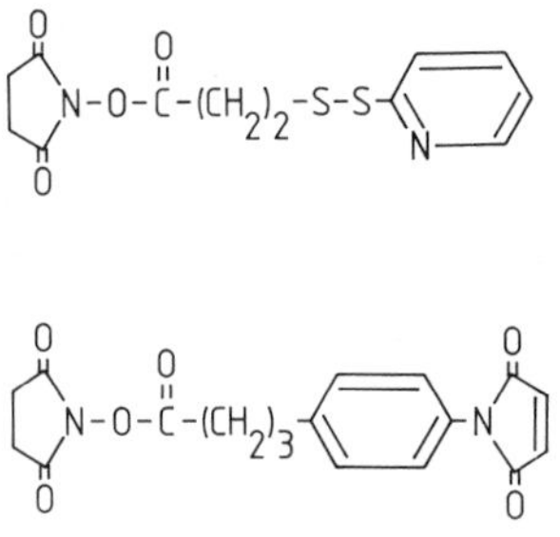

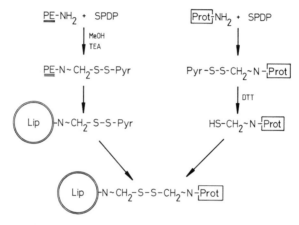

Figure 1. Structures of the two compounds most widely used for introducing thiol groups into moieties containing amino groups. SPDP, *N*-succinimidyl pyridyl dithiopropionate; SMPB, *N*-succinimidyl (-4-[*p*-maleimidophenyl) butyrate.

Figure 2. Use of SPDP for conjugation of proteins to liposomes. In this method, a potential sulphydryl-(SH-) forming residue is introduced into each of the components to be linked together by reaction of both of the individual components separately with SPDP. In the case of the protein, this can be carried out in aqueous solution at neutral pH; the lipid reaction takes place in organic solvent at high pH. Both derivatized products can be stored for long periods of time. In order to link the protein to liposomes, liposomes are prepared which contain the lipid derivative (PDP-PE) in the membrane. At the same time, the disulphide bond (-S-S-) in the derivatized protein (PDP-protein) is reduced by reaction with dithiothreitol (DTT) at pH 5.5 for 30 min. After removal of the DTT (e.g. by column chromatography) the liposomes and the protein solution are mixed together at neutral pH for 18 h, during which time the less stable aromatic pyridyl sulphydryl group (-S-Pyr) is displaced from the disulphide bond by the aliphatic -SH group of the reduced protein, resulting in linkage of the protein to liposomes via a disulphide bridge. Because the two components to be conjugated contain sulphur residues in two different oxidation states (i.e. oxidized as a disulphide bond, or reduced as a sulphydryl group), there is little chance of cross-linking of the homologous reagents with themselves, so that protein–protein and liposome–liposome cross-linking does not occur.

with thiols to produce thioether bonds. All of these reactions occur rapidly at neutral pH, making them ideal for use with proteins, since denaturation can be avoided. A series of heterobifunctional reagents based on these reactions have been developed (13–15), and proved most efficient for protein conjugation to liposomes (16–18).

In this chapter, we will begin by described two of the most promising and efficient methods for the conjugation of proteins to liposomes. Both procedures rely upon the reaction of thiol-reactive phospholipid derivatives included in the liposome with thiol groups of the proteins to be conjugated. The thiol-reactive lipids are synthesized using the reagents *N*-succinimidyl pyridyl dithio propionate (SPDP) and *N*-succinimidyl (-4-[*p*-maleimidophenyl]) butyrate (SMPB) (*Figure 1*). The former approach results in the reversible coupling of protein via a disulphide bond while the latter produces an irreversible thioether linkage. These reagents have been used by several laboratories for the conjugation of proteins to liposomes. Because both methods are similar or identical in a number of steps, only the SPDP method will be given in detail, and modifications relevant to SMPB will be noted at the end. The basis of the methods is outlined in *Figure 2*.

2. SPDP METHOD OF LINKING PROTEINS TO LIPOSOMES

2.1 **Reagents**

(i)   Phosphatidyl ethanolamine (PE). It is best to use lipid of the highest purity available, preferably a saturated synthetic product or a *trans*-esterified product derived from egg phosphatidyl choline. Egg phosphatidyl ethanolamine can be used, but is highly susceptible to peroxidation.

(ii) *N*-succinimidyl pyridyl dithiopropionate (SPDP). This reagent can be purchased from Sigma Chemical Co., Pierce or Pharmacia-LKB Ltd.

(iii) Triethylamine (TEA). This reagent is available in a sufficiently pure form from Pierce.

2.2 **Procedure**

2.2.1 *Synthesis of PPD-PE*

(i)   Dry down 15 mg of PE (approx. 20 μmol) in a 5 ml glass bottle.

(ii)   Redissolve in 2 ml of dry chloroform (dried over a molecular sieve type 3a—see Appendix I).

(iii) Add 30 μmol of TEA (3 mg), followed by 30 μmol of SPDP (approx. 10 mg) in 1 ml of dried methanol.

(iv)   Stir the mixture at room temperature under nitrogen for $1-2$ h until the reaction is complete.

(v)   While the reaction is proceeding, prepare a chromatography column of 2 g of silicic acid in chloroform. (Dissolve the silicic acid in 10 ml of chloroform and pour the contents into a 10 ml plastic syringe barrel plugged with glass fibre. Allow the surplus to drain out and fit the syringe barrel with a plastic disposable three-way tap.)

(vi)   Check the progress of the reaction by TLC using silica gel plates developed with chloroform-methanol-water (65:25:4 by vol). The derivative runs faster than free PE. Visualize the spots using phosphomolybdate or iodine (see Chapter 3, Section 2.2).

(vii) If the reaction has not gone to completion, add another 1 mg of TEA and continue stirring.

(viii) When the reaction is complete (i.e. no more free PE on the TLC plate), dry down the mixture on a rotary evaporator.

(ix)   Resuspend the dried lipids in chloroform, and apply immediately to the top of the silicic acid column. (Note: it is not advisable to leave the lipids in contact with TEA for a long period of time, since the TEA appears to cause hydrolysis

165

of ester linkages, leading to the formation of lyso- compounds. If it is necessary to wait before removing the TEA chromatographically, connect the flask of dried lipids to a lyophilizer, and maintain under high vacuum overnight).

(x)   Wash the column with 4 ml of chloroform, then elute with 4 ml portions of a series of chloroform:methanol mixtures, first 4:0.25 (v/v) followed by 4:0.5 (v/v), 4:0.75 (v/v) and finally 4:1 (v/v). Collect 2 ml fractions and locate the pure derivative by TLC.

(xi)   Pool the fractions containing the desired product and concentrate by evaporation at reduced pressure in a rotary evaporator.

(xii) Recheck for purity by TLC. The purity can also be confirmed by determining whether the phosphate/2-thiopyridine ratio is unity. [Measure phosphorus by the Bartlett assay (see Chapter 3, Section 2.11); 2-thiopyridine may be measured spectrophotometrically after reduction with dithiothreitol (the molar extinction coefficient of 2-thiopyridine at 343 nm is 8300).]

Store the product at $-20°C$ or below under nitrogen in a chloroform solution, preferably sealed in glass ampoules. Completion of reaction, and purity of the product can also be checked by staining with ninhydrin spray after TLC. The spray is made up by dissolving 0.25 g of ninhydrin in 100 ml of acetone/lutidine (9:1 v/v-lutidine is dimethyl pyridine). Lipids such as PE, which contain free amino groups, stain blue after $5-10$ min although weaker spots may take longer; PDP-PE gives no coloration at all. Ninhydrin spray seems to interfere with iodine staining, so it is advisable to run two samples side by side and mask one lane during spraying.

2.2.2 *Protein thiolation (illustrated using IgG)*

(i)   Prepare 5 ml of IgG solution in 0.1 M phosphate buffer (pH 7.5) at a concentration of 5 mg ml^{-1}.

(ii)   Prepare SPDP solution at a concentration of 20 μmol ml^{-1} (6 mg ml^{-1}) in ethanol.

(iii) Add 150 μl of SPDP solution slowly to 5 ml of the stirred protein solution with a Hamilton syringe to give a molar ratio of SPDP to protein of 15:1. The ethanol concentration should not exceed 5% or protein denaturation may occur. Allow the mixture to react for 30 min at room temperature (appox. 20°C).

(iv)   Separate the protein from reactants by gel chromatography on Sephadex G-50 equilibrated with 0.05 M sodium citrate ($Na_3C_6H_5O_7.2H_2O$, 19.7 g l^{-1}), 0.05 M sodium phosphate ($Na_2HPO_4.7H_2O$, 13.4 g l^{-1}), 0.05 M sodium chloride (2.9 g l^{-1}) pH 7.0.

The product is characterized by measuring the number of pyridylthiols per molecule as described earlier, in which a sample of the protein is treated with 50 mM DTT to release free thiopyridone, whose absorbance is measured at 343 nm. Typically, these conditions yield $3-5$ pyridyl thiols per IgG molecule. If measuring the concentration of protein by absorbance at 280 nm, allowance must be made for the fact that bound thiopyridone also absorbs at this wavelength (see ref. 13, p. 726). The product can be stored for long periods at 4°C provided it is sterilized by filtration, or by the addition of azide (10 μg ml^{-1}). Note: the examples given here are typical thiolation procedures for monoclonal mouse IgG and rabbit Fab' fragments. It should be remembered that if proteins other than immunoglobulins (or their binding fragments) are to be used, the number of thiols required may be different. An example of this is given in Shek and Heath (7), where bovine serum albumin required 15 thiols per molecule for efficient conjugation to liposomes.

2.2.3 *Effecting the coupling reaction*

(a) *Liposome preparation*

(i)   Prepare liposomes suspended in pH 7.0 buffer by any of the methods described in Chapter 2. A suitable lipid composition (see later) is PC:Chol:PG:PDP-PE molar ratio 8:10:1:1. The concentration range of PDP-PE can be $1-10$ mole per 100 mole total lipid.

(b) *Reduction of derivatized protein*

(ii)   Titrate the pyridyl dithio-protein solution in citrate-phosphate buffer to pH 5.5 by addition of 1 M HCl.

(iii) Make up a solution of 2.5 M dithiothreitol (DTT 380 mg ml^{-1}) in 0.2 M acetate buffer pH 5.5 (165 mg of sodium acetate in 10 ml).

(iv)   Add 10 μl of DTT solution for each millilitre of protein solution.

(v)   Allow to stand for 30 min.

(vi)   Separate the protein from the DTT by chromatography on a Sephadex G-50 column equilibrated with a buffer at pH 7.0. Note: care should be taken to exclude oxygen from the reaction mixture since thiols are rather unstable and prone to oxidation. Therefore, bubble nitrogen through all buffers to remove oxygen, and collect the protein fractions under nitrogen.

(c) *Conjugation of components*

(vii) Mix liposomes with protein, and allow to stand overnight, with stirring, at room temperature. Separation of liposomes from free protein may be carried out using the flotation methods described in Chapter 2, Section 5.3.2.

Aromatic dithiols are much more susceptible to cleavage by free sulphydryl groups than are aliphatic disulphide bridges. Consequently, in the presence of reduced protein, the PDP moiety is readily broken down, the thiopyridine being replaced by the protein thiol forming a disulphide bridge which is resistant to further attack. If desired, unreacted residues may be blocked or destroyed. Thiols may be blocked with Ellman's reagent (20), or may be reacted with alkylating reagents such as *N*-ethylmaleimide or iodoacetamide. Maleimide residues (see Section 3) may be reacted with thiol-containing reagents, such as DTT, mercaptoethanol, or cysteine. Residual pyridyldithiols could theoretically be eliminated by reduction at pH 4.5, with subsequent alkylation of the thiols generated. To date, no investigators have considered this process necessary.

2.2.4 *Use of endogenous thiol residues*

In the method described above, SPDP has been used to introduce extra reactive groups into the protein. In certain cases, however, it is possible to use as precursors for free thiol groups the disulphide bridges already present in the protein, which after reduction

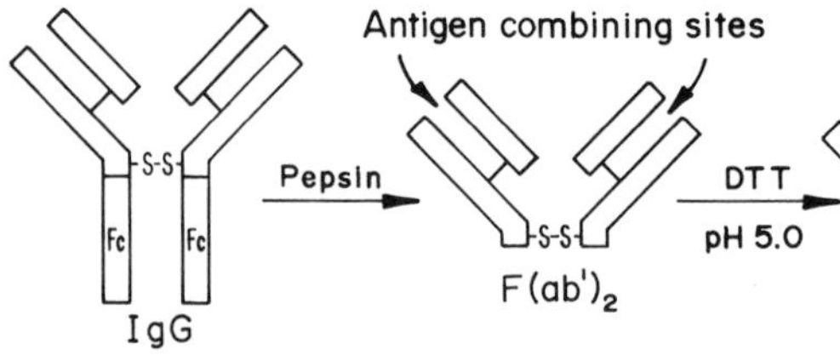

Figure 3. Conversion of IgG to Fab′ fragments containing free endogenous thiol groups. The Fab′ fragments formed in this way contain free sulphydryl groups which can react directly with liposomes containing PDP-PE, without the need for introduction of PDP residues into the protein.

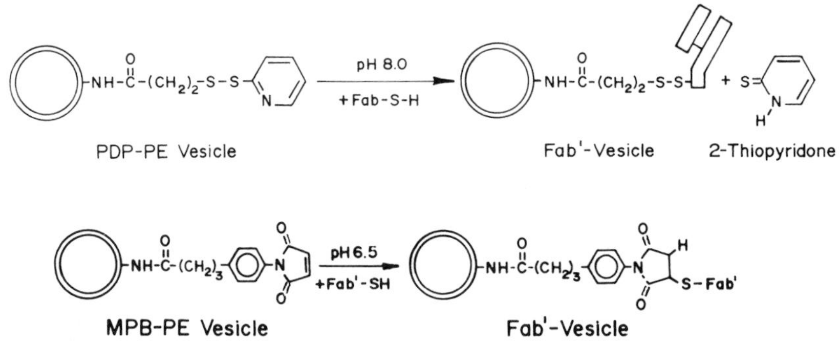

Figure 4. Comparison of SMPB and SPDP. This figure shows the difference in reaction conditions and the type of linkage finally obtained when using MPB-PE rather than PDP-PE as the liposomal conjugating agent. In both cases, the reaction is illustrated using Fab′ protein subunits bearing endogenous thiol residues. The pH tolerance for the PDP- reaction (pH 6.0−8.0) is much wider than for the MPB- reaction (pH 6.0−6.8). The product of linkage via the PDP- group is a disulphide bridge, which may be easily cleaved under certain conditions, whereas the MPB- method gives rise to a very stable thioether linkage.

by DTT, are converted to cysteine residues which are then available for linking to the derivatized PE on the liposome surface. Such an approach is particularly convenient for IgG which may be converted to F(ab′)$_2$ fragments by pepsin digestion, followed by reduction with 20 mM DTT (final concentration) at pH 5 for an hour (*Figure 3*). DTT removal and handling of protein are carried out exactly as before.

3. MODIFICATIONS OF THE SPDP METHOD WHEN USING SMPB

With regard to the protein component of conjugation, the method using SMPB is identical to that described above. Free thiol groups are introduced into the protein either by reaction with SPDP, or by conversion of endogenous disulphide linkages. The difference between the two methods is in the derivatization of the PE, where SPDP is replaced by SMPB (*Figure 4*), but otherwise exactly the same protocol is followed. The product of derivatization, 4-(p-maleimidophenyl)butyryl phosphatidyl ethanolamine (MPB-PE) has an R_f of 0.52 in the TLC solvent system given above. In the final conjugation step (*Figure 4*) one should be aware that the maleimide residue of MPB-PE is rather unstable, particularly at pH values greater than pH 7.0 (17). Liposomes containing MPB-PE should be prepared at pH 6.0−6.8 not more than a few hours before their conjugation to the protein. The stability of MPB-PE is sufficient to allow gel filtration and other

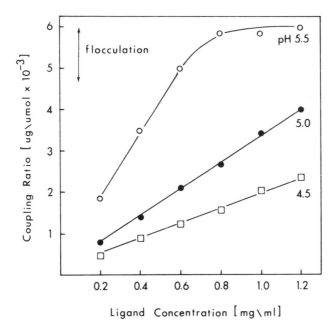

Figure 5. Relationship between Fab′ concentration and coupling ratio (expressed as μg Fab′ per μmol lipid). Relationship between protein concentration (Fab′) and coupling ratio (μg protein per μmol lipid). The reduction step in which Fab′ was generated from F(ab)′$_2$ was carried out at three different pH values.

manipulations prior to conjugation. The pH range for conjugation of MPB-PE (pH 6.5 − 6.8) is considerably narrower than that permissible for SPDP (pH 6.0 − 8.0) to allow reaction while avoiding excessively rapid degradation of the maleimide residue. Often the reaction is allowed to proceed overnight (12 − 16 h) with stirring, although 6 h is sufficient time for protein conjugation to MPB-PE liposomes.

4. FACTORS AFFECTING CONJUGATION EFFICIENCY

4.1 Concentration of reduced protein and lipid

Conjugation is most easily regulated by adjustment of the protein and the liposome concentration. For the most part, the protein concentration in the conjugation mixture determines the protein-to-lipid ratio of the product (*Figure 5*). The authors have observed a linear relationship between the two parameters, and have been able to control the products accordingly. The lipid concentration does not appear to affect appreciably the protein − lipid ratio of the product, but must nonetheless be controlled for other reasons (see Section 4.3). In the case of rabbit Fab′ fragments, the pH at which the F(ab′)$_2$ fragment is reduced also affects the efficiency of conjugation. As shown in *Figure 6*, the degree of reduction of the F(ab′)$_2$ fragment by DTT is a function of the pH and ranges from 0.9 thiols per monomer at pH 4.0 to 1.2 thiols at pH 5.5. As expected, the degree of conjugation of these fragments to liposomes containing thiol-reactive groups increases with increasing number of thiols per fragment. *Figure 6*, also demonstrates that coupling efficiency improves from about 15% for fragments reduced at pH 4.0 to 75% at pH 5.5.

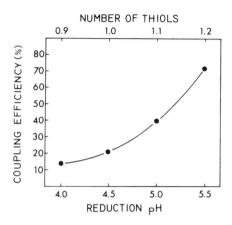

Figure 6. Relationship between reduction pH (number of thiols) and coupling efficiency, expressed as a percentage of the total added Fab' fragments coupled to liposomes.

Table 1. Effect of cholesterol on ligand conjugation.

Liposome composition			
PC	*Cholesterol*	*MPB-BE*	*Coupling ratio ($\mu g\ \mu mol^{-1}$)*
10.0	0	1.0	<25
10.0	1.0	1.0	62
10.0	2.5	1.0	146
10.0	5.0	1.0	168
10.0	10.0	1.0	174

4.2 Presence of cholesterol in the liposome membrane

As shown in *Table 1*, cholesterol appears to be required for efficient coupling of Fab' fragments to MPB-PE containing liposomes. Liposomes containing no cholesterol fail to couple the fragments above control levels. The inclusion of up to 30 mol% of cholesterol produces a dramatic increase in the coupling efficiency.

4.3 Aggregation during conjugation

In some circumstances, protein can induce aggregation of liposomes. Such aggregation can be particularly prevalent with small liposomes, and is best controlled by minimizing the liposome concentration and keeping the protein concentration to a level which does not induce aggregation (11,21). Aggregation may be due either to the covalent cross-linking of liposomes via a protein bridge, or to the clumping of the protein-bearing liposomes as a result of reduction in the number of mutually repulsive surface changes. This phenomenon has not been extensively investigated, although a recent report has suggested that it may be controlled by minimizing the number of thiols on the protein molecules (22). Inclusion of charged lipids such as PG or PS has also been used to reduce aggregation by providing electrostatic repulsive forces among the suspended liposomes, limiting close approach and thereby inhibiting aggregate formation (23). As shown in *Figure 7*, the ionic strength of the suspending medium also affects

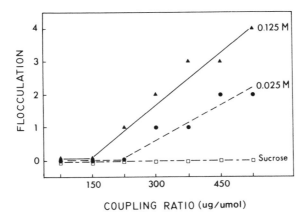

Figure 7. Effect of ionic strength of the medium and coupling ratio on flocculation (aggregation) behaviour or liposomes. Liposomes were suspended either on 0.3 M sucrose solution, in 0.025 M NaCl, or in isotonic saline (0.125 M). The extent of aggregation (flocculation) was assessed qualitatively on a score of 1 to 4 by microscipc examination.

aggregation behaviour. In general a balance must be struck between the tendency of the protein-conjugated liposomes to aggregate (manifesting visibly as flocculation), and the electrostatic repulsive forces engineered into the system.

5. COMPARISON OF PROTEIN THIOLATION METHODS

The method employed for thiolating proteins depends to a large extent on the protein one wishes to use. If the protein has endogenous disulphide bridges which can be broken without destroying essential enzymic, binding or immunogenic properties of the molecule, then this approach is to be preferred, since it appears that endogenous free thiols are often more reactive than those introduced as thiopropionyl derivatives, so that one may expect the rate of reaction, and the efficiency of conjugation to be increased.

While the most widely used method for introduction of exogenous thiols employs SPDP, one disadvantage of using this reagent is the large number of purifications which have to be carried out at each stage—one to separate unreacted SPDP from protein, and one to remove DTT, with concomitant loss of protein on columns, etc. Hashimoto *et al.* (24) have pointed out that one separation step can be avoided by using the reagent SAMSA instead (S-acetylmercaptosuccinic anhydride) to introduce thiol groups (again employing an *N*-hydroxy succinimide ester). In this case, the sulphur atom is introduced not as part of a disulphide bridge, but as a thioester, which is easily unblocked with hydroxylamine. Since the presence of hydroxylamine (unlike DTT) does not compete, or otherwise interfere, with the subsequent formation of either disulphide or thioether linkages, the two reactions (unblocking and conjugation) can be carried out in the same vessel without the need for removal of the hydroxylamine before addition of the protein to liposomes (see *Figure 8*). One difficulty with SAMSA is that upon binding to the amino group of the protein, a free carboxyl group is revealed, which is retained as part of the linker molecule, thus changing the net charge of the protein radically after only a few substitutions. Such alterations, while sometimes being innocuous, may easily lead to increased aggregation, or non-specific binding to cells. This problem is overcome

171

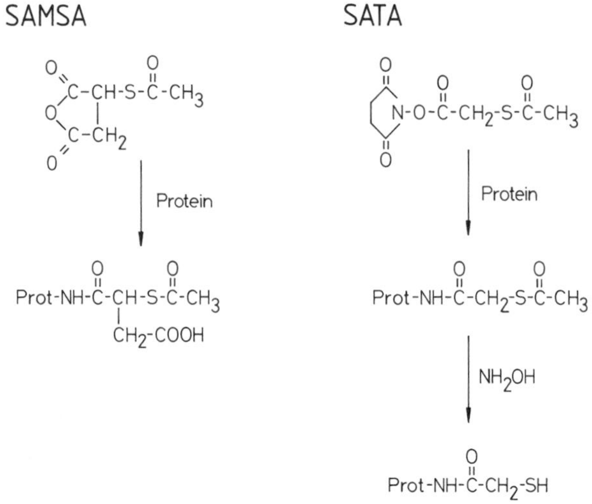

SAMSA   SATA

Figure 8. Introduction of thioacetate groups into proteins. SAMSA, S-acetyl mercaptosuccinic anhydride; SATA, Succinimidyl-S-acetyl thioacetate. Reaction of SAMSA with proteins results in the introduction of an extra carboxy function, in addition to the thioacetate group. Use of SATA avoids this complication. Proteins derivatized with either agent can yield a free sulphydryl group upon incubation with hydroxylamine, as shown here for the SATA derivative.

with SATA (succinimidyl-S-acetylthioacetate) which works in the same way as SAMSA, but releases the free carboxy moiety into the bulk medium (see *Figure 8*). The protocol adopted for using these reagents is as follows [after Rector *et al.* (15) and Derksen and Scherphof (25)].

5.1 SAMSA/SATA modification of proteins

(i)   Prepare the buffer containing 145 mM NaCl, 2 mM EDTA, 10 mM Hepes-NaOH (pH 7.6) by dissolving 2.4 g of Hepes, 8.4 g of NaCl and 0.75 g of EDTA disodium salt in a litre of distilled water. Bring the pH to pH 7.4 with about 1 ml of 5 M NaOH.

(ii) Prepare 2 ml of IgG solution (5 mg ml^{-1} in phosphate buffered saline).

(iii) To the stirred solution of IgG add 40 μl of SAMSA/SATA (1 mg) at a concentration of 25 mg ml^{-1} in dimethyl formamide.

(iv) Incubate at room temperature under nitrogen for 30 min.

(v)   Remove unreacted reagent by dialysis or column chromatography. Equilibrate with Hepes:saline:EDTA buffer. The protein may be concentrated and stored at -20°C until required.

(vi) To 100 μl of the protein solution (20 mg ml^{-1} in Hepes:saline:EDTA) add 10 μl of 0.1 M hydroxylamine (7 mg ml^{-1} NH$_2$OH.HCl) in a sealed plastic microcentrifuge tube.

(vii) Incubate at room temperature for 30 min under nitrogen.

(viii) Add 100 μl of derivatized liposomes (containing about 3 mg of lipid) to the above mixture, and continue incubation, under nitrogen, at room temperature overnight.

(ix) Remove unbound protein from liposomes by column chromatography or centrifugal flotation.

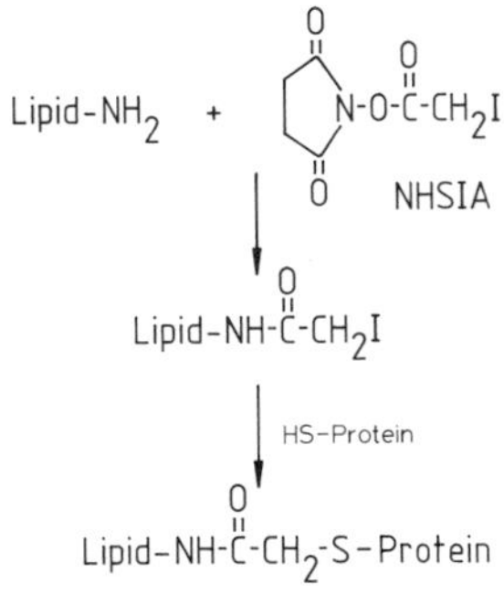

Figure 9. Conjugation of iodoacetate to PE using NHSIA.

6. COMPARISON OF LIPOSOME THIOLATION METHODS

The choice between use of SPDP and SMPB as the derivatizing agent for liposomes is determined by the use to which the liposomes will be put. PDP-PE binds rapidly and efficiently to thiolated proteins, and the reaction can tolerate a wider pH range than can conjugation via MPB-PE (defined in *Figure 4*). The disulphide linkage formed, however, is a reversible one, and can easily be broken in the presence of thiols such as glutathione, which is present in high concentration in biological tissues. For *in vivo* use, therefore, SPDP derivatization may be inadvisable, except in cases where dissociation of the protein from liposomes is particularly desired. The MPB reaction, on the other hand, results in a thioether linkage which is very stable in biological environment (see *Figure 4*).

One objection to the use of SMPB is the possibility that the large maleimido benzoyl residue may act as an immunogenic determinant in its own right (especially if conjugated with small proteins or haptens), and may interfere with the stimulation of immune responses by liposomes, or create difficulties in interpretation of the results of immunological experiments. It has been proposed (24) that the introduction of iodoacetate residues onto the liposome surface (via *N*-hydroxysuccinimido-iodoacetate, NHSIA) may be a suitable alternative (*Figure 9*). However, one group of workers found that the iodoacetate resulted in a very low efficiency of binding of protein (26) while others found that use of a spacer group was necessary to overcome steric hindrance (24). MPB is already large enough not to need an extra spacer group.

Other approaches to effecting the coupling reaction between protein and liposomes have been to employ sandwich techniques in which the methods described above are used to bind staphylococcal protein A (2), avidin (27) or biotin to the liposomes (*Figure 10*), which are then able to bind a whole range of IgG molecules of varying specificities, or biotinylated proteins. Avidin precipitation of biotin liposomes has also been used as a means of purification of liposomes from unassociated protein (24).

7. DERIVATIZATION OF LIPOSOMES *IN SITU*

7.1 **Thio-derivatives**

In the methods described so far, lipophilic thio-derivatives have been synthesized prior to their incorporation during formation of liposomes, into the bilayer membrane. Under these circumstances, the thiol precursors are distributed on both sides of the membrane,

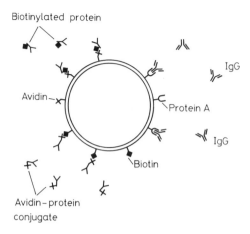

Figure 10. Sandwich techniques for binding proteins to 'multi-purpose' liposomes. The use of non-covalent methods for linking proteins to liposomes opens up the possibility of attaching several different types of protein to the same liposome, or of producing different populations of liposomes, each bearing a different protein ligand, but otherwise absolutely identical with respect to membrane composition and aqueous contents. The concept is illustrated here for avidin and biotin (× and ♦ respectively) and for protein A binding of IgG.

and come into direct contact with solvents and entrapped solute. In cases where the entrapped material is sensitive to the action of these precursors, the presence of such thio-derivatives in the membrane during liposome manufacture is clearly undesirable. Prior synthesis of the derivatives is also tedious and time-consuming, and their use in this way is uneconomic, since only a proportion of the ligand molecules are exposed on the outer surface of the liposome membrane, available for coupling to proteins.

An alternative approach which overcomes these problems is to derivatize the PE or other lipid after the liposomes have been formed. In this way, only the residues on the outer membrane are transformed, and no reagents come into contact with entrapped materials. The method adopted for labelling is exactly the same as is employed for protein molecules—that is, the reagent (SPDP, SMPB, NHSIA) is dissolved in concentrated form in an organic solvent such as dioxane, dimethyl formamide or ethanol, and is added with stirring to the liposome suspension. In the case of liposomes containing 50 mol% cholesterol, the concentration of solvent added to the suspension can be tolerated up to a value of 5% without causing leakage of entrapped solutes. Good buffering capacity is required to ensure that the pH does not decrease upon hydrolysis of the reagents to the free acid. Aminolysis proceeds at a much faster rate than hydrolysis, but any unreacted reagent will have been hydrolysed after half an hour. It is also possible to add the liposome suspension to the dry solid reagents (to give a final concentration of about 1 mg ml^{-1}), with a molar excess of reagent to PE of between 10- and 20-fold. Separation can be carried out by dialysis or column chromatography at the same time as the protein is being prepared for conjugation.

7.2 Formation of carbonyl groups

As mentioned in the introduction to this chapter, the coupling of proteins to liposomes can also be carried out utilizing the reaction of a Schiff base with primary or secondary amino groups (*Figure 11*). Since the endogenous amino groups of the protein can be

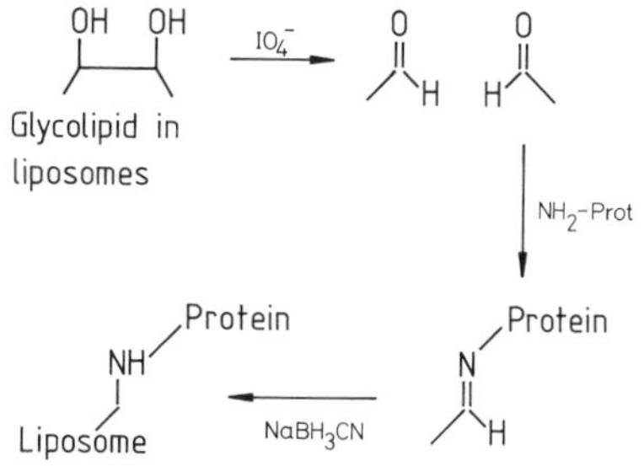

Figure 11. Linkage of liposome-bound sugars to proteins via Schiff base. Glycolipids such as cerebrosides or gangliosides incorporated into the liposome membrane can be used to link liposomes to proteins. Membrane-bound sugars containing two adjacent (vicinal) hydroxyl groups can be oxidized by periodate to aldehyde functional groups, which then react readily with the amino groups in proteins. Unreacted aldehyde functions, and the newly formed imine linkage, can be converted to unreactive species by reduction with sodium cyanoborohydride.

used without modification, and liposome-bound glycolipids can be oxidized *in situ* without destroying the integrity of the liposomes, conjugation can be performed in a simple two-step reaction. Liposomes containing 50 mol% cholesterol and 10 mol% lactosyl ceramide, 20 mol% galactose cerebroside or 5% gangliosides have been employed successfully. Use of sucrose stearate-palmitate (28) and phosphatidyl inositol (8) has also been reported. A protocol [after Heath and co-workers (12)] for periodate coupling of liposomes is given below.

7.2.1 *Periodate coupling of liposomes*

(i)   Prepare borate-buffered saline containing 20 mM borate ($Na_2B_4O_7.10H_2O$ 7.6 mg ml^{-1}) and 120 mM NaCl (7 mg ml^{-1}) adjusted to pH 8.4.

(ii)   To 1 ml of glycolipid liposome suspension, containing approximately 5 mg of total lipid ml^{-1} borate-buffered saline, add 200 μl of 0.6 M sodium periodate ($NaIO_4$ 128 mg ml^{-1} distilled water) with stirring.

(iii)   Incubate at room temperature in the dark for 30 min.

(iv)   Remove free periodate either by dialysis overnight, or by minicolumn chromatography.

(v)   To 1 ml of oxidized liposomes, add 0.5 ml of the protein solution, containing 10−30 mg of IgG in borate-buffered NaCl solution (pH 8.4). Incubate for 2 h at room temperature.

(vi)   Add 10 μl of 2 M sodium cyanoborohydride (125 mg ml^{-1}) per ml reaction mixture and leave overnight at 4°C.

(vii)   Separate free protein from liposomes by column chromatography or centrifugal flotation.

Although the optimal pH for periodate oxidation is pH 5.5, this reaction is carried out at higher pH values here to avoid increase in the permeability of the liposome membrane, and resultant loss of contents due to leakage at low pH. Also, at low pH, neutral liposomes are permeable to periodate, leading to oxidation of internal contents as well as surface glycolipids. Negatively-charged liposomes, however, are impermeable to periodate at low pH. If desired, periodate oxidation may be carried out at low pH provided the final periodate concentration is reduced to 10 mM or less. The incubation mixture may then be left overnight.

Sodium cyanoborohydride is less reactive than sodium borohydride, but is used in preference since it is stable around pH 8.0 or lower (i.e. it does not require pH 9 as when NaBH$_4$ is used) and will not attack the disulphide bridges of proteins. The protein:lipid ratio is increased with increasing protein concentration, while the total protein bound is higher when the lipid concentration is higher. Under optimal conditions, approximately 20% initial protein can be bound, and about 40% of the theoretical available space on the liposome surface can be filled. With small sonicated liposomes, coupling at high concentration of both lipid and protein can lead to aggregation by cross-linking of adjacent liposomes with proteins; this behaviour is not displayed with larger liposome (e.g. REVs). It is important to note that with glycolipids such as galactocerebroside, good coupling is not observed at concentrations less than 20 mol%, presumably because only one sugar residue is available for oxidation per molecule. Linkage with PG appears to be difficult, possibly because of steric considerations.

8. ALTERNATIVE CONJUGATION PROCEDURES

8.1 Employing liposomal carboxyl groups

Phosphatidyl ethanolamine can be derivatized in such a way that the lipid moiety bears an active carboxyl group which, after incorporation into liposomes, is able to bind directly to amino groups on proteins and other molecules. This is achieved via the intermediary of a bifunctional straight-chain α-ω dicarboxylic acid, which acts as a bridge between the PE and the protein (*Figure 12*). By choosing carboxylic acids containing different numbers of carbons, the length of this spacer group can be varied to suit one's purposes. A distance of 6−8 carbons in between the amide linkages has been found to be optimal in terms of the efficiency of linkage and minimization of non-specific, non-covalent binding of proteins.

The lipid moiety containing the carboxylic acid spacer can be prepared in two ways. In the first, the dicarboxylic acid is presented to the PE in the form of the di-*N*-hydroxy succinimide derivative (29), so that both the carboxy functions are activated: one of these reacts with the PE amino group, to form a phospholipid containing a free carboxyl group also activated with NHS. This compound is incorporated into liposomes, which will then react spontaneously with proteins, binding to them via the protein amino groups. The only problem with this method is that the PE-spacer-CO.NHS derivative, while moderately stable in organic solvents, breaks down rapidly in aqueous media, so that it can only be used in cases where liposomes are prepared by a fairly rapid procedure, and little subsequent processing is required.

In the second method (30), the dicarboxylic acid is converted to an internal acid anhydride by reaction with carbodiimide (CDI); the anhydride then reacts with PE to give a derivative containing a free carboxyl group at the end of a spacer chain. This compound is very stable in both organic and aqueous media. Upon incorporation in liposomes, the carboxyl function is activated with a carbodiimide at low pH. After excess free CDI has been removed, protein is introduced into the liposome suspension which binds rapidly and efficiently as the pH is raised. Up to 60% binding efficiency has been reported using this method. Procedures for the preparation of both PE derivatives are given below.

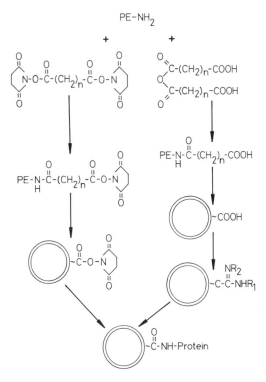

Figure 12. Two alternative methods of conjugation using dicarboxylic acids as spacer groups. This figure shows reaction schemes for the use of disuccinimidyl suberate (**left**) or the dicarboxylic acid anhydride (**right**) to introduce into liposome membranes spacers of varying lengths terminated with a carboxyl functional group.

8.1.1 *PE derivatization with suberimidate*

(i)   Dissolve 10 mg of PE in 10 ml of '*bone dry*' ethanol (see method in Appendix I on purification of solvents.

(ii)   Add a 10-molar excess of disuccinimidyl suberate (50 mg) and an equimolar quantity of triethylamine (1.5 μl).

(iii) Let the reaction proceed at room temperature for several hours, following the conversion of PE to the NHS-derivative by TLC (chloroform – methanol – water 65:25:4 v/v). If the reaction stalls (i.e. if the PE is not completely consumed), add a little more TEA or more disuccinimidyl suberate until full conversion is achieved.

Note: do not let the reaction continue for more than a few hours—certainly not overnight. The TEA will eventually cause the formation of a lyso form so it is important to isolate the product from the reaction mixture as soon as possible.

(iv)   Following completion of the reaction, reduce the volume of ethanol to a minimum while keeping everything in solution.

(v)   Add a 20-fold volume excess of methylene chloride to precipitate any unreacted disuccinimidyl suberate.

(vi)   Centrifuge at low speed for 10 min in a bench centrifuge at 20°C, decant the supernatant containing the product, and discard the pellet.

(vii)   Remove the methylene chloride from the supernatant collected by drying down under vacuum.

(viii) Dissolve the solid in a minimal volume of ethanol and repeat the precipitation procedure in Steps (v)–(vii) twice more.

(ix)   Dry the product down in a weighed flask on a rotary evaporator, and place it under vacuum in a lyophilizer overnight (to remove traces of TEA which is fairly volatile).

(x)   After removal of the TEA weigh the flask to determine the product weight.

(xi)   Dissolve the product in a suitable solvent (e.g. chloroform) to the desired working concentration and combine with other lipids in the usual way for preparation of liposomes.

Do not try to store the NHS-PE for longer periods—it breaks down even at low temperature (-70°C) in anhydrous solvents or dry form. It is best to use fresh NHS-PE each time in order to achieve a reproducible and efficient conjugation.

8.2.1 *PE derivatization with acid anhydride (after Kung and Redeman)*

(a) *Formation of sebacic anhydride*

(i)   Dissolve 16.2 mg of sebacic acid (0.08 mmol) and 8.7 mg (0.042 mmol) of dicyclohexyl carbodiimide (DCCI) in 2 ml of methylene chloride in a $15-20$ ml glass screw-capped tube.

(ii)   Flush with nitrogen, cap tightly, and incubate the mixture for 48 h at room temperature with stirring. The anhydride formed may be used without further isolation.

(b) *Conjugation of PE*

(iii)   To the above reaction mixture add 2 ml of chloroform containing 28 mg (0.038 mmol) of PE and 15 μl of triethylamine (0.108 mmol).

(iv)   Allow the reaction to proceed for a few hours at room temperature, following its progress by TLC (chloroform–methanol–water 65:25:4 v/v).

(v)   To stop the reaction, add 5 ml of chloroform, and 4 ml of 0.02 M phosphate, 0.02 M citrate buffer ($Na_2HPO_4.7H_2O$ 5.4 mg ml^{-1}, $Na_3C_6H_6O_7.2H_2O$ 5.9 mg ml^{-1}) (pH 5.5), and shake vigorously.

(vi)   Separate the two phases by low speed centrifugation (1000 g for 10 min at room temperature).

(vii)   Discard the upper aqueous phase, and dry the organic phase over anhydrous sodium sulphate.

(c) *Purification of derivative*

(viii) Run the dried chloroform solution on to a 1 × 20 cm silica gel column (Kiesel gel 60).

(ix)   Follow with 50 ml of eluant solutions of the following composition:

   (1) 50 ml chloroform—0% methanol v/v
   (2) 45 ml chloroform and 5 ml methanol—10% methanol v/v

(3) 40 ml chloroform and 10 ml methanol—20% methanol v/v
(4) 35 ml chloroform and 15 ml methanol—30% methanol v/v
(5) 25 ml chloroform and 25 ml methanol—50% methanol v/v

(x)   Analyse the fractions eluted by TLC. The product should have an R_f value of
        0.42. PE may be distinguished from the product by use of a ninhydrin spray
        reagent. Most of the product should elute in the later fractions.

The product may be stored, or used directly for incorporation into liposomes. Protein
is bound to liposomes containing the derivative by activating the liposomes for an hour
with EDCI (1-ethyl-3-(dimethyl aminopropyl)-carbodiimide, 2.5 mg ml^{-1}) in pH 5
phosphate-buffered saline (concentration of derivative 0.05 μmol ml^{-1}). Add 50 μl of
protein solution (e.g. IgG 10 mg ml^{-1}) to each millilitre of liposome suspension.
Increase the ionic strength by addition of 50 μl of 1 M NaCl and adjust the pH to pH 8.
The reaction is carried out overnight at 4°C before separating the reactants from the
product by usual methods.

Recently, a method has been reported (31) in which linkage of proteins to a synthetic
lipid in the liposome membrane is brought about by diazotization. The lipid is *N*-(*p*-
aminophenyl) stearylamide (APSA) and the aromatic groups in the protein react with
the APSA (activated by treatment with cold sodium nitrite) to give a diazo linkage.
One advantage of this procedure is that it is possible to gauge the success or otherwise
of the final conjugation step by observing the colour change which takes place as a
result of formation of the diazo bond, to give liposomes with a distinctive 'tan' coloration.
The efficiency of conjugation can vary considerably, depending upon the protein used.
Apart from the slight explosion hazard involved in diazotization if the temperature rises
too high, the most severe drawback of this method is the requirement for the conjugation
reaction to be carried out at pH 10, which could damage some sensitive proteins. Details
of the one-step synthesis of APSA and the conjugation procedure, can be found elsewhere
(31).

8.2 **Employment of protein-bound carboxyl groups**

The methods described previously have involved using the amino groups on the surface
of proteins to conjugate them to linkage groups, and subsequently to liposomes. This
is because the amino group (together with sulphydryl groups) is one of the few functional
groups capable of forming covalent linkages under physiological conditions without
the need for activation. Amide linkages can be formed by reaction with activated carboxy
groups (e.g. *N*-succinimido ester), where the latter reagents have been synthesized prior
to coupling, using much harsher chemical conditions. Circumstances will arise, however,
where linkage to proteins or polypeptides via amino-groups is unsatisfactory because
the biological activity of the molecule is reduced, and methods need to be found for
conjugating proteins using other functional groups. The approach using chemical
modification of oligosaccharide moieties has already been mentioned at the beginning
of this chapter. Another possibility is to make use of the free carboxyl groups on proteins,
and this can be accomplished by employing carbodiimide (CDI) or analogues to catalyse
the formation of an amide bond between the protein carboxyl group and the amino group
of PE or of a linking reagent. Unfortunately, since proteins almost always possess
numerous carboxyl and amino groups on each molecule, extensive cross-linking will

Table 2. Summary of conjugation procedures.

Lipid-reactive agent	Lipid moiety	Linkage	Protein functional group	Protein-reactive agent
(A) SPDP	PE−NH$_2$	−S−S−	NH$_2$-protein	SPDP + DTT or
			HS-protein	SATA/SAMSA + NH$_2$OH
SMPB/NHSIA	PE−NH$_2$	−S−	NH$_2$-protein	SPDP + DTT or
				SATA/SAMSA + NH$_2$OH
			HS-protein	−
(B) Activated dicarboxylic acids	PE−NH$_2$	−CO−(CH$_2$)$_n$−CO−	NH$_2$-protein	−
(C) −	PE−NH$_2$	Peptide	HOOC-protein	NH$_2$ block, then Carbodiimide
(D) −	PE−NH$_2$	2ry amine	Glycoprotein	Periodate/borohydride
Periodate/borohydride	Glycolipid	2ry amine	NH$_2$-protein	−
(E) Nitrite	APSA	−N=N−	Aromatic side groups	−

Five basic types of conjugation procedure are outlined here, classified according to the type of linkage formed. (A) Linkages involving sulphur atoms. (B) Ester linkages. (C) Amide linkages. (D) Linkages via a secondary amine. (E) Diazo linkages.

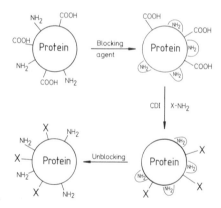

Figure 13. Use of protein carboxyl groups: Blocking followed by CDI. In order to employ the carboxyl functions on proteins as linking groups for conjugation to liposomes, the endogenous amino groups must first be inactivated, to prevent protein−protein cross-linking from taking place after the carboxyl groups have been activated so that they may react with exogenous amino groups. The activating agents most commonly used are water-soluble forms of carbodiimide (CDI). A convenient reversible blocking agent is citraconylate.

occur if CDI is used on its own. Thus before carbodiimide is used, it is necessary to block the endogenous amino groups of the protein, to prevent their participating in the coupling reaction (*Figure 13*).

One of the most satisfactory blocking agents which can be used for this purpose is probably citraconic acid, since this can be removed under conditions (dialysis against pH 4.4 buffer) which may not be too deleterious for the protein. This method has been used successfully for conjugating antibodies to PE (32), which is then inserted into membranes during formation of liposomes by detergent dialysis. If one wished to prepare

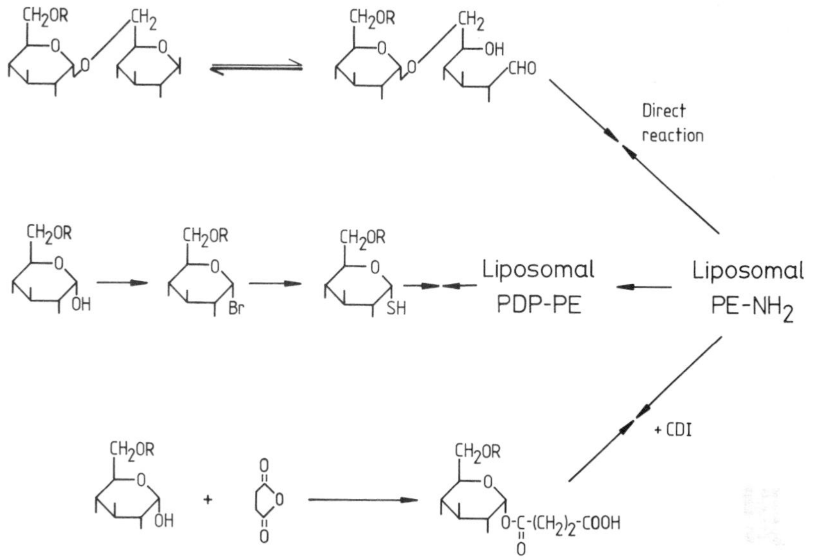

Figure 14. Conjugation of sugars to liposomes. This figure illustrates three different methods of conjugation of sugars to liposomes. In each case, a different functional group has been associated with the sugar, which can then react either with the amino group of liposomal PE, or the PDP-derivative of this molecule. **Top**, Certain sugars possess endogenous aldehyde groups, present in low concentration in the chain form of the sugar, which exists as an automer in equilibrium with the more stable ring form. An imine linkage is formed by direct reaction of these aldehyde functions with amino groups. **Centre**, Introduction of a thiol residue into the sugar permits linkage with PDP- derivatized lipids to occur. **Bottom**, Free hydroxyl groups on sugars can be converted to carboxyl-bearing species by reaction with succinic anhydride. The hemisuccinate thus formed can bind to liposomal PE via the carbodiimide reaction.

liposomes by other methods, and thus attach the protein/peptide to pre-formed membranes, the same procedure could be employed—that is, reaction of the blocked protein with liposomes in the presence of CDI. One would then be faced with the task, however, of unblocking the protein at pH 4.4 in the presence of liposomes, which can cause some undesirable leakage or hydrolysis. This problem could be avoided by conjugating the protein first to a linking group (e.g. an aminothiol, or aminothioester) which can then be unblocked separately, before conjugating it to liposomes, as described earlier, using the maleimido-derivative of PE. Those readers who may have the necessity to adopt this approach are referred elsewhere (33) for basic methods for citraconylation, CDI conjugation and unblocking. A summary of all the different methods of conjugation discussed in this chapter is given in *Table 2*.

9. NON-PROTEIN CONJUGATES

The methods described in this chapter are all suitable for use in conjugation of compounds other than proteins to liposome membranes. In particular, mono- and oligosaccharides have been linked to liposomes and macromolecules in a variety of ways, using either the thiol analogues of the sugar (34), succinyl derivatives formed by action of succinic anhydride (35), or the endogenous aldehyde group revealed upon automerization of some sugars from the ring to the chain form (36). The reaction pathways for these approaches are given in *Figure 14*. The conjugation of biotin, either as the *N*-hydroxysuccinimide ester (37), or as the hydrazide, has also been accomplished (38).

10. ANCHOR GROUPS OTHER THAN PE

The behaviour of ligands when bound to the liposome membrane via different anchor molecules is a very new and unexplored field. Numerous lipids are available, both natural and synthetic, which can be used for conjugation—amino cholesterol, phosphatidyl serine, stearylamine, phosphatidyl glycerol, phosphatidyl inositol, gangliosides etc. It has been found that PS conjugated with SPDP incorporates poorly into liposomes, preferring to form micellar structures. This problem does not appear to arise with biotinylated PS (BPS), and it has been reported that liposomes containing BPS are more susceptible to fusion with cell membranes than are those containing the PE derivatives (37). Such phenomena are worthy of further study in regard to the use of proteins to effect targeting and delivery of liposomal contents to cells *in vitro* and *in vivo*.

11. REFERENCES

1. Heath,T.D., Montgomery,J.A., Piper,J.R. and Papahadjopoulos,D. (1983) *Proc. Natl. Acad. Sci. USA*, **80**, 1377.
2. Leserman,L.D., Machy,P. and Barbet,J. (1981) *Nature*, **293**, 226.
3. O'Connel,J., Campbell,R., Fleming,B., Mercolino,T., Johnson,M. and McLaurin,D. (1985) *Clin. Chem.*, **31**, 1424.
4. Kung,V.T., Maxim,P., Veltri,R. and Martin,F. (1985) *Biochim. Biophys. Acta*, **839**, 105.
5. Kung,V.T., Vollmer,Y. and Martin,F. (1986) *J. Immunol. Methods*, **90**, 189.
6. Martin,F. and Kung,V.T. unpublished results.
7. Shek,P.S. and Heath,T.D. (1983) *Immunology*, **50**, 101.
8. Torchilin,V.P., Goldmacher,V.S. and Smirnov,V.M. (1978) *Biochim. Biophys. Res. Commun.*, **85**, 983.
9. Dunnick,J.K., McDougall,R., Aragon,S., Goris,M. and Kriss,J. (1975) *J. Nucl. Med.*, **16**, 483.
10. Heath,T.D. and Martin,F.J. (1986) *Chem. Phys. Lipids*, **40**, 347.
11. Heath,T.D., Robertson,D., Birbeck,M.S.C. and Davies,A.J.S. (1980) *Biochim. Biophys. Acta*, **599**, 42.
12. Heath,T.D., Macher,B.A. and Paphadjopoulos,D. (1981) *Biochim. Biophys. Acta*, **640**, 66.
13. Carlsson,J., Drevin,H. and Axen,R. (1978) *J. Biochem.*, **173**, 723.
14. Kitagawa,T. and Aikawa,T. (1976) *J. Biochem.*, **79**, 233.
15. Rector,E.S., Schwenck,R.J., Tse,K.S. and Sehon,A.H. (1978) *J. Immunol. Methods*, **24**, 321.
16. Martin,F.J., Hubbell,W.L. and Papahadjopoulos,D. (1981) *Biochemistry*, **20**, 286.
17. Martin,F.J. and Papahadjopoulos,D. (1982) *J. Biol. Chem.*, **257**, 286.
18. Leserman,L.D., Barbet,J., Kourilsky,F. and Weinstein,J.N. (1980) *Nature*, **288**, 602.
19. Barlett,G.R. (1959) *J. Biol. Chem.*, **234**, 466.
20. Ellman,G.L. (1959) *Arch. Biochim. Biophys.*, **82**, 70.
21. Matthay,K.K., Heath,T.D. and Papahadjopoulos,D. (1984) *Cancer Res.*, **44**, 1880.
22. Jou,Y.H., Jarlinski,S., Mayhew,E. and Bankert,R.B. (1984) *Fed. Proc.*, **43**, 1971, #3218.
23. Martin,F. and Kung,V.T. (1985) *Ann. NY Acad. Sci.*, **446**, 443.
24. Hashimoto,K., Loader,J.E. and Kinsky,S.C. (1986) *Biochim. Biophys. Acta*, **856**, 556.
25. Derksen,J.T.P. and Scherphof,G.L. (1985) *Biochim. Biophys. Acta*, **841**, 151.
26. Wolff,B. and Gregoriadis,G. (1984) *Biochim. Biophys. Acta*, **802**, 259.
27. Urdal,D.L. and Hakomori,S. (1980) *J. Biol. Chem.*, **255**, 10509.
28. Bogdanov,A.A., Klibanov,A.L. and Torchilin,V.P. (1984) *FEBS Lett.*, **175**, 178.
29. Kinsky,S.C., Loader,J.E. and Benson,A.L. (1983) *J. Immunol. Meth.*, **65**, 245.
30. Kung,V.T. and Redeman,C.T. (1986) *Biochim. Biophys. Acta*, **862**, 435.
31. Snyder,S.L. and Vannier,W.E. (1983) *Biochim. Biophys. Acta*, **772**, 288.
32. Jansons,V.K. and Mallett,P.L. (1980) *Anal. Biochem.*, **111**, 54.
33. Jansons,V.K. (1984) In *Liposome Technology*, Gregoriadis,G. (ed.), CRC Press Inc, Boca Raton, Florida, Vol 1, p63.
34. Chabala,J.C. and ShenT.Y. (1978) *Carbohydr. Res.*, **67**, 55.
35. Pittman,R.C. and Steinberg,D. (1978) *Biochim. Biophys. Res. Comm.*, **81**, 1254.
36. Ghosh,P., Bachhawat,B.K. and Surolia,A. (1981) *Arch. Biochim. Biophys. Acta*, **206**, 454.
37. Bayer,E.A., Rivnay,B. and Skutelsky,E. (1974) *Biochim. Biophys. Acta*, **550**, 464.
38. Spiegel,S., Skutelsky,E., Bayer,E.A. and Wilchek,M. (1982) *Biochim. Biophys. Acta*, **687**, 27.

CHAPTER 5

Physical methods of study

G.R.JONES and A.R.COSSINS

1. INTRODUCTION

Of the various methods available for measuring the physical properties of liposomes and interactions between liposomes, many lie outside the scope of this book. This is because they are sophisticated techniques requiring complex and specialized instruments which are usually only found in laboratories dedicated to the exploitation of that particular technique. Furthermore, the theoretical sophistication underlying these techniques and the equipment used is often inversely proportional to the practical expertise required in performing the measurements. Although the data gained from these methods may be inaccessible by any other means, the range of different types of information obtainable by using a given technique is often rather limited. Mention has already been made of differential scanning calorimetry, which is the method of choice for the direct measurement of thermotropic lipid phase transitions. Mixing of lipids as a result of liposome fusion may be inferred from this technique by observation of discontinuous changes in phase transition behaviour of a mixed population of vesicles. Applications of NMR have already been described for the measurement of entrapped volume, and of lamellarity, and further application will be described in this chapter.

Fluorescence studies have been singled out here for detailed discussion, firstly, because of the relative simplicity of the equipment required, secondly, because of the wide variety of techniques which can be employed to give information on different aspects of liposome behaviour, and thirdly, because of the relative ease of learning the skills and techniques necessary to carry out such measurements. The instrumentation used is inexpensive and usually available in most well-established biochemical laboratories.

Fluorescence is a very sensitive technique and for most measurements, the micromolar concentrations of fluorescent probe required do not noticeably perturb the bilayer properties of liposomes. Because the emission of fluorescent light by fluorophores is sensitive to their immediate microenvironments, information can be obtained on the state of the liposome without having first to separate it from the bulk medium, as has already been described in Chapter 3 for the use of carboxyfluorescein to monitor leakage. Another advantage is that environmentally-induced changes in fluorescence properties are virtually instantaneous, and measurements are rapid so that parameters which change quickly can be conveniently monitored.

This chapter is intended to give an introduction to the use of fluorescence techniques in liposome research, concentrating on the more practical aspects. Examples are taken from the literature to illustrate the sort of conclusions that can be drawn from the experimentation.

2. FLUORESCENCE TECHNIQUES IN LIPOSOME RESEARCH

2.1 **Fusion of liposomes**

Fluorescence is the technique of choice for monitoring the fusion of liposomes. Several ingenious methods have been developed that are generally based upon the mixing of two markers originally present in two different sets of liposomes. The interaction of the two molecules leads to changes in fluorescence which can be used as an assay of the extent of fusion. The probes such as those shown in *Figure 1* can be placed either within the phospholipid bilayers or in the aqueous lumen of the liposomes to determine separately the mixing of hydrophobic or aqueous compartments. In that the techniques depend upon a mixing of compartments they largely overcome the problem of distinguishing between the aggregation of liposomes and the actual fusion and mixing of contents, which other techniques based on the analysis of liposome size, such as light scattering and gel filtration, do not.

Perhaps the most useful and extensively used technique is the terbium/dipicolinate assay in which terbium citrate is encapsulated in the aqueous space of one population of liposomes and sodium dipicolinate in another (1). Intermixing of the contents as a result of fusion leads to the formation of a terbium−picolinate complex ($Tb(picolinate)^{33-}$) which has a 10^4-fold greater fluorescence intensity than the dissociated ions. The leakage or release of liposome contents is not registered provided that calcium ions and EDTA are included in the external medium to prevent the stable formation of the fluorescent complex (see *Figure 2*).

Kendall and Macdonald (2) have introduced a similar method in which the fluorescent compound calcein (2′,7′-{[bis(carboxymethyl)amino]methyl}fluorescein), is combined with Co^{2+} to form a non-fluorescent chelate. EDTA is a stronger chelator of Co^{2+} and so it releases the intensely fluorescent free dye. By incorporating the Co^{2+}− calcein complex in one set of liposomes and EDTA solution in a second set, fusion of vesicles from the two populations can be monitored by following the progressive increase in fluorescence. Leakage from liposomes can be quantified by titrating the external medium with Co^{2+} or EDTA. Because calcein fluoresces at visible wavelengths this technique also enables the direct observation of fusion events by fluorescence microscopy.

Fusion of liposomal membranes may be detected by measuring the fluorescence resonance energy transfer (RET) between two lipid analogues originally placed in separate vesicle preparations. This technique depends upon an overlap in the emission spectrum of one fluorophore (the donor) and the excitation spectrum of a second fluorophore (the acceptor); see *Figure 3* and Section 4.8 for a detailed description. Excitation of acceptor molecules can then occur by direct transfer of energy from the excited donor molecule and this is manifest as a reduction in the emission intensity of the donor fluorescence and a corresponding increase in emission intensity of the acceptor fluorescence. Because RET decreases as the sixth power of the distance between donor and acceptor molecules, it occurs only when the two molecules come into close contact, as is the case when they both occupy the same bilayer.

By incorporating donor and acceptor molecules into different liposome populations, fusion will be detected by the onset of RET (*Figure 4a*). The relative change of RET on fusion is linearly related to the extent of fusion. Clearly, this technique is dependent on demonstrating that the fluorophores are not able to migrate from one bilayer to another

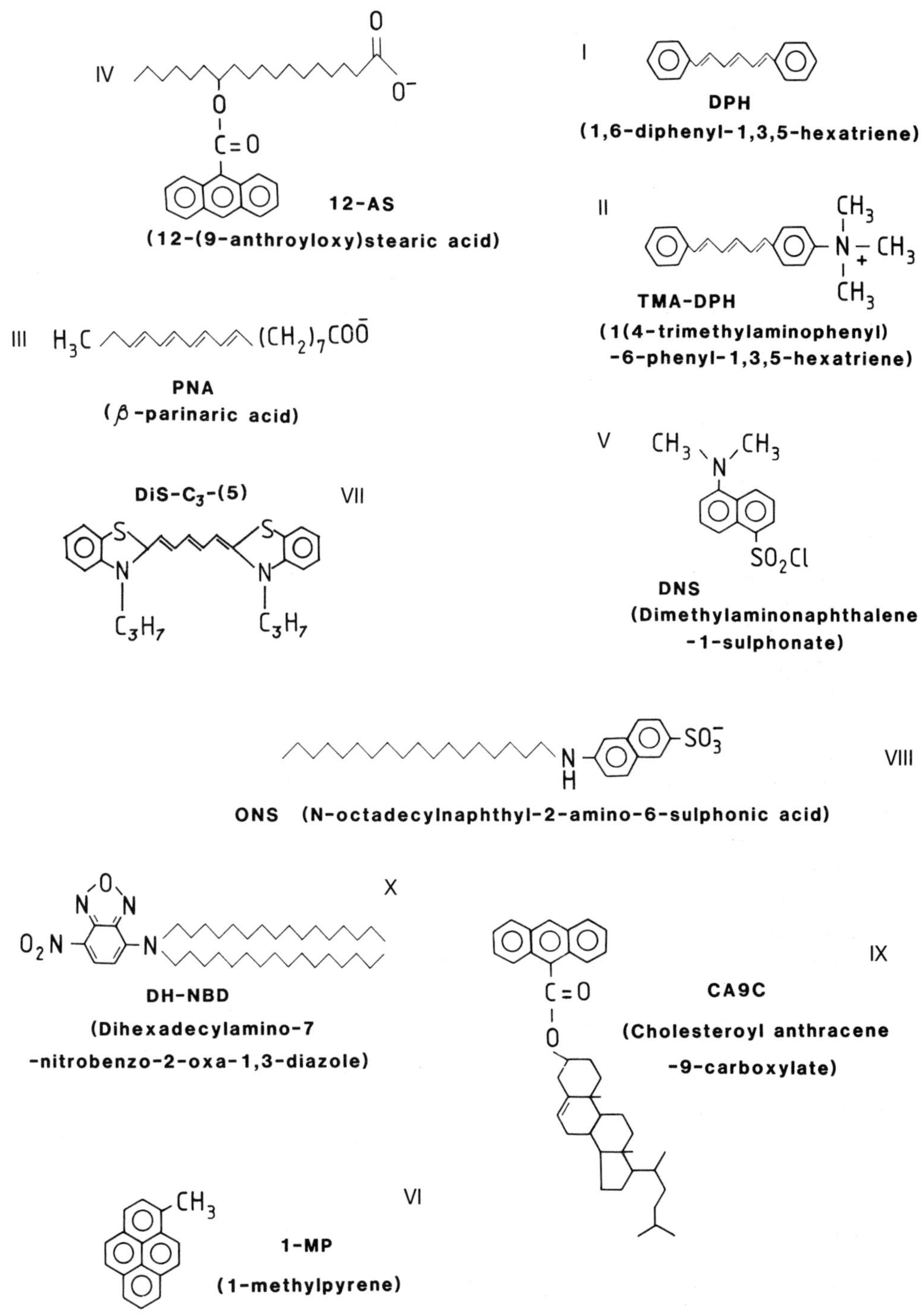

Figure 1. Chemical structures of some fluorescent probes mentioned in the text.

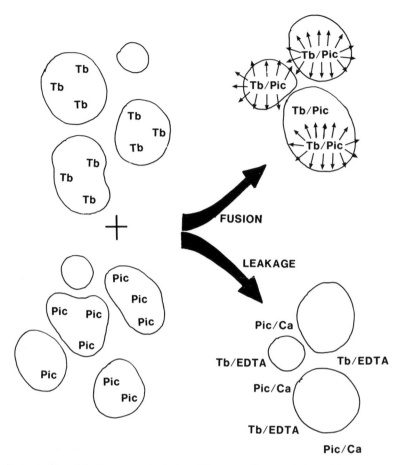

Figure 2. The terbium dipicolinate assay for fusion of liposomes. Two populations of liposomes containing terbium citrate and dipicolinate, respectively, are mixed. Fusion of liposomes from one population with those from the other leads to mixing of their contents and to the formation of a highly fluorescent terbium−dipicolinate complex. Addition of the chelator EDTA and Ca^{2+} to the external medium prevents formation of the fluorescent complex through leakage from the liposomes.

without fusion, which, in the case of phospholipid analogues, is true (3,4). A variation on the RET method is to pre-mix the donor and acceptor in the same liposomes and to induce fusion in the presence of an excess of unlabelled liposomes (*Figure 4b*). The dilution of the fluorophores as they disperse through the unlabelled membranes then leads to a decrease in RET.

2.2 Membrane fluidity

The physical structure of the bilayer is of obvious importance in determining the properties of liposomes. Because liposomes are usually prepared using pure or nearly pure phospholipids they display particularly marked thermotropic behaviour; thus transitions between liquid−crystalline and gel phases occur in a cooperative manner

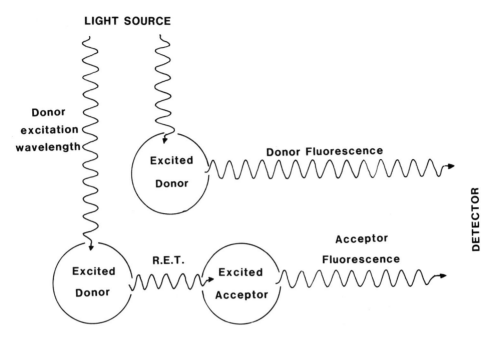

Figure 3. The principle of resonance energy transfer (RET). The donor fluorophore exhibits fluorescences at short wavelengths. The close proximity of a second fluorophore with an absorption band which overlaps with the emission band of the donor leads to excitation of the acceptor by the donor. The resulting fluorescence is now from the longer wavelength spectrum of the acceptor whilst shorter wavelength fluorescence is correspondingly reduced.

and over a restricted range of temperatures. Although membrane phospholipids are anisotropically arranged in ordered semi-crystalline arrays there may be, nevertheless, a significant degree of molecular motion and disorder, especially in the liquid−crystalline state. This motion may be of several distinct types ranging from the flexing of hydrocarbon chains by rotation of carbon−carbon bonds to the wobbling motion of entire molecules and the lateral displacement of molecules. The resulting condition is commonly described as being fluid, though it must be recognized that the fluidity of this compartment is quite distinct from the fluidity of bulk hydrocarbon solvents such as paraffin. Information on the degree of membrane order or fluidity is provided by several spectroscopic techniques, such as nuclear magnetic resonance (NMR) spectroscopy, electron spin resonance (ESR) spectroscopy and fluorescence spectroscopy. Because of differences in the nature of these techniques and the time-domains over which they are sensitive they emphasize rather different aspects of the fluid condition.

Perhaps the most widely used technique is the measurement of fluorescence polarization (see Section 4.9 and *Figure 5*) in which the emission polarization (or anisotropy) of membrane-bound fluorophores, such as the rigid rod-shaped probe 1,6-diphenyl-1,3,5-hexatriene (DPH, see *Figure 1*), is used as a measure of the extent of probe wobbling motion during its excited lifetime (approximately 10^{-8} sec). In that this wobbling motion is limited by the motional properties and degree of ordering of

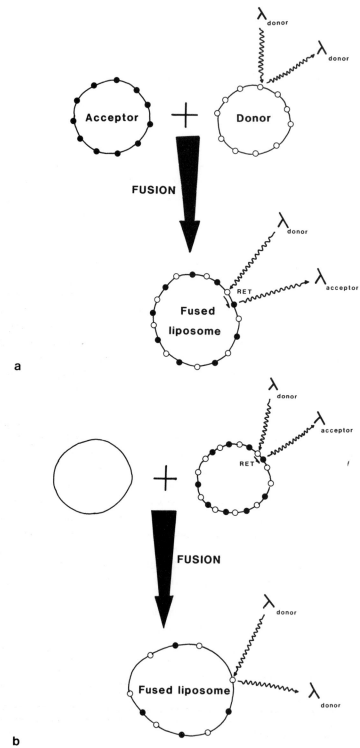

the surrounding hydrocarbon chains the probe obviously provides a measure of hydrocarbon order (5). Thermotropic phase transitions in liposomes are particularly well demonstrated as dramatic changes in the emission intensity and polarization of, for example, DPH (*Figure 6*) at the transition temperature (T_c), together with some pre-transitional changes at lower temperatures.

Understanding the precise significance of DPH polarization in fluid membranes is less straightforward. To appreciate fully the meaning of such information it is necessary to define carefully the membrane characteristics which limit the depolarizing rotations of the probe and this requires a complex time-resolved analysis of anisotropy which is described in Section 5.2. In the case of DPH the depolarization observed is made up of two components, one caused by the rate of wobbling motion and the other component by the hindrance placed on free rotational motion of the probe by the cage-like structure of the hydrocarbon chains (*Figure 7*). These components can only be separated by time-resolved measurements of anisotropy, which requires a more sophisticated apparatus and access to considerable computing power (see for example refs 6–8). In membranes which contain cholesterol and which are relatively ordered the second component predominates. Thus, steady state DPH polarization provides information primarily on membrane order in these membranes rather than on a dynamic aspect of fluidity as implied by the terms 'microviscosity' or fluidity, and this measure of order can be used to calculate an order parameter. In most circumstances where, for example, comparative experiments are performed, steady state measurements of anisotropy are probably sufficient to demonstrate a difference in fluidity. It is necessary to perform time-resolved measurements only when the nature of the difference requires demonstration but even then some limited interpretation of steady state data in terms of order parameters is possible (9).

DPH has been the most widely used probe mainly because it becomes highly fluorescent when introduced into bilayers and has some very convenient photophysical properties for polarization measurements. However, it does suffer from one major problem in that, because it is an entirely hydrophobic molecule, it does not occupy a single, well defined position in the bilayer. Recent work suggests that it may occupy several discrete sites such as between the hydrocarbon chains within a monolayer, or between the phospholipid monolayers (10). Other probes overcome this problem by the incorporation of a charged group onto the molecules so that it becomes 'tethered' to the interfacial area of the bilayer. Various derivatives of DPH have been synthesized for this purpose including TMA-DPH (see *Figure 1*) and a phospholipid—DPH analogue (11). In these cases, the probe is influenced by only half of the monolayer, the half near the headgroup.

Figure 4. (a) The use of resonance energy transfer (RET) to demonstrate the fusion of liposomal membranes. Each population of liposomes contains a different membrane-bound fluorescent probe. On fusion the donor and acceptor fluorophores come into close contact within the same bilayer. Illumination of light at the excitation wavelength of the donor leads to emission of fluorescence at the wavelengths characteristic of the acceptor. (b) The dilution technique using RET to demonstrate fusion of liposomes. Here one population of liposomes contains both donor and acceptor such that RET occurs. Fusion of these liposomes with liposomes not containing fluorophores leads to a dilution of the donor and acceptor such that collisional interactions between them are less frequent. This reduces RET so that acceptor fluorescence is reduced whilst donor fluorescence becomes more evident.

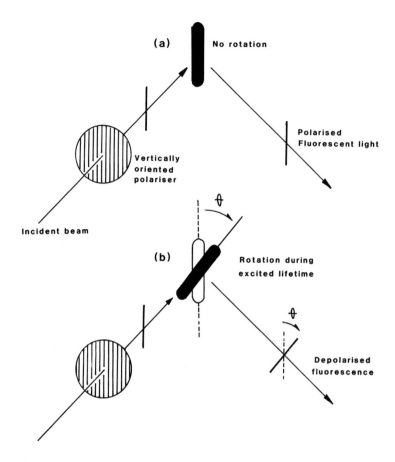

Figure 5. The principle of fluorescence polarization spectroscopy. Polarized light preferentially excites those fluorophores whose molecular axes are orientated in a particular direction with respect to the plane of polarization. If the fluorophore remains immobile during its excited lifetime (10^{-8} sec), then the fluorescent light will also be highly polarized (**a**). If, however, the fluorophore rotates during its fluorescent lifetime then the resulting fluorescence will become less highly polarized. The polarization of fluorescence thus act as a convenient index of the extent of molecular rotation during its excited lifetime.

A second major consideration is the disturbance of the bilayer properties by the inclusion of bulky groups such as the anthracene molecule or phenyl groups. This is particularly important because probes are sensitive to their immediate microenvironments and it is precisely this that is disturbed by the inclusion of ill-fitting groups. Parinaric acid (β-PNA, see *Figure 1*) is a conjugated polyene fatty acid which does not possess bulky groups and does not suffer from this problem. This and the fact that its headgroup is 'tethered' make β-PNA popular as a membrane probe. The *trans* isomer is thought to partition naturally into gel phase bilayers and, therefore, is a useful probe of phase separations. However, β-PNA is extremely sensitive to oxidation and this makes it less easy to use than DPH or its analogues.

Another potential difficulty in liposome systems is the effect of sample turbidity upon polarization measurements. Scattering of the excitation beam into the emission signal

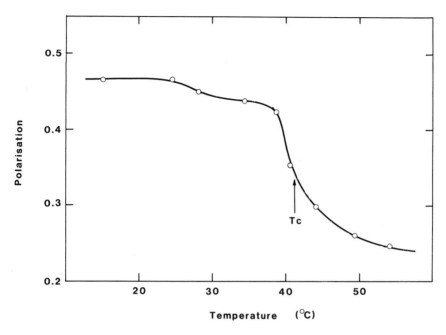

Figure 6. The effects of a thermotropic phase transition of phospholipid liposomes upon the fluorescence polarization of the membrane-bound probe, DPH. At low temperatures the polarization is high indicating only limited probe rotation during its excited lifetime. At approximately 41 °C the gel phase is dramatically transformed into the more mobile liquid−crystalline condition which permits a far greater degree of molecular rotation and a corresponding reduction in polarization (see arrow). (Data of M.Behan.)

obviously alters the measured polarization and it is sometimes necessary either to reduce this effect by the inclusion of emission filters in addition to the monochromator, or by determining the magnitude of this effect and making the appropriate correction (12). The former method is preferred but the latter is simply achieved by determining the relationship between polarization and sample turbidity and then extrapolating to zero turbidity. In that the relationship between polarization and turbidity varies according to the chemical composition and physical state of the membrane, as well as the method of membrane preparation, the correction factors should be determined for each series of polarization measurements. Another solution to this problem is to use small unilamellar liposomes. However, the high surface curvature of extremely small vesicles may perturb the properties of the bilayer interior making them a rather poor model of bilayer structure in general.

The so-called 'fluidity gradient' can be investigated using the *n*-(9-anthroyloxy) fatty acids where the carboxyl terminal group is located at the interfacial region of the membrane and the fluorescent anthracene group is attached by an ester linkage at different positions along the fatty acid chain (*Figure 1*). This provides labelling at a graded series of depths in the bilayer so that determination of polarization provides a relatively easy means of establishing the depth-dependence of fluidity. There is some uncertainty concerning the packing problems created by the bulky anthracene group, but since this technique provides information which is in very good agreement with that provided

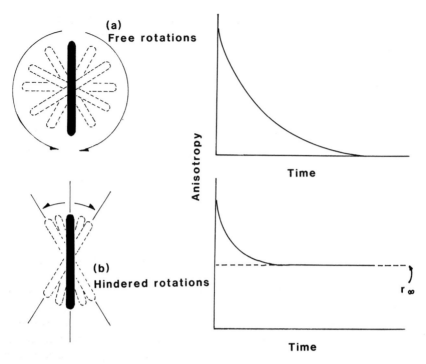

Figure 7. The use of time-resolved analysis of anisotropy to determine the restriction on rotational behaviour of a rod-shaped membrane-bound fluorescent probe, such as DPH. In (**a**) the probe is able to rotate freely so that anisotropy declines progressively to zero as the probes rotate to random orientations before they fluoresce. In (**b**) the rotations of the probe are constrained by the cage-like structure of the neighbouring hydrocarbon chains. Anisotropy decays to a non-zero value because of this constraint; the resulting lowest value of time-resolved anisotropy (r_∞) is related to the degree of hindrance offered by the bilayer interior. The initial slope of the time-resolved graph provides information on the rate of rotation.

by NMR spectroscopy this effect is probably not too serious. Anilinonaphthalene sulphonate (ANS) (and its derivative DNS—structure V in *Figure 1*) has been widely used as a probe of the interfacial region of phospholipid bilayers where it shows saturable binding and a sensitivity to membrane composition, medium composition, pH gradient and membrane potential and surface charge (13).

Finally, pyrene (*Figure 1* structure VI) has frequently been used to determine the magnitude of another aspect of the fluid condition, namely lateral diffusion. This probe displays an emission peak at longer wavelengths than the normal emission which is due to the formation of excited state dimers or 'excimers' (see Section 4.7). It was originally thought that the formation of these dimers was a diffusion-controlled reaction such that the rate of formation and hence the intensity of excimer emission was a function of the diffusion coefficient for the probe and the viscosity of the solvent, which in this case means the hindrance to lateral diffusion offered by the hydrophobic portion of the bilayer. However, subsequent studies (14) have shown that in phospholipid bilayers, at least, the fluorescence kinetics of pyrene do not satisfy the criteria expected for a diffusion-controlled mechanism and instead it has been suggested that excimers are created from an aggregated form of pyrene. This means that great caution must be exercised in the interpretation of pyrene fluorescence as it may not provide a rigorous measure of lateral diffusion.

2.3 pH of the liposome lumen

The measurements of pH gradients across vesicular membranes by following the distribution of weakly acidic or basic radioactive probes is well established (15). Although this technique is sensitive it is not suitable for rapid kinetic analysis and this has led to the introduction of several methods based upon the use of fluorescent probes. One technique is based upon the quenching of fluorescent amines, such as 9-aminoacridine, as the probes redistribute from the external medium into the intravesicular space to an extent depending on the existing pH gradient (16). Although the mechanism by which quenching occurs is not well characterized, Deamer *et al.* (17) have shown that it is closely correlated with the magnitude of the pH gradient. These authors also describe several methods for creating pH gradients across liposomal membranes.

An alternative technique for the direct measurement of internal pH (pH_i) uses the pH-dependent shift of emission spectra of fluorescent weak bases such as quinine and acridine (18). As these probes distribute into the liposome they emit fluorescence at longer wavelengths than probes that remain in the more alkaline external medium. By measuring the decrease in the alkaline peak and the enhancement of the acidic peak, and by comparison with a standard curve of fluorescence against pH, the internal pH can be deduced. This method has the important advantages firstly of not requiring the measurement of the internal H^+ space, which in the case of liposomes with rather small internal compartments is difficult, and secondly it requires only one hundredth of the probe concentration of earlier techniques.

Pyranine is a useful pH-sensitive probe for the continuous measurement of pH_i in liposomes which contain small internal spaces (19). Because of its polyanionic character it does not interact significantly with liposomes having a net anionic surface charge. Liposomes are formed in the presence of pyranine and external probe is removed by gel filtration. Leakage is minimal and the properties of the entrapped probe closely resemble that of a bulk solution of pyranine.

A popular fluorophore for studies on live cells and which may be useful for liposome studies is 2,7-biscarboxyethyl-5(6)-carboxyfluorescein (BCECF). This probe offers excellent time resolution, a very high sensitivity to small changes in pH_i and the ability to make continuous recordings on small samples (20). It can be incorporated into the internal aqueous space of liposomes and again the external probe removed by gel filtration. At the end of the experiment calibration can be achieved by solubilizing the liposomes and releasing the dye into the medium. The response of the dye can then be measured at different known pHs by titrating with concentrated buffer solutions. BCECF does suffer from the disadvantage of a slow but definite leakage through lipid bilayers and from small spectral shifts in living cells. Quene 1 is a newer probe with a slightly higher pK_a of pH 7.3, which exhibits a much reduced leakage rate ($<10\%$ h^{-1}) and with no spectral shifts (21).

Finally, intravesicular pH can be determined using a novel membrane-bound probe prepared by covalently linking fluorescein to phosphatidyl ethanolamine. Under appropriate conditions, such as when the external medium is highly buffered, this probe is able to report directly on the pH of the internal compartment (22).

2.4 **Membrane potential**

A variety of fluorescent techniques are available for estimating the transmembrane potential in cells, subcellular organelles and liposomes. In many cases the precise mechanism by which the fluorescence is altered is not known and the interpretation of these responses to changes in potential is frequently not without some controversy. In some of these cases the responses are related to changes in partitioning into or binding onto the bilayer or in the transbilayer distribution of the fluorophore. This indirectly causes changes in emission characteristics by the formation of dimers, or, perhaps, by the light-induced isomerization of adsorbed fluorophore (for a review, see ref. 23).

Because these probes work by a redistribution mechanism, their ability to respond to potential changes depends upon their polarity and partitioning into the bilayer. Therefore it is usual to examine several analogues to determine which has the most favourable characteristics. These may vary in the length of the alkyl group and hence in lipophilicity. The most widely used fluorophores are the carbocyanine dyes, such as Di-S-C_3(5) (*Figure 1*, structure VII) and Di-O-C_5(3), and the oxonol dyes. Because of the redistribution mechanism, these probes react rather slowly to changes in potential (i.e. several seconds) though the size of the response may be large ($8-10\%$ per 10 mV). The measurements are technically simple and allow for continuous recording (see, for example, ref. 24). Other types of dye are available, some with responses in the millisecond or microsecond time-scale. The changes in fluorescence seen are 10^2-10^5 times smaller than for the 'redistribution' dyes and their detection requires more specialized apparatus (23). The reader is referred to Haugland (25) for a recent overview of the available fluorophores and the more recent literature.

Although the changes in fluorescence result from changes in membrane potential, other factors may contribute to the changes in fluorophore binding and distribution. For example, the chemical composition of the bilayer (e.g. cholesterol content and headgroup composition) affects the amount of adsorbed dye and hence interferes with a straightforward relationship between fluorescence and membrane potential (26). Furthermore, the high surface curvature of small unilamellar liposomes alters the distribution of adsorbed dye (27). Clearly, these difficulties make it impossible to predict the magnitude of the response and independent calibration is necessary. This may be achieved by setting the membrane potential at a series of known values and determining the resulting fluorescence intensity. Generally, this is achieved by using the potassium ionophore, valinomycin, to induce a potassium diffusion potential, the magnitude of which can be altered by varying the concentration of potassium in the medium as predicted by the Nernst equation (*Figure 8*). Obviously the success of this procedure depends upon the applicability of the Nernst equation to the liposomes under study. The presence of uncharacterized conductances leads to unknown deviations from the Nernst equation, though in liposomes this is unlikely to be the case. The practical aspects of probe selection, fluorescence methodology, and calibration are dealt with in detail elsewhere (28).

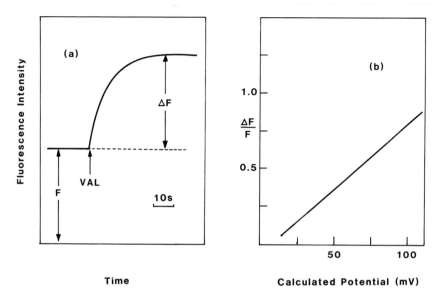

Figure 8. The use of a fluorescent probe as an indicator of membrane potential in liposomes. The potential was generated by adding the potassium ionophore, valinomycin (VAL), to liposomes in which a transmembrane gradient of potassium ions has been established. The potential was calculated according to the Nernst equation. Graph (**a**) shows the changes in fluorescence intensity of a berberine-derivative resulting as the potential is created by addition of VAL. Graph (**b**) displays the linear dependence of the change in fluorescence upon the calculated potential. (Modified after ref. 29.)

3. PRINCIPLES BEHIND THE USE OF FLUORESCENCE TECHNIQUES

A full description of the origins of luminescence is clearly impossible in an article of this size when the main theme is based on the practical aspects of the subject. However, a brief introductory description of the absorption of a photon and the subsequent fluorescence is given here to provide a background for the experimentation that follows. For a more complete theoretical treatment of fluorescence spectroscopy the reader should consult specialized reviews (30–33).

Fluorescence occurs when a compound containing conjugated double bonds absorbs light of a certain energy to form the short-lived excited state of the molecule, which in returning to the ground state releases energy in the form of light. The fate of the energy absorbed by a polyatomic molecule that remains intact is best illustrated by the Jablonski diagram (*Figure 9*). Electronic excitation is inevitably accompanied by an increase in the vibrational and rotational activity of the molecule. If the absorption process results in the excitation of an electron to an orbital which retains a paired spin, that is the same spin orientation as that of the ground state, then the transition is termed singlet-singlet and is efficient (i.e. allowed). The absorption process results in the redistribution of an electron, producing the excited state of the molecule, which because of its greater anti-bonding character, has properties different from that of the ground state of the molecule. Emission from the single manifold (i.e. from one of a number of singlet electronic states with differing vibrational energies) is termed fluorescence. When absorption results in the change of spin orientation the electron enters the triplet

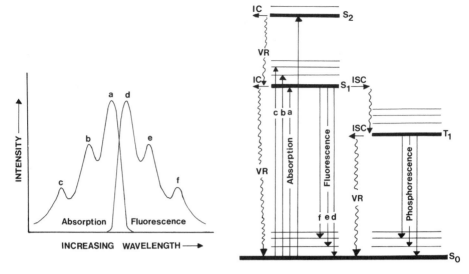

Figure 9. Idealized absorption and fluorescence spectrum of an organic molecule with its corresponding Jablonski diagram.

manifold by the process of intersystem crossing (see *Figure 9*). This transition is normally inefficient (i.e. forbidden) but under suitable conditions, such as low temperature, the resulting emission occurs as phosphorescence.

The absorption process occurs in about 10^{-15} sec. This is too rapid for an adjustment in nuclear geometry (the Franck–Condon principle) and gives a situation which leads to vibrational structure in the absorption spectrum. For an efficient transition between the ground and excited state (i.e. a large decadic molar extinction coefficient) two other conditions or selection rules have to be fulfilled, besides that of the spin condition. These are, firstly, that electronic transitions must occur between states of unlike electronic symmetry, and secondly, a transition must be favoured when there is a high degree of spatial overlap between the molecular orbitals of the two states. For example, a transition from a non-bonding orbital to an anti-bonding π orbital ($n \rightarrow \pi^*$) has little spatial overlap and consequently gives rise to molar extinction coefficients of 10^4 or less, whereas a $\pi \rightarrow \pi^*$ transition has a greater degree of electronic overlap making molar extinction coefficients of $\sim 10^5$ dm^3 mol^{-1} cm^{-1} possible.

Although there are one or two notable exceptions, fluorescence only occurs from the first excited state, S1 (Kasha's Rule). Energy losses occur by vibrational relaxation to S1, which is a faster process than internal conversion, or fluorescence to the ground state, and this results in the so-called Stokes' loss. Energy is also lost during solvent relaxation (see Section 4.5) and this accounts for the wavelength difference, in *Figure 9*, between vibrational bands a and d; that is, the maximum wavelength of excitation and the minimum wavelength of the fluorescence.

There are two further properties of fluorescence that distinguish it from other types of emission, or scattering processes. First, the vibrational structure of the fluorescence spectrum is, in most cases, the mirror image of that of the absorption, or excitation spectrum (*Figure 9*). Secondly, as absorption and fluorescence are specific properties of the molecular orbitals of a particular compound, the shape and position of the fluorescence spectrum is independent of the wavelength of excitation. This latter characteristic can be used to distinguish weak fluorescence from Raman scattering, which is quite often observed at longer wavelengths than the excitation wavelength. By changing the excitation wavelength one would expect a corresponding shift in the position of the Raman scattering peak, but not in the fluorescence maximum.

It can be seen from the Jablonski diagram that fluorescence is not the only path by which excited state is deactivated to the ground state. Therefore, not all photons absorbed by a fluorescent molecule are emitted as fluorescence. The ratio of photons emitted to those absorbed is termed the quantum yield of fluorescence (ϕ_f), which can be almost unity for efficient fluorophores. Although quantum yield is a simple concept, it is a difficult parameter to determine experimentally (30). Fluorescence lifetime (τ_f) and quantum yield are related by the following equation:

$$k_f = \phi_f / \tau_f$$

where k_f is the rate at which fluorescence photons are emitted. Thus lifetime is a useful indicator of changes in quantum yield. The fluorescence lifetime can be defined as that time after excitation when the intensity of fluorescence has decayed to $1/e$ of its maximum intensity. Fluorescence lifetimes for organic molecules are normally in the range from 0.1 to 20 nanoseconds (10^{-9} sec). The relationship between k_f, ϕ_f and τ_f can be used, for example, to distinguish between dynamic and static quenching processes (see Section 4.6).

4. BASIC TECHNIQUES AND THEIR APPLICATIONS

4.1 Instrumentation

The instrument used in the measurement of fluorescence is the spectrofluorimeter and a good understanding of its internal workings is essential in order to obtain unambiguous results. The spectrofluorimeter should be sited in a clean part of the laboratory (in an instrument room if possible) away from bright light sources, such as windows, with adequate ventilation to avoid the build-up of ozone which may be generated by the lamp. A schematic drawing of the essential components of a spectrofluorimeter is shown in *Figure 10*.

The most widely used and versatile light source is a high pressure xenon arc lamp, which emits a continuum of useful wavelengths spanning from the ultraviolet (~ 240 nm), through the visible region, to the infra-red (~ 900 nm,). The stability of the lamp output is ensured by the lamp supply unit; any lamp output fluctuations, or lamp 'striking' problems usually necessitate the replacement of the lamp.

Light from the lamp is collimated onto the entrance slit of the excitation monochromator and then falls on to the grating. The function of this grating is to split the white light of the lamp into its spectrum so that a particular wavelength can be selected for excitation of the sample. As the grating is rotated about its axis different wavelengths pass through the exit slit. It is important to realize that it is the slits that control the band-pass of the monochromator; the wider the slits are set, the greater the range of wavelengths that are allowed to fall on the sample. In some less expensive

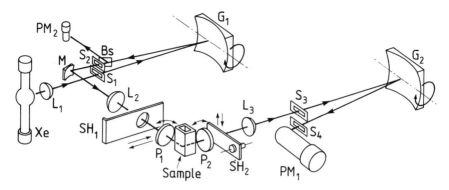

Figure 10. Optical layout of a typical spectrofluorimeter, simplified and drawn from the Perkin-Elmer model 650-40 manual; where Xe is the lamp, L_1, L_2 and L_3 are lenses, S_1, S_2, S_3 and S_4 are monochromator slits, G_1 and G_2 are the excitation and emission monochromator gratings respectively, Bs is a beam splitter, SH_1 and SH_2 are shutters, and PM_1 and PM_2 are the signal and reference photomultipliers respectively. When polarizers P_1 and P_2 are fitted they are positioned immediately before and after the sample (see Section 4.9.1).

spectrofluorimeters interference filters are used instead of an excitation monochromator. The disadvantages of these are that the excitation of the sample is limited by the pass-wavelength of the filters available, their band-pass, and their lifetime in the beam. Before the light from the monochromator falls on the sample it is split into two beams, one of which is diverted to a reference photomultiplier and which monitors the intensity of the excitation light, while the other passes through the beam-splitter to the sample.

Fluorescence from the sample is usually monitored at 90° to the excitation beam; the more intense excitation beam passes straight through the sample to a light trap. Fluorescence is then collected by a lens and passes through a slit into the emission monochromator where it falls on to a second rotatable grating. This scans the fluorescence across the exit slit, behind which the signal photomultiplier is situated. The signal from this photomultiplier tube is then divided by the signal from the reference photomultiplier and this ratio is displayed on a recorder, or fed to a microprocessor. The reason why the ratio of signal to reference output is taken, as opposed to just the signal, is that in doing so the fluorimeter is then correct for minor fluctuations in lamp intensity and changes in the spectral efficiency of the lamp and excitation monochromator as a function of wavelength. The latter occurs because a grating is more efficient at a particular wavelength corresponding to its blaze angle. As fluorescence occurs at higher wavelengths than the excitation the blaze angle of the emission grating is selected to be more efficient at higher wavelengths (e.g. 450 nm) than the excitation grating (e.g. 300 nm). The most sophisticated spectrofluorimeters have provision for the correction of the emission response for emission monochromator and photomultiplier wavelength sensitivity. Without this facility emission spectra may be distorted or even shifted (30).

4.2 Cuvette care

It is important to keep fluorescence cuvettes scrupulously clean by using a cleaning solution such as described in *Table 1*. Stubborn deposits on quartz may require harsher treatment such as washing in chromic acid. In extreme cases a solution of 20%

Table 1. Cleaning solutions for fluorescence cuvettes.

A. 4 M hydrochloric acid, diluted from conc. with distilled H_2O.
B. Absolute ethanol.

Solutions A and B are carefully mixed and stored in a capped brown bottle. These can be stored indefinitely at room temperature.

---

hydrofluoric acid can be used but cuvettes should not be exposed to this solution for more than 2 min. Also, it must be remembered that this is an extremely dangerous reagent that must be used in a fume cupboard. As quartz has a particularly high melting temperature, cells can be baked-out in a furnace at 560°C.

The inner surface of the cell can be made hydrophobic in order to reduce the deposition of liposomes, by coating with a silating reagent using the following procedure.

(i)   Clean the cell thoroughly then wash with distilled water.
(ii)   Flick the water out of the cell to leave a film of distilled water on the inner surface of the cell.
(iii)   In a fume cupboard, pour the silating agent [this is best purchased as a dilute solution of dimethyldichlorosilane (2%) in 1,1,1-trichloroethane] into the cell until it is approximately half full.
(iv)   Seal the cell with its stopper, or a convenient clean plug and, using a gloved hand, shake the cell vigorously, but carefully, as the silating agent is harmful.
(v)   After shaking for a few seconds, pour out the silating agent into a suitable waste container.
(vi)   Dry the cell in a ventilated oven at 110°C overnight. Be careful when opening the oven as the fumes from the silating agent may still be present.
(vii)   Wash the cell thoroughly in distilled water, to remove acid deposits formed by the silating process.

4.3 Fluorescent probes

In general, for biological systems, there are two types of fluorescent probes; (i) intrinsic probes are fluorescent species that are natural components of the system, and (ii) extrinsic probes that are fluorescent molecules added to the system to act as reporter groups.

4.3.1 Extrinsic probes

These are usually large, highly conjugated aromatic molecules. The points to consider when choosing a fluorescent probe are listed below.

(i)   The probe must be highly fluorescent and its absorption and fluorescence bands separate from those of the membrane system under study.
(ii)   The probe must partition mainly into, or bind mostly to the liposome. Many potentially useful probes fail because the fluorescence of the probe in the solvent 'swamps-out' the fluorescence of the probe which is partitioned into the liposome. There are some molecules that have little, or no fluorescence in water [e.g. 1,6-diphenyl-1,3,5-hexatriene (DPH) and parinaric acid] ensuring their popularity as probes in membrane/water systems.

(iii)   Obviously, the probe has to be sensitive to the membrane property under investigation. For example, DPH is well suited for probing membrane fluidity in liposomes, while 3-hexadecanoyl-7-hydroxycoumarin is better suited for probing pH at the surface of liposomes. Each fluorescent probe has its own unique properties of fluorescence lifetime and quantum yield, solvent-dependent spectral positions, pK_a, and membrane binding groups.

(iv)   Addition of the probe to the liposome suspension must not cause changes in the structure of the bilayer, or aggregate, or fuse the membranes. Perturbation of the liposome membrane is minimized by keeping the ratio of lipid to probe greater than 50:1.

(v)   The purity of the probe must be as high as possible. Contamination of the probe with other fluorescent species can lead to inexplicable spectral shifts, or energy transfer causing, for example, depolarization of polarized fluorescence.

(vi)   When possible, probes should be chosen that do not show photochemical changes on excitation, such as photodegradation or photodimerization. In any case, it is advisable to protect the probe from the light whenever possible, by storing the probe in the dark. When not taking measurements the shutter of the fluorimeter should be closed to reduce the exposure time of the sample to light. Photodegradation can be greatly reduced, for example, in the case of parinaric acid, by excluding oxygen from solutions by purging with argon (in membrane systems it is inadvisable to put solutions through freeze-thaw cycles).

(vii)   It is usually desirable for the fluorescent probe to resemble, or be covalently attached to, a membrane component (e.g. fatty acid) so that the region of the bilayer that it is reporting from is known.

The chemical structures of some of the fluorescence probes referred to in this chapter are shown in *Figure 1*.

Fluorescent probes are best stored desiccated, in the dark, at $-20°C$. Purchase of fluorescent probes is best done immediately prior to their use and from a reputable company which quotes purity. Specialist fluorescence probe companies which provide informative catalogues, such as Molecular Probes (PO Box 22010, Eugene, OR 97402, USA) and Lambda Probes and Diagnostics (Grottenhof Str. 3, A-8053 Graz, Austria), are highly recommended.

4.3.2 Incorporation of the fluorescent probe into liposomes

Normally the probe is mixed with the membrane lipid (assuming that they are soluble in the same solvents) prior to hydration. However, when harsh treatment is used in the formation of the liposomes (e.g. sonication) or in the case of natural membranes, usually it is best to add the probe dissolved in ethanol to the suspension of pre-formed liposomes, or biological membranes, with gentle but thorough mixing. In the experience of the authors a final concentration of ethanol of $1-2\%$ has little effect on most systems studied. In the case of liposomes with a high phase transition temperature, such as dipalmitoyl phosphatidyl choline, incorporation of the probe is assisted by heating the liposome suspension containing the probe so that the membrane lipids are in the liquid−crystalline phase.

4.3.3 *Intrinsic probes*

Most lipids do not fluoresce strongly and, therefore, are not used as intrinsic fluorescence probes. Indeed, when lipid preparations do show fluorescence it is generally indicative of their peroxidation. On the other hand, proteins and polypeptides containing the amino acid residues tryptophan and tyrosine are fluorescent in the UV spectrum and can be used as intrinsic probes. In such cases, the protein must naturally associate with membranes and be available at high purity (e.g. melittin, ref. 34).

4.4 **Reagents and lipids**

It is essential that reagents, solvents (including water), and lipid are all of the highest purity, showing no fluorescence when excited at the wavelengths to be used for fluorescence measurements. It is usual at some stage to scan the solutions, at high photomultiplier sensitivity, in the absence of probe, not only to establish the purity of the reagents used, but to assess the effect of light scattering of the liposomes and the Raman scattering of the solution on the fluorescence signal.

4.5 **Spectral changes**

Many of the fluorescent compounds that are used as membrane probes have emission wavelengths, fluorescence quantum yields and lifetimes which are dependent on the polarity of their environments. This is because the dipole moment of the probe molecule in the excited state is usually larger than in the ground state. As a result, the environment of the probe is perturbed and, in turn, it has to reorganize (solvent dipolar relaxation) in order to accommodate the excited state of the probe. To achieve this redistribution of solvent dipoles, excitation energy is lost, on a time scale of 10^{-11} sec and a spectral shift appears as part of the Stokes' loss.

Generally, fluorescence which arises from a $n \to \pi^*$ transition (e.g. 5-dimethylamino-naphthalene-1-sulphonate, DNS, *Figure 1*), where a non-bonding electron from the nitrogen atom is excited to a π anti-bonding orbital of the naphthalene ring, shows a blue shift in emission maximum with decreasing solvent polarity. In this case, as for anilinonaphthalene sulphonate, ANS (13), decreasing the solvent polarity also increases the quantum yield and fluorescence lifetime. For fluorescence species where a $\pi \to \pi^*$ transition is involved, decreasing solvent polarity results in a red shift in the emission maximum.

By linking polarity-dependent probes to membrane components such as fatty acids, cholesterol, phospholipids or proteins, the position of the probe within the bilayer can be 'pin-pointed' by comparison of the emission maximum with that of the probe in various solvents. For example, from a comparison of the wavelength of the emission maximum of the probe *N*-octadecylnaphthyl-2-amino-6-sulphonic acid, ONS (*Figure 1*) (35), in liposomes and in several solvents, it was found that this probe adopts a position at the lipid—water interface near the surface of the bilayer. Similarly, PRODAN appears to occupy two distinct sites, one at the lipid—water interface and the other within the bilayer interior (36).

Although measurements of this kind under continuous illumination are useful for the identification of the location of various probes within the membrane, when used in conjunction with fluorescence lifetime measurements using a pulsed excitation source, a powerful method for investigating membrane structure results.

4.6 **Quenching**

Fluorescence quenching is a process of deactivation of the excited state which competes with fluorescence and results in a decrease in fluorescence intensity. Quenching can occur by two types of process; by collisional encounters of the fluorophore with the quenching species (dynamic, or collisional quenching), or by the complexation of fluorophore and quencher (static quenching). Many molecules quench fluorescence. The most common of these is molecular oxygen which quenches most fluorophores to some extent. Other collisional quenchers include xenon, halogens, acrylamide, nitrous oxide and chlorinated hydrocarbons. Quenching is also observed in solutions where the solvent interacts with the excited fluorophore, for example, by charge-transfer. Electron scavengers, such as cupric ions, quench by accepting the electron from the fluorophore.

Figure 11 shows the effect of the addition of iodide ions (the quencher) to a solution of small unilamellar liposomes containing 1(4-trimethylaminophenyl)-6-phenyl-1,3,5-hexatriene, TMA-DPH (*Figure 1*) (11). In this case, significant quenching indicates that the probe is situated close to the water–lipid interface, as expected from

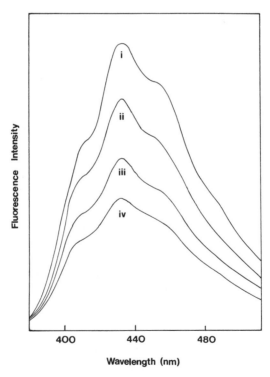

Figure 11. Effect of an increasing concentration of a quencher (sodium iodide) on the fluorescence of TMA-DPH in small unilamellar liposomes, where (**i**) is in the absence of quencher, (**ii**) 0.75, (**iii**) 2.0 and (**iv**) 3.0 M sodium iodide.

its amphipathic character. In this case iodide is a collisional quencher, deactivating the excited state of DPH by spin orbit coupling of the electrons of the fluorophore and quencher, resulting in the formation of an excited triplet state from which fluorescence cannot occur.

Chalpin and Kleinfeld (37) have used acrylamide, copper sulphate, and potassium iodide to investigate the probe depths of a series of *n*-(9-anthroyloxy)stearic acid probes (*Figure 1*) in egg lecithin liposomes and membranes derived from erythrocytes. They found that only iodide penetrated the bilayer to any reasonable depth forming a concentration gradient into the bilayer.

4.7 Excimer formation

As previously mentioned, the properties of the excited molecule can be quite different from those of the ground state and in some cases probe molecules that would not normally show strong binding interactions with other species in the ground state may do so when excited. Excited state dimers are called excimers when the molecules are the same and heteroexcimers (or exciplexes) when different. Their formation is indicated by the quenching of the normal, monomer fluorescence and the corresponding appearance of a broad structureless fluorescence spectrum at longer wavelengths. This occurs most strongly with increasing fluorophore concentration as the probability of collisional interactions increases. In solution the quenching of monomer fluorescence obeys the Stern−Volmer equation:

$$F_0/F = 1 + K[C]$$

where F_0 and F are the fluorescence intensities before and after adding the quencher, respectively, $[C]$ is the fluorophore concentration and K is the Stern−Volmer constant. The application of excimer forming molecules, such as pyrene and its derivatives possessing acyl chains, as fluorescent probes in liposome systems, has been exploited by Sackmann (38) to give information on lateral mobility, lateral phase separations in mixed lipid liposomes, and the kinetics of lipid exchange between liposomes. Excimer formation is usually measured as the ratio of excimer fluorescence intensity to monomer intensity. When a ratio of 10−20 mol% of pyrene to lipid is used, efficient excimer formation is observed on changing from the liquid−crystalline phase to the gel phase, due to the precipitation of the probe as probe clusters (*Figure 12*).

Van den Zegel *et al.* (39) found that in the case of 1-methylpyrene (1-MP, *Figure 1*) in small unilamellar vesicles no excimer was apparent at a probe to lipid molar ratio of ~1:2000, but excimer formation was observed at a ratio of 1:200. These authors have also proposed that in such systems, excimer formation is compatible with that in isotropic media. Chong and Thompson (40) have used the effect of oxygen quenching on the apparent excimer formation constant of a pyrene−sphingomyelin and pyrene−phosphatidyl choline adduct to show that in multilamellar liposomes of 1-palmitoyl-2-oleoyl-L-α-phosphatidyl choline the former showed marked phase separation, whereas the latter did not.

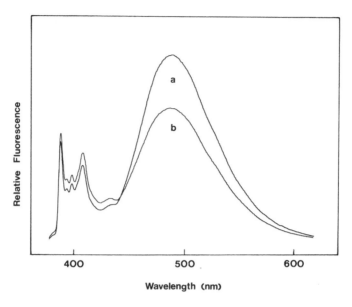

Figure 12. Change in the fluorescence spectrum of pyrene (12 mol%) in small unilamellar liposomes of dipalmitoyl phosphatidyl choline on changing from the gel phase (**a**) to the liquid−crystalline phase (**b**). The highly structured band system at 400 nm is the pyrene monomer fluorescence and the broad featureless fluorescence at 490 nm the pyrene excimer fluorescence. (Data of G.R.Jones.)

4.8 **Energy transfer**

The term 'spectroscopic ruler' has been used to describe the information that can be obtained from resonance energy transfer, RET (31−33). This technique relies on the non-radiative transfer of excited state energy from a fluorescent donor to an acceptor. The rate of energy transfer depends on:

(i)   the fluorescence lifetime of the donor;

(ii)   the overlap of the fluorescence spectrum of the donor and the absorption spectrum of the acceptor;

(iii)   the orientation of their transition dipoles; and

(iv)   most importantly on the distance between the donor and acceptor groups (the inverse sixth power of their separation). In other words, the smaller the distance between donor and acceptor the greater the amount of RET that would be expected. As the distance that RET can occur over can be as great as 5 nm, it has been used to great effect in investigating certain membrane processes, for example, protein binding and membrane fusion.

Donor−acceptor pairs suitable for RET should ideally have:

(i)   a large amount of spectral overlap between donor emission and acceptor excitation;

(ii)   high extinction coefficients for both donor and acceptor; and

(iii)   a donor with a high quantum yield.

RET can be used to monitor liposome fusion, by measuring the reduction in the donor fluorescence when two populations of liposomes are mixed, one containing the donor

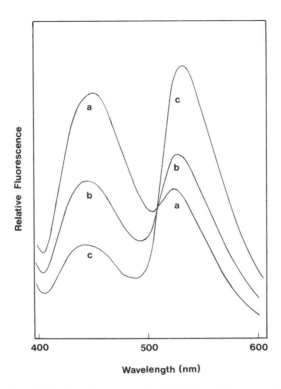

Figure 13. Increase in the RET observed accompanying the fusion of liposomes of egg lecithin, 50% of which contained CA9C (1 mol%) and 50% containing DH-NBD (3 mol%). (**a**) No fusion, (**b**) approximately 50% fusion, and (**c**) 100% fusion. Note how the mixing of the two fluorophores through fusion leads to the appearance of longer wavelength fluorescence and the disappearance of the shorter wavelength fluorescence. The degree of fusion was linearly related to the fluorescence in the long wavelength region. (Modified after ref. 3.)

and the other the acceptor probe. *Figure 4* explains the principles which underlie the technique and *Figure 13* shows an example where the increase in RET between two lipid-bound probes, cholesteroyl anthracene-9-carboxylate (CA9C) and dihexadecyl-amino-7-nitrobenzo-2-oxa-1,3-diazole (DH-NBD, see *Figure 1*), by the method developed by Uster and Deamer (3). As the amount of lipid mixing increases (liposome fusion), there is a decrease in the fluorescence of CA9C with an increase in the fluorescence of DH-NBD. One problem with this particular method is the difficulty found in labelling pre-formed liposomes with DH-NBD.

RET can also be used to monitor the binding of proteins to liposomes. For example, Shaklai *et al.* (41) have used RET to follow the binding of haemoglobin (the acceptor) to erythrocyte ghosts containing 12-(9-anthroyloxy)stearic acid (the donor). In this case the emission spectrum of the latter probe overlaps the Soret absorption band of haemoglobin.

4.9 Polarization of fluorescence

When a small fluorophore, which has been immobilized in a solid, or highly viscous medium, is excited with plane polarized light, the fluorescence emitted is also polarized.

This is due to the fact that only those fluorophores which are aligned parallel, or nearly so, to the plane of polarization of the incident beam become excited and to the fact that both the electric dipole oscillator which absorbs the light, as well as the emission oscillator, have fixed orientations on the fluorophore (31,32). As the viscosity of the medium is decreased the fluorophore will begin to rotate, or tumble, due to Brownian motion so that the molecular axis rotates during the excited lifetime (see *Figure 5*). This results in the loss of polarization of fluorescence. As the viscosity of the medium is further reduced, the fluorescence will appear more and more depolarized because molecular rotation is more extensive. If the rotational diffusion of the fluorophore is much faster than the decay of its fluorescence, then the fluorescence will be completely unpolarized. If the rotational diffusion of the fluorophore is much slower than the fluorescence decay, then the emitted light will be highly polarized. In certain cases, when the fluorophore is attached to a large carrier, such as a protein, or is embedded in a phospholipid bilayer, the now restricted rotational diffusion of the fluorescent probe can be of the same time-scale as the fluorescence decay. Under these circumstances, quite subtle changes in, for example, the rigidity of the matrix of the liposome which forms the environment of the probe, will produce changes in the polarization of fluorescence. This means that fluorescent probes can report on the microviscosity, or 'fluidity', of that region of the membrane which it occupies.

In the simplest case of a sphere of volume V_0, rotating in solution and having a fluorescence lifetime of τ, its polarization, p, is related to the rotational relaxation time, ϱ, and the microviscosity, η, by the Perrin equation (31−33):

$$\left(\frac{1}{p} - \frac{1}{3}\right) = \left(\frac{1}{p_0} - \frac{1}{3}\right)\left(1 + \frac{3\tau}{\varrho}\right) = \left(\frac{1}{p_0} - \frac{1}{3}\right)\left(1 + \frac{RT\tau}{\eta V_0}\right)$$

where p_0 is the intrinsic, or limiting polarization (the polarization in the absence of Brownian motion), R is the universal gas constant, and T is the absolute temperature. From the Perrin equation it can be seen that a change in the rotational relaxation time caused by a change in the volume of rotating probe, or the viscosity of its environment, will result in a change in the fluorescence polarization. Changes in the fluorescence lifetime, as indicated by the fluorescence intensity (expressed as $k_f = \phi/\tau$) will also produce a change in the polarization and this will be discussed later (Section 5). When the measurement takes the form of fluorescence anisotropy, r, a term related to the fluorescence polarization, the Perrin equation takes the following form:

$$r_0/r = 1 + 3\tau/\varrho$$

where r_0 is the anisotropy in the absence of rotation, the so-called 'limiting anisotropy'.

Where steady state fluorescence polarization, or anisotropy is used to measure the liposome fluidity it is not usual to calculate the microviscosity using the Perrin equation, but rather to use the polarization measurements themselves in comparison with those undertaken in solvents of known viscosity. The processes which affect the polarization of fluorescent probes in liposomes are summarized in *Table 2* (derived from refs 31 and 43).

Table 2. Conditions affecting fluorescence polarization.

Condition	Effect on polarization or anisotropy	Cause
Dilute probe in highly viscous or rigid medium	Maximum value for given wavelength	Electronic structure of molecule
Conc. probe in highly viscous or rigid medium	Decrease in value from that above	Energy migration due to electromagnetic coupling of probe molecules
Dilute probe in moderately rigid medium	Decrease in polarization or anisotropy	Brownian motion or rotations controlled by liposome matrix
Dilute probe in low viscosity medium (solution)	Little or no polarization or anisotropy	Depolarization of fluorescence by Brownian motion
Light scattering (Rayleigh or Tyndall)	Values greater than theoretical maximum	Poor choice of filters or incorrect monochromator setting
Light scattering (multiple-scattering)	Decrease in polarization or anisotropy	Loss of orientation of exciting and emitted light
Reabsorption of emitted light	Decrease in polarization or anisotropy	Loss of orientation by transfer to non-parallel dipoles
Energy transfer	Decrease in polarization or anisotropy	Loss of orientation by transfer to non-parallel dipoles

4.9.1 *Instrumentation*

Most spectrofluorimeters have the facility to place rotatable polarizers in the light path immediately before and after the sample (see *Figure 10*); the latter is sometimes called the analyser. Plane polarizers preferentially select one electronic vector of light (allow one polarization of light to pass through it), the rest being absorbed, or reflected, depending on the type of polarizer. There are two types of polarizer commonly used in spectrofluorimeters. First, some polarizers polarize light by double refraction (e.g. calcite crystal prism polarizers such as Glan-Thomson or Glan-Foucault), while others use the dichroism of iodine in a sheet of polyvinyl alcohol film which has been stretched to produce a very fine banded arrangement of molecules (Polaroid film). The suitability of a polarizer for a particular purpose is judged by its transmission of the wavelength of light used in the experiment, and the extinction ratio, which is the ratio of light transmitted through a pair of polarizers with crossed axes to that with parallel axes as a function of wavelength. Calcite prisms have good ultraviolet transmission and extinction, but have small apertures making them more suitable for excitation where the beam is more collimated. Film polarizers are less expensive than prisms and have larger apertures, making them more suitable as analysers.

The most common optical arrangement for the measurement of fluorescence polarization is the 'L' format of the spectrofluorimeter (*Figure 10*). The sample is excited with vertically polarized light and fluorescence intensities are recorded with the analysing polarizer orientated parallel ($I_{\parallel}$) and perpendicular (I_{L}) to the excitation polarizer. From these values the steady state polarization, p, or anisotropy, r, can be calculated from:

$$p = \frac{I_{\parallel} - I_{\text{L}}}{I_{\parallel} + I_{\text{L}}} \; ; \; r = \frac{I_{\parallel} - I_{\text{L}}}{I_{\parallel} + 2I_{\text{L}}}$$

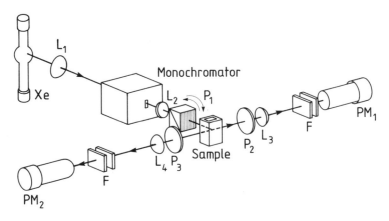

Figure 14. Optical layout of a typical 'T' format polarization fluorimeter where L is a lens, P_1 is a rotatable prism polarizer, P_2 and P_3 are film polarizers which are vertically and horizontally orientated, respectively. Components labelled F are 'cut-off' filters and PM_1 and PM_2 are both photomultipliers. The liposome suspension is contained in a quartz cuvette.

One problem encountered when using the 'L' format with an emission monochromator, is that if the grating is not set at the wavelength that corresponds to its blaze angle, it will pass polarized fluorescent light of different orientations with different efficiencies. This means that a correction factor (G-factor) has to be introduced into the above equations to account for this. The G-factor is determined by setting the excitation polarizer to the horizontal orientation and recording the fluorescence for parallel ($I_{\parallel'}$) and perpendicular ($I_{L'}$) analyser orientations (the terms parallel and perpendicular refer to the position of the analyser with respect to the excitation polarizer). This correction is made by multiplying I_L in the above equations by G, where:

$$G = \frac{I_{L'}}{I_{\parallel'}}$$

The G-factor need only be measured once for a set of experiments which require no alteration to the emission monochromator wavelength. One way of eliminating the necessity of determining the G-factor in a 'L' format fluorimeter is to replace the emission monochromator with a cut-off filter, or an interference filter.

The 'T' format fluorimeter (*Figure 14*) simultaneously measures the parallel and perpendicular components of the fluorescence by utilizing two analysers, mounted on each detection limb, oriented orthogonally to each other, and each with its own photomultiplier tube. The instrument has to be set up by adjusting the excitation polarizer to pass horizontally polarized light on to the sample and then adjusting the gain of the photomultiplier, which monitors the parallel fluorescence, so that the ratio of signals from the two photomultiplier tubes (x/y) is unity. The excitation polarizer is then set to the vertical and the ratio x/y recorded. With the 'T' format:

$$p = \frac{1 - (x/y)}{1 + (x/y)} \ ; \ r = \frac{1 - (x/y)}{1 + 2(x/y)}$$

The advantages of this format are 2-fold; firstly, by using a ratiometer to give the value of x/y, the fluorescence polarization, or anisotropy can be obtained from one measurement, which may be essential if the probe photobleaches. Also, if monochromators are used to monitor the emission, then their G-factors are 'balanced-out' by the set-up procedure.

4.9.2 *Setting the orientation of the polarizers in the fluorimeter*

In cases where the orientations of the polarizers are unknown they are best set up in the fluorimeter in the following manner, using a dilute solution of 'Ludox' (Dupont), or soluble glycogen, made up to an absorbance of ~ 0.01 at 400 nm.

(i)   With the polarizer held in the hand, rotate it until light reflected from a horizontal surface can be seen to be extinguished. By this method a rough vertical orientation is obtained.
(ii) Place the polarizers in the fluorimeter, so that their orientations are both vertical (from above).
(iii) Remove the filters from the fluorimeter, if applicable, or set both monochromators to ~ 400 nm (or zero order), so that the intensity of scattered light is measured.
(iv) Rotate the analysing polarizer to give a maximum reading for the light scattering.
(v)   Rotate the excitation polarizer to give a maximum reading for the light scattering.
(vi) Repeat Step (iv).

When the horizontal orientation of the polarizers has to be set, the same procedure is used to give minimum readings for the light scattering when the excitation polarizer is in the vertical position. Scattered light should give polarization values approaching unity, with values below 0.96 considered poor. This may be due to the scattering agent being too concentrated or too dilute, so it is wise to try a few concentrations. The theoretical maximum values for fluorescence polarization, as opposed to scattered light, are 0.5 and -0.333 (41), respectively, when excitation is with polarized light and 0.333 and -0.143 with unpolarized light. In the latter cases, it is possible to record polarizations using only an analysing polarizer by a combination of the photoselection process and the geometry of the instrument.

4.9.3 *Fluorescence polarization and anisotropy in liposomes*

Steady state polarization and anisotropy can be used to report on the viscosity of liposomes with changes in lipid composition (43), with the addition of fluidizing substances such as ethanol and anaesthetics (44), and with changes in temperature (44) and pressure (45). *Figure 15* shows the effect of elevating the hydrostatic pressure on the main transition temperature of TMA-DPH in small unilamellar liposomes of dimyristoyl phosphatidyl choline. At high pressure and low temperature the fluorescence polarization is high and representative of the gel phase. At low pressures and high temperatures the polarization is lower and characteristic of the liquid−crystalline phase. *Figure 15* shows that a pressure of 1000 atmospheres has shifted the main transition temperature from $\sim 20°C$ to $\sim 40°C$.

The relationship between fluorescence polarization, microviscosity, and fluidity has been exploited by several authors to show changes in biomembrane composition. For example, Inbar and Shinitzky (46) and later Johnson (47) have attempted, with some

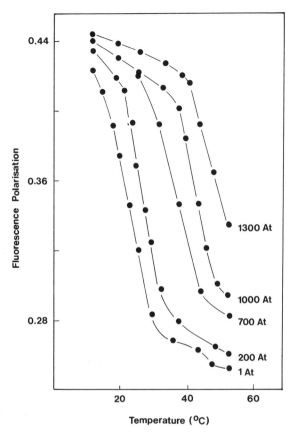

Figure 15. Effects of temperature and hydrostatic pressure on the fluorescence polarization of liposomes of dimyristoyl phosphatidyl choline containing TMA-DPH (0.3 mol%). At atmospheric pressure the liposomes undergo a transition at approximately 23°C but the application of hydrostatic pressure (200–1300 atmospheres) increases the temperature of the transition. (Data of G.R.Jones.)

success, to distinguish between cancer and normal cells, of the same type, by using the difference in the fluorescence polarization of DPH added to their suspensions. These differences are associated with the different lipid compositions of normal and oncogenically transformed cells. Cossins (48), again using DPH, has shown that brain synaptosomes of fish which have been acclimatized at various temperatures exhibit marked adaptive differences in membrane fluidity.

Unfortunately, in whole cell systems DPH does not confine itself to plasma membranes, but enters all membranes as well as lipid droplets, giving rise to ambiguities. Mély-Goubert and Freedman (49) have challenged the usefulness of DPH (and by implication all other untethered probes) as a probe of fluidity in membrane systems, mainly because the exact distribution and extent of interaction with proteins is unknown. These criticisms are circumvented to a large extent by the use of probes which are anchored in the bilayer, such as TMA-DPH.

5. ADVANCED TECHNIQUES

5.1 **Time-resolved fluorescence** (50)

When using fluorescence techniques such as polarization, quenching, energy transfer, or excimer formation, a knowledge of the accompanying changes in the excited-state lifetime(s) of the probe can be of considerable value. For example, it can be seen from the Perrin equation in Section 4.9 that a change in fluorescence polarization may result from either a change in the rigidity of the medium surrounding the probe, or a change in the lifetime of fluorescence, or both. A combination of polarization and lifetime measurements can therefore lead to more reliable rotational information. In quenching experiments, the mode of quenching, that is whether collisional, where a decrease in the lifetime occurs, or complex quenching, where there is no change in lifetime, establishes the form of interaction between probe and quenching species. From knowledge of the excited state lifetime of the donor the rate of energy transfer can be determined. In certain energy transfer experiments, where the donor−acceptor separation changes during the fluorescence lifetime of the donor, knowledge of the lifetime of the donor can be used to calculate the mean distance between donor and acceptor molecules.

It is well within the reach of modern electronics to directly measure fluorescence lifetimes of around 1 nanosecond. Two different techniques are employed; the first of these uses the shift in the phase of the fluorescence compared with that of the sinusoidally modulated exciting beam, and the second uses the single photon counting technique, which relies on a pulsed excitation source, producing a decay profile of the fluorescence. Although phase-modulation was the first technique to be developed, it has been limited in the past because it relies on a match between the modulation frequency and the fluorescence lifetime in order to produce a substantial phase-shift, and also it can only be used to quantify monoexponential fluorescence decays. There have been a number of technical advances such as the use of lasers as the light source, with multi- or continuous variability of modulation from 5 to >80 MHz, and the introduction of dual beam heterodyned instruments. Also, the application of data analysis techniques, such as fast Fourier transformations, has meant that this technique has become more applicable to the 'less than perfect' conditions which prevail when measuring fluorescence lifetimes of probes in liposomes. When the phase fluorimeter is fitted with polarizers in the 'T' format between the parallel and perpendicular components, the differential tangent gives information on the rotational rate of the fluorophore.

Although the single photon counting method is limited by the width of the excitation pulse, and the rise-time of the detection system, direct measurement of the fluorescence decay can be obtained. This means that a theoretical exponential equation of the form:

$$I_t = \Sigma_i A_i e^{-t/\tau_i}$$

where I_t is the time-resolved fluorescence intensity and A_i and τ_i are the pre-exponential and lifetime terms for the ith exponential, can be used to fit the fluorescence decay profile, after the finite width of the instrument response (sometimes called the pump, or excitation pulse) has been taken into account. When this instrument is fitted with polarizers, the rotational correlation time(s) of the probe is directly obtainable from the time-resolved fluorescence anisotropy, which is constructed from the parallel and perpendicular components of the decay.

The following sections will deal with the single photon counting method only.

5.2 Measurement of fluorescence lifetimes and time-resolved fluorescence anisotropy (51)

5.2.1 *Instrumentation*

There are now several suppliers of complete single photon counting systems (e.g. Edinburgh Instruments Ltd, Edinburgh; and Applied Photophysics Ltd, Leatherhead, UK). Instruments are usually based on the now familiar 'L' format though some 'T' format fluorimeters are becoming available. They are provided with a pulsed excitation source, which is usually a flash lamp or, more recently, pulsed laser systems, or even synchrotron radiation sources.

The flash lamp, which is by far the least expensive source, consists of two electrodes in a gas-filled housing, to which a high voltage is applied through a charging resistor. On dielectric breakdown of the gas, a flash of light is emitted from between the electrodes, where the electric field is greatest. Lamps can be ungated (free-running), or gated with a fast on/off switching device, usually a thyratron, which allows variable repetition rates up to 200 kHz. Flash lamps are designed to be filled with either high (~ 15 bar), or low pressure (~ 1 bar, or lower) gas. Stable operation of the flash lamp is crucial to the measurement and so diagnostic warning devices, which are triggered when the lamp misfires, or free-runs, are of value. The width and intensity of the pulse of light produced by a lamp depends on the electrode gap, voltage applied, condition of the electrodes, repetition rate, the gas and its pressure. The wavelengths of light produced depend on the gas. For example, low pressure hydrogen (and the more intense deuterium) emits a continuum of wavelengths, the intensity of which rapidly decreases above 300 nm, whereas nitrogen mainly emits light in the form of spectral lines from ~ 300 nm to ~ 400 nm.

A laser is a rather expensive upgrade on a flash lamp. To obtain short pulses at useful wavelengths a continuous wave laser source, such as an argon ion laser, has to be mode-locked, pump a dye laser, have a cavity dumping device and have a doubling crystal to produce light of half the wavelength of the output of the dye laser. Such systems produce a high intensity, high frequency pulse-train of narrow pulse-width (~ 10 picoseconds), appropriate for single photon counting. The maintenance of an effective laser system that is useful as an excitation source in single photon counting requires skilled research personnel. However, there are central facilities [e.g. the Rutherford Laboratory (UK) and the Center for Fast Kinetics, Austin, Texas] which make the equipment and expertise available to investigators for approved experiments.

Synchrotron radiation at central facilities, such as Daresbury Laboratory, UK; Stanford, Brookhaven (USA); Orsay (France); and Frascati (Italy) provide an attractive alternative to the purchase of expensive equipment. Light from this type of storage-ring is eminently suitable for single photon counting techniques in that it is pulsed (each pulse being, for example, ~ 300 picoseconds in 'width'), of a high repetition rate (e.g. ~ 3 MHz), intense, Gaussian in shape, having a high stability with time and tuneable in terms of wavelength.

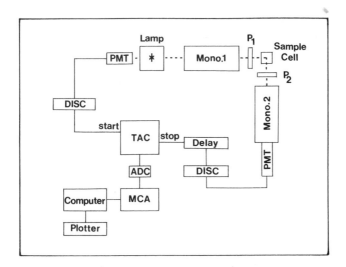

Figure 16. Schematic diagram of a single photon counting fluorimeter and its associated electronics as used for time-resolved analysis. Optical paths are designated by the broken line. Polarizers P_1 and P_2 are used in time-resolved anisotropy measurements. TAC, time to amplitude converter; ADC, analogue to digital converter; MCA, multichannel analyser; PMT, photomultiplier; DISC, discriminator.

The single photon counting instrument has the familiar 'L' format layout (*Figure 16*) with the excitation beam wavelength selected by a monochromator, or interference filter (these are of course unnecessary when laser excitation is used). The fluorescence is monitored by a fast photomultiplier tube at right angles to the excitation beam. Any contamination of the fluorescence signal, with for example scattered light, is eliminated with an emission monochromator, or optical 'cut-off' filter. When fluorescence lifetimes of solutions which have measurable anisotropies, such as liposomes, are to be determined, it is essential that a polarizer is positioned between the sample and emission monochromator at an angle of 54.6° to the vertical (the 'magic angle'), to avoid the effects of the time-dependent anisotropy the lifetime data, which causes problems in the analysis.

The requirements for the fluorescence detector in single photon counting experiments are that it has a fast rise-time (low transit time spread) and a high gain (10^7). Photomultiplier tubes, such as the Philips XP2020Q, fill these requirements but exhibit 'after-pulsing' which can be reduced, but not entirely removed. Recent developments in the production of fast rise-time devices called microchannel plate photomultipliers mean that the instrumental response times of the single photon counting instruments, with a laser or synchrotron source, will be considerably reduced. Previously, these types of detectors have not been employed because of their low gain ($< 10^6$) and prominent after-pulsing. However, developments have eliminated after-pulsing and, given two- and three-stage microchannel plate detectors, an acceptable gain for use with a laser or synchrotron source.

The principle behind this instrument is to record the time delay between the flash of excitation light and the first photon that is detected at the photomultiplier tube, over many repetitions, so that a histogram of the fluorescence decay is built up. Each time the excitation source pulses, a detector sends a synchronization signal to a leading edge, or constant fraction discriminator, which in turn provides a suitable pulse for the time to amplitude convertor (TAC). This 'start' pulse initiates a 'voltage ramp' in the TAC, the voltage of which is linear with time. This voltage ramp continues until a 'stop' pulse is received from the photomultiplier tube, after shaping by a second constant fraction discriminator, or the reset time of the TAC has elapsed. When a stop pulse is detected by the TAC, the ramp (capacitor) output pulse is first digitized by an analogue to digital converter (ADC) and subsequently stored as one count in a multichannel analyser operating in pulse height analysis mode (PHA). When taking single photon measurements, it is necessary to limit the ratio of stop to start pulses to ~2% by means of an iris diaphragm or other attenuating device positioned between the light source and the sample, in order that anomalous decay profiles, due to the double photon events, are prevented. Computer programs are now available to correct for this phenomenon, allowing much higher ratios of stops to starts, thereby decreasing the counting time for highly fluorescent samples.

As well as obtaining detailed information on the fluorescence lifetimes of probes in liposomes it is sometimes useful to record the emission spectrum as a function of the time domains of the decay. This is achieved by setting a 'time window' within the TAC (perhaps from 0 to 5 nanosec, or from 10 to 30 nanosec) and scanning the fluorescence with an emission monochromator. When recording time-resolved emission spectra the TCA is operated in single channel analyser mode and the MCA in multichannel scaler mode.

The introduction of an excitation polarizer immediately before the sample and a remotely controlled rotatable analysing polarizer positioned immediately after the sample (*Figure 16*), replacing the magic angle polarizer, facilitates the collection of parallel and perpendicular fluorescence decay data from which the time-resolved fluorescence anisotropy can be calculated.

5.2.2 *Data analysis* (50,51)

At present the most popular method used to extract fluorescence lifetime parameters from decay data is the non-linear least squares analysis, incorporating the Marquardt minimization algorithm. Here the response function of the instrument is convoluted with a synthesized mono-, bi- or tri-exponential decay of trial parameters and then compared with the experimentally derived fluorescence data set. A reduced chi-squared (χ^2) is calculated from the weighted residuals, which is a measure of how good the estimates were. New trial decay parameters are chosen by the Marquardt algorithm in an effort to reduce the χ^2 value, the process repeating (iterating) until a minimum value is obtained. In single photon counting measurements a good fit is indicated by a residual χ^2 close to unity. In experiments on liposomes it is usual to find that the fluorescence decays give a poor fit with the mono-exponential law, but better fits with the double, or triple exponential laws. This occurs because the dye usually occupies more than one environment within the bilayer. Also, scattered light from liposome solutions cause fitting problems, although computer programs are in existence which account for this (51).

5.2.3 *Application to lifetime measurements in liposomes*

In the case of collisional quenching, the reduction in the fluorescence lifetime of the fluorophore gives an absolute measure of the amount of quenching, whereas complex quenching has the effect of completely 'switching-off' the fluorescence and therefore has little effect on the observed fluorescence lifetime. In solution, collisional quenching follows the Stern−Volmer law:

$$\tau_0/\tau = 1 + K_D[Q]$$

This is probably not strictly the case where the fluorescent probe is incorporated in a liposome and the quencher partitions mostly into the solution, or when both quencher and probe are partitioned in the liposome. However, semi-quantitative information can be extracted from the sort of lifetime data shown in *Figure 17*. For example, in an experiment where sodium iodide was used to assess the difference in the depths of different probes, based on the same fluorophore group (11), the high concentrations of iodide necessary cause liposome aggregation. This in turn decreases the total amount of fluorescence observed from the sample, by processes other than quenching. Fluorescence lifetime measurements are unaffected by these artefacts. A time dependency has been observed for the emission spectrum of TNS (6-[*p*-toluidinyl]naphthalene-2-sulphonate) in egg lecithin liposomes (52), indicating the occurrence of an environmental relaxation process in the nanosecond time-scale in the region of the bilayer close to the water/lipid interface.

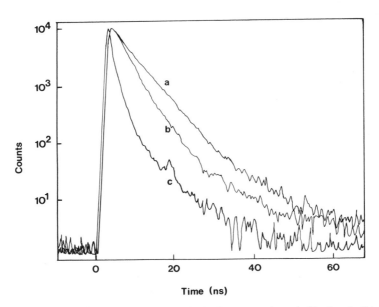

Figure 17. Effect of the addition of quencher (iodide) on the fluorescence decay profile of egg lecithin liposomes containing TMA-DPH (0.3 mol%), where (**a**) is the decay in the absence of quencher, (**b**) with 1.5 M sodium iodide, and (**c**) is the instrument response. (Data of G.R.Jones.)

215

5.3.4 *Application to time-resolved anisotropy measurements in liposomes*

The determination of the rotational correlation time from the time-resolved anisotropy has two main advantages over measurements under continuous illumination (steady state measurements). First, the measurement is usually independent of the fluorescence lifetime; in´ fact the total fluorescence can be calculated from the parallel and perpendicular components and the lifetimes determined independently from the same data. Secondly, if the rotational dynamics of the probe are complex (i.e. if the fluorophore is not a spherical rotor in an isotropic environment, which is the case for probes in liposomes) then more detailed information on its environment can be obtained from the multiexponential (or complex) nature of the anisotropy decay kinetics, described by the equation

$$r(t) = r_0 \Sigma_i g_i e^{-t/\phi_i}$$

where $r(t)$ is the decay of anisotropy, r_0 is the initial, or limiting anisotropy, g_i a pre-exponential term, and ϕ_i the rotation correlation times. For a monoexponential profile $\phi = 1/3(\varrho)$ (Section 3) or $1/6R$, where R is the rotational rate.

In the case of the probe DPH in dioleoyl phosphatidyl choline/cholesterol liposomes (*Figure 18*) there are two distinguishable rotational correlation times, one quite short and the other long or even infinite (r_∞). Whether the latter is best modelled as an

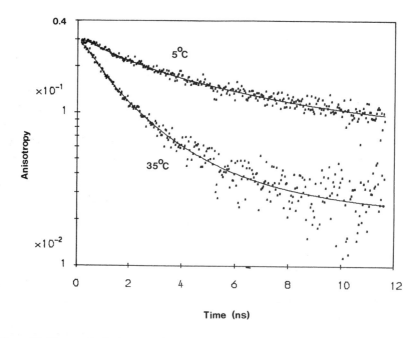

Figure 18. Time-resolved anisotropy of DPH in dioleoyl phosphatidyl choline/cholesterol (30%) liposomes at 5°C and 35°C. Dots show the decay data and solid lines show the bi-exponential fits to the decay with rotational correlation time of 4 and 40 nsec at 5°C and 2 and 22 nsec at 35°C. (Data of G.R.Jones.)

infinite decay or not, r_∞ is still subject to argument (53,54), but in any case the time-resolved anisotropy profile characterizes a rather hindered rotation, as well as a rapid rotation for the probe. Decays which have a r_∞ are described by

$$r(t) = r_\infty + (r_0 - r_\infty) \Sigma_i g_i e^{-t/\phi_i}$$

where r_∞ does not decay.

Using time-resolved fluorescence anisotropy of n-(9-anthroyloxy) fatty acids in dipalmitoyl phosphatidyl choline liposomes, Vincent *et al.* (55) showed that the addition of cholesterol markedly increased the r_∞ and correlation time of the probe when at the $7-9$ carbon region, but decreased the r_∞ value in the $12-16$ carbon region.

5.3 Measurement of liposome diffusion processes in the microsecond to second time-scale

The rotational diffusion of proteins in liposomes is in the microsecond time-scale, which is incompatible with the nanosecond decay times of fluorophores. Phosphorescent probes, however, have much longer lifetimes ($10^{-5}-1$ sec) and can therefore be used to study such systems, using depolarization measurements. On a longer time-scale again, fluorescence photobleaching has been used to follow the lateral diffusion of membrane proteins labelled with suitable probes. These experimental techniques require the use of lasers and optical microscopy and are adequately reviewed elsewhere (56).

6. LIGHT SCATTERING BY LIPOSOMES

6.1 Apparent absorption

Perhaps the simplest method of using light scattering in liposome research in order to monitor changes in the size of liposomes and membrane phase, is not to monitor light scattering at all, but rather to measure the intensity of the transmitted beam using a spectrophotometer. Measuring the apparent absorption, or turbidity of a solution has some advantages over the light scattering measurement *per se*, in that there is no angular dependence involved. This angular dependence of light scattering occurs for solutions of liposomes with diameters greater than ~ 20 nm ($\sim 1/20$th of the wavelength of incident beam). The concentration of lipid suitable for turbidometric measurements is about 1 mg ml^{-1}, and for large multilamellar liposomes, the medium should contain a density neutralizing agent, such as sucrose (16.7%), to avoid the liposomes settling out when lengthy measurement periods are necessary. The intensity of scattering increases by the inverse fourth power of the wavelength of the incident light and so there is an advantage in choosing as low a wavelength as possible, without encroaching on the electronic absorption bands of the sample. Multiple scattering must be avoided in all light scattering measurements and can be demonstrated experimentally by the departure from the linear relationship between apparent absorption, or the intensity of scattering and liposome concentration.

Bangham *et al.* (57) and Yoshikawa *et al.* (58) have shown that the apparent absorption of multilamellar liposomes is proportional to the two-thirds power of the total volume, the latter workers having used this relationship to study the behaviour of osmotically

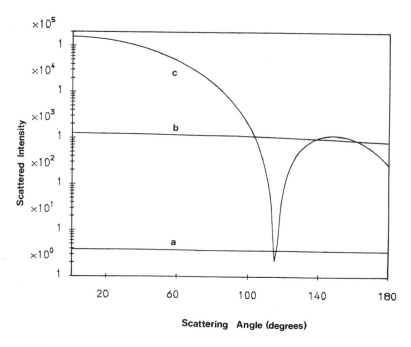

Figure 19. Computer generated curves for the angular dependence of light scattering at a wavelength of 500 nm, with solid spheres of diameters; 50 nm (**a**), 100 nm (**b**) and 500 nm (**c**). (Data of G.R.Jones.)

active liposomes. Rowe (59) has used turbidity measurements (at 400 nm) to show that the addition of ethanol decreases the phase transition temperature of dipalmitoyl phosphatidyl choline multilamellar liposomes.

6.2 Scattered light

Gibson and Strauss (60) have followed the fusion of synthetic, small unilamellar liposomes by monitoring the scattered light intensity at 90° to the incident beam. These authors, together with Kerker (61), show that for liposomes up to a diameter of 1/20th of the wavelength of the incident light the scattered intensity is proportional to the volume of the particle, or the cube power of the particle diameter. For particles greater than 1/20th to twice the incident wavelength the intensity of scattering is roughly proportional to the square of the diameter. *Figure 19* shows computer generated curves, from the Debye equation, showing the effect of the increase of diameter of a solid sphere (an approximation to a liposome) on the angular dependence of light scattering at 500 nm. As the diameter increases there is a greater increase in the scattering at lower angles than at higher angles. Eventually, the effect of the form factor is imposed on the angular scattering to give optical diffraction. The minimum in the scattering intensity at ~120° in *Figure 19* would not be as pronounced in large liposomes which have a variation in shape and size distribution. Nevertheless, this figure does make the point that 90° may not always be the most experimentally appropriate angle of measurement. Boni *et al.* (62), for example, when studying the fusion of small unilamellar synthetic liposomes induced with polyethylene glycol of various molecular weights, used a scattering angle of 60°.

6.3 **Instrumentation**

Turbidity measurements are best performed in a spectrophotometer. Scattering intensities at 90° to the incident beam can be measured in a spectrofluorimeter, with the excitation and emission monochromators set at a suitable but identical wavelength. For light scattering measurements at other angles it is necessary to use a light scattering photometer. The light source is usually a mercury lamp with interference filters which pass light of 436 nm or 546 nm. The photomultiplier is mounted on a turntable, with the sample cell at its centre. The light scattering measurement is usually taken as the ratio of scattered intensity to transmitted light and as the latter is far more intense than the former, a set of neutral density filters of known transmissions or electronic circuitry can be used to attenuate the signal.

Cylindrical cuvettes (Burchard cuvettes) are usually used with this type of instrument when the photomultiplier is set to measure scattering at angles other than 90°. The cells are by necessity voluminous in order to avoid reflections which would give anomalous scattering intensities, especially at low angles. It is of paramount importance that the cells are cleaned thoroughly, using the solutions and apparatus described previously (Section 4.2), as the scattering of dust and sample cannot be distinguished by the instrument. Solvent and solutions should be filtered prior to the experiment with 0.22-micron cellulose acetate membrane filters. Solvents can also be freed of dust by subjecting them to several hours of high speed centrifugation; only the solvent in the top half of the centrifuge tube should be used. Avoid using buffer salts which have not been filtered, as even analytical grade reagents contain insoluble matter as contaminants. If liposomes have to be prepared from biological sources, work in a dust-free environment as often as possible, filter buffers, and keep sample and buffers under cover. Dust is sometimes impossible to remove after making up a solution without significantly reducing the sample concentration.

7. REFERENCES

1. Wilschut,J. and Papahadjopoulos,D. (1979) *Nature*, **281**, 690.
2. Kendall,D.A. and Macdonald,R.C. (1982) *J. Biol. Chem.*, **257**, 13892.
3. Uster,P.S. and Deamer,D.W. (1981) *Arch. Biochem. Biophys.*, **209**, 385.
4. Morgan,C.G., Thomas,E.W. and Yianni,Y.P. (1983) *Biochim. Biophys. Acta*, **728**, 356.
5. Shinitzky,M. and Barenholz,Y. (1978) *Biochim. Biophys. Acta*, **515**, 367.
6. Kawato,S., Kinosita,K., Jr and Ikegami,A. (1977) *Biochemistry*, **16**, 2319.
7. Lakowicz,J.R., Prendergast,F.G. and Hogan,D. (1979) *Biochemistry*, **18**, 508.
8. Kinosita,K. and Ikegami,A. (1984) *Biochim. Biophys. Acta*, **769**, 523.
9. Van der Meer,B.W., Van Hoeven,R.P. and Van Blitterswijk,W.J. (1986) *Biochim. Biophys. Acta*, **854**, 38.
10. Van Langen,H., Engelen,D., Van Ginkel,G. and Levine,Y.K. (1987) *Chem. Phys. Lett.*, **138**, 99.
11. Cranney,M., Cundall,R.B., Jones,G.R., Richards,J.T. and Thomas,E.W. (1983) *Biochim. Biophys. Acta*, **735**, 418.
12. Lentz,B.R., Moore,B.M. and Barrow,D.A. (1979) *Biophys. J.*, **25**, 489.
13. Slavik,J. (1982) *Biochim. Biophys. Acta*, **694**, 1.
14. Blackwell,M.F., Gounaris,K. and Barber,J. (1986) *Biochim. Biophys. Acta*, **858**, 221.
15. Rottenberg,H. (1979) In *Methods in Enzymology*, S.Colowick and N.Kaplan (eds), Academic Press Inc., New York. Vol. **55F**, p. 547.

16. Schuldiner,S., Rottenberg,H. and Avron,M. (1972) *Eur. J. Biochem.*, **25**, 64.
17. Deamer,D.W., Prince,R.C. and Crofts,A.R. (1972) *Biochim. Biophys. Acta*, **274**, 323.
18. Lee,H.C. and Forte,J.G. (1980) *Biochim. Biophys. Acta*, **601**, 152.
19. Clement,N.R. and Gould,J.M. (1981) *Biochemistry*, **20**, 1534.
20. Rink,T.J., Tsien,R.Y. and Pozzan,T. (1982) *J. Cell Biol.*, **95**, 189.
21. Rogers,J., Hesketh,T.R., Smith,G.A. and Metcalfe,J.C. (1983) *J. Biol. Chem.*, **258**, 5994.
22. Thelen,M., Petrone,G., O'Shea,P.S. and Azzi,A. (1984) *Biochim. Biophys. Acta*, **766**, 161.
23. Waggoner,A.S. (1979) *Annu. Rev. Biophys. Bioeng.*, **8**, 47.
24. Sims,P.J., Waggoner,A.S., Wang,C.-H. and Hoffman,J.H. (1974) *Biochemistry*, **13**, 3315.
25. Haugland,R.P. (1985) *Handbook of Fluorescent Probes and Research Chemicals*. Molecular Probes, 9849 Pitchford Ave., Eugene, Oregon 97402, USA.
26. Deleers,M., Servais,J.-P., de Laveleye,F. and Wulert,E. (1984) *Biochem. Biophys. Res. Commun.*, **123**, 178.
27. Baker,J.A., Duchek,J.R., Hooper,R.J., Koftan,R.J. and Huebner,J.S. (1979) *Biochim. Biophys. Acta*, **553**, 1.
28. Bashford,C.L. and Smith,J.C. (1979) In *Methods in Enzymology*, S.Colowick and N.Kaplan (eds), Academic Press Inc., New York, Vol. **55F**, p. 569.
29. Mikes,M. and Kovar,J. (1981) *Biochim. Biophys. Acta*, **640**, 341.
30. Parker,C.A. (1968) *Photoluminescence of Solutions*. Elsevier Publishing Co., London.
31. Pesce,A.J., Rosen,C.-G. and Pasby,T.L. (1971) *Fluorescence Spectroscopy*. Marcel Dekker, New York.
32. Lakowicz,J.R. (1983) *Principles of Fluorescence Spectroscopy*. Plenum Press, New York and London.
33. Guilbault,G.G. (1973) *Practical Fluorescence: Theory, Methods, and Techniques*. Marcel Dekker, New York.
34. Georghiou,S., Thompson,M. and Mukhopadhay,A.K. (1982) *Biochim. Biophys. Acta*, **688**, 441.
35. Waggoner,A.S. and Stryer,L. (1970) *Proc. Natl. Acad. Sci. USA*, **67**, 579.
36. Chong,P.L.-G. (1988) *Biochemistry*, **27**, 399.
37. Chalpin,D.B. and Kleinfeld,A.M. (1983) *Biochim. Biophys. Acta*, **731**, 465.
38. Sackmann,E. (1976) *Z. Physik. Chem.*, **101**, 391.
39. Van den Zegel,M., Boens,N. and De Schryver,F.C. (1984) *Biophys. Chem.*, **20**, 333.
40. Chong,P.L.-G. and Thompson,T.E. (1985) *Biophys. J.*, **47**, 613.
41. Shaklai,N., Yguerabide,J. and Ranney,H.M. (1977) *Biochemistry*, **16**, 5585.
42. Weber,G. (1966) In *Fluorescence and Phosphorescence Analysis*. Hercules,D.M. (ed.), Wiley-Interscience, New York.
43. Smutzer,G. and Yeagle,P.L. (1985) *Biochim. Biophys. Acta*, **814**, 274.
44. Harris,R.A. and Schroeder,F. (1982) *J. Pharm. Exp. Ther.*, **223**, 424.
45. Chong,P.L.-G. and Cossins,A.R. (1984) *Biochim. Biophys. Acta*, **772**, 197.
46. Inbar,M. and Shinitzky,M. (1974) *Proc. Natl. Acad. Sci. USA*, **71**, 4229.
47. Johnson,S.M. (1981) In *Fluorescent Probes*. Beddard,G.S. and West,M.A. (eds), Academic Press, London, pp. 143−58.
48. Cossins,A.R. *ibid.*, pp. 39−80.
49. Mély-Goubert,B. and Freedman,M.H. (1980) *Biochim. Biophys. Acta*, **601**, 315.
50. Cundall,R.B. and Dale,R.E. (eds) (1983) *Time-Resolved Fluorescence Spectroscopy in Biochemistry and Biology*. NATO ASI Series A **69**, Plenum Press, New York and London.
51. O'Connor,D.V. and Phillips,D. (1984) *Time Correlated Single Photon Counting*. Academic Press, London.
52. Easter,J.H., De Toma,R.P. and Brand,L. (1976) *Biophys. J.*, **16**, 571.
53. Kinosita,K., Kawato,S. and Ikegami,A. (1977) *Biophys. J.*, **20**, 289.
54. Lakowicz,J.R., Cherek,H., Maliwal,B.P. and Gratton,E. (1985) *Biochemistry*, **24**, 376.
55. Vincent,M., de Foresta,B., Gallay,J. and Alfsen,J. (1982) *Biochem. Biophys. Res. Commun.*, **107**, 914.
56. Hoffmann,W. and Restall,C.J. (1983) In *Biomembrane Structure and Function*. Chapman,D. (ed.), Macmillan Press, London, Vol. **4**, p. 257.
57. Bangham,A.D., Gier,J.D. and Greville,G.D. (1967) *Chem. Phys. Lipids*, **1**, 225.
58. Yoshikawa,W., Akutsu,H. and Yoshimasa,G.D. (1983) *Biochim. Biophys. Acta*, **735**, 397.
59. Rowe,E. (1982) *Mol. Pharmacol.*, **22**, 133.
60. Gibson,S. and Strauss,G. (1984) *Biochim. Biophys. Acta*, **769**, 531.
61. Kerker,M. (1971) In *Chemistry and Physics of Interfaces II*. Ross,S. (ed.), Am. Chem. Soc., Washington.
62. Boni,L.T., Hah,J.S., Hui,S.W., Mukherjee,P., Ho,J.T. and Jung,C.Y. (1984) *Biochim. Biophys. Acta*, **775**, 409.

CHAPTER 6

Liposomes in biological systems

ROGER R.C.NEW, CHRISTOPHER D.V.BLACK, ROBERT J.PARKER,
ANU PURI, and GERRIT L.SCHERPHOF

1. INTRODUCTION

This chapter covers techniques which have been found useful in studying the behaviour and fate of liposomes in biological systems—that is, their interactions with cells and organs. *In vitro* and *in vivo* systems have been taken together here since many of the considerations with regard to methodology are common to both situations. The chapter starts with a discussion, first, of the way in which the disposition of liposomes is dependent upon membrane composition, and secondly of how the composition can be varied in such a way as to alter the fate of the liposomal contents. The second section deals with markers which can be utilized to give useful information about these two processes, employing the techniques described in the final part, which covers ways in which the components of biological systems can be dissected and analysed for these markers.

2. INTERACTIONS OF LIPOSOMES WITH CELLS

Liposomes can interact with cells in many ways to cause liposomal components to become associated with those cells. These are summarized in the following section.

2.1 Cellular interactions

2.1.1 *Intermembrane transfer*

Intermembrane transfer of lipid components can take place upon close approach of the two phospholipid bilayers without the need for disruption of the liposome or prejudice of the membrane integrity (*Figure 1a*). Indeed it is possible for such transfer to occur (often in both directions) with complete retention of the contents of the liposome's aqueous compartment.

Exchange of phospholipids with cells is thought to occur possibly via the intermediary of a specific cell surface exchange protein, since only certain phospholipids exchange (PC and PE), and the process is slowed down after trypsin treatment (1). Fluorescent derivatives of PE conjugated through the amino group will not exchange, while those with the fluorophore incorporated in the fatty acyl chains do so readily.

Similar interactions can take place between liposomes and lipoproteins (particularly high density lipoproteins), and in certain circumstances proceed to the extent of destroying the liposome altogether. The degree to which destruction occurs depends on the ratio of lipoprotein to liposomes, and the system is saturable at high concentrations of liposomes. At low lipid concentrations, both *in vivo* and *in vitro*, small

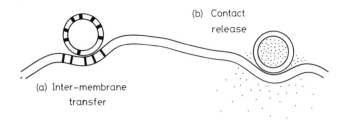

Figure 1. Contact-mediated transfer mechanisms. This drawing illustrates two ways in which liposomal contents can be incorporated into cells without any need for internalization of the liposome itself. (**a**) Intermembrane transfer. Lipophilic materials situated in the liposome membrane can insert themselves into other membranes with which that liposome comes into contact provided that the distance between the two membranes is small enough to prevent the materials being transferred from coming fully into contact with the bulk water phase during the process of transfer. (**b**) Contact release. This can occur when the membranes of the cell and the liposome, upon being brought together, experience perturbations as a result of that contact which increase the permeability of the liposome membrane (and perhaps of the cell membrane as well) to the solute molecules entrapped in the aqueous compartment of the liposome. This brings about leakage of solute from the liposome directly into the cell.

unilamellar vesicles composed only of egg lecithin (no cholesterol) can rapidly disintegrate in plasma, the lipid components being redistributed among the lipoproteins (2). Incorporation of cholesterol into the egg lecithin membrane partially inhibits disintegration, and sphingomyelin SUVs containing 30 mol% cholesterol or more remain intact with complete retention of aqueous contents. On the other hand, for large liposomes, both multi- and unilamellar, transfer of phospholipids to plasma lipoproteins is low, irrespective of cholesterol content.

Cholesterol itself transfers very readily between bilayers in such a way as to equilibrate, so that its molar concentration is equalized throughout all membranes present. Thus, liposomes have been used as a method of supplementing or depleting cell membranes of cholesterol *in vitro*, simply by incubating the cells in the presence of liposomes containing low or high levels of sterol (3). Such manipulation can have profound effects on cell function.

2.1.2 *Contact-release*

'Contact-release' of aqueous contents of liposomes occurs by a poorly understood mechanism in which contact with the cell causes an increase in permeability of the liposome membrane. This leads to release of water-soluble solutes in high concentration in the close vicinity of the cell membrane, through which these solutes may, under certain circumstances, then pass (*Figure 1b*). Curiously, cell-induced leakage of solutes has been observed to be greater in membranes with cholesterol concentrations above 30 mol% (4).

The above two phenomena can provide very effective means for introducing materials into specific cells without the need for ingestion of the whole liposome, and would be of particular value for cells which are not actively phagocytic. The methods will work best under conditions where flow and turbulence of the medium surrounding the cells is reduced, and where physical interactions between liposomes and cells are strengthened by means of receptor/ligand binding between the two membranes. Whether these processes take place to a significant extent must depend upon membrane composition as well as the nature of the compounds themselves.

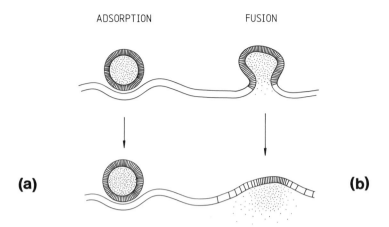

Figure 2. Liposome binding to cells with and without uptake. Not all liposomes which become associated with cells are subsequently taken up. In some cases, the liposomes may remain passively adsorbed on the cell surface indefinitely (**a**) with complete retention of aqueous and lipid contents within the liposome, separate from the cell. In other cases, after the initial adsorption stage, fusion may take place (**b**) with incorporation of the liposomal lipids into the cell membrane, and introduction of aqueous contents into the cytoplasm.

2.1.3 *Adsorption*

Adsorption of liposomes to the cell surface can often occur with little or no internalization of either aqueous or lipid components (*Figure 2a*). It may take place either as a result of physical attractive forces, or as a result of binding by specific receptors to ligands on the vesicle membrane. It is thought that physical adsorption of liposomes may occur through binding to a specific cell surface protein (5). Surprisingly, uptake is greatest at or below the phase transition temperature of the liposome membrane (i.e. when lipids are in the gel phase), presumably because the binding sites involved are membrane defects which are more stable at low temperature. Adsorption is an essential prerequisite for ingestion of the liposome by cells, but what factors determine whether or not a liposome is consumed thereafter by pinocytosis or phagocytosis is not yet fully understood. Studies by Leserman *et al.* (6) have shown that attachment of liposomes to cell membranes via certain surface proteins, but not others, can result in rapid uptake into the cell.

2.1.4 *Fusion*

Close approach of liposome and cell membranes can lead to fusion of the two, resulting in complete mixing of liposomal lipids with those of the plasma membrane of the cell, and the release of liposomal contents into the cytoplasm (*Figure 2b*). In the case of a multilamellar vesicle, this will involve introduction of the internal membrane lamellae of the liposome intact into the cytoplasm, so that interactions such as those cited above may come into play all over again between liposomes and subcellular organelles. In *in vitro* situations, fusion may be brought about quite readily by incorporation of fusogens (e.g. lysolecithin, PS, detergents, surfactants) into the membrane, although these materials run the risk of being quite toxic to cells, presumably because they continue to manifest their non-specific membrane perturbing effects after incorporation within

PHAGOCYTOSIS OF LIPOSOMES

(i)

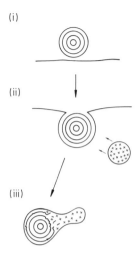

(ii)

(iii)

Figure 3. Uptake of liposomes by phagocytosis. In certain types of cell, such as macrophages, the adsorption stage (**i**) is rapidly followed by invagination of the cell membrane and envelopment of the cell by plasma membrane (**ii**), to form an endocytic vacuole, in which the liposome is brought into contact with lysosomal enzymes which can break down the liposome membrane (**iii**). In contrast to the process of fusion (see *Figure 2b*), phagocytosis results in all the liposomal lipid being completely internalized (not inserted into the plasma membrane), and the aqueous contents of the liposome initially being restricted to the lysosomal compartment, rather than the cell cytoplasm.

the cell. To date, the most effective method for *in vitro* delivery of liposomal materials into cells by fusion is use of Sendai virus fusion proteins (see Section 4.3 of Chapter 2), in which not only can water-soluble components enter into the cytoplasm, but functional membrane proteins can be inserted into the cell plasma membrane. Other methods, which may involve fusion, have been described also, using PE and oleic acid or positively-charged lipids (7).

In vivo, however, it appears that the process of fusion, as a means of inducing cellular uptake, takes second place to phagocytosis. Under most circumstances, liposomes are cleared far too rapidly from the bloodstream by phagocytic cells for fusion events to occur to any significant extent. (However, see Section 2.2 for a discussion of 'long-lived' liposomes incorporating gangliosides in the membrane.) More favourable settings for fusion processes would be localized sites with low involvement of the reticular endothelial system (RES), such as the aqueous humor of the eye, cerebrospinal fluid, or following passive adsorption to the walls of capillaries.

2.1.5 *Phagocytosis/endocytosis*

Cells with phagocytic activity take liposomes up into endosomes, subcellular vacuoles formed by invagination of the plasma membrane, which, following the action of membrane-bound proton pumps, have a pH of $5-5.5$. These endosomes then fuse with lysosomes to form secondary lysosomes where cellular digestion takes place in a milieu of approximately pH 4.5 (*Figure 3*). Lysosomal enzymes break open the liposomes,

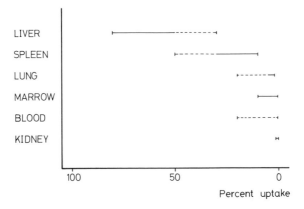

Figure 4. Liposomal localization 24 h after intravenous injection. The solid lines indicate the normal ranges for uptake of liposomes into given organs, the variation resulting from differences in liposome size, dose, charge etc. The dotted lines indicate values which may be obtained under special circumstances, usually as a result of incorporation of particular ligand-binding lipids in the membrane (see text). Retention of liposomes in the blood stream is especially variable, with a $t_{1/2}$ for plasma clearance time ranging from 30 min for large negatively-charged MLVs bearing aminomannose residues, to 12 h or more for small sphingomyelin liposomes containing gangliosides.

the phospholipids being hydrolysed to fatty acids, which can then be recycled, and reincorporated into host phospholipid. The fate of liposomal fatty acids can be traced using labelled DMPC, since the short chain myristic acid is easily distinguishable from the longer host fatty acid chains. In mammalian liver, cholesterol is metabolized to bile salts. Certain anionic lipids and drugs can inhibit lysosomal breakdown and produce a pseudo-storage disease in the lysosomes (8).

During the process of breakdown of liposomes in lysosomes, the contents of the aqueous compartment are released, after which they will either remain sequestered in the lysosomes until exocytosis, particularly if they are highly charged at low pH, or they will slowly leak out of the lysosome and gain access to the rest of the cell. Ways of manipulating the subcellular fate of liposomal contents will be discussed in later sections of this chapter. In addition to phagocytosis, liposomes may also be taken up by receptor-mediated endocytosis; if liposomes are coated with low density lipoproteins (LDL) or transferrin, they will bind to the cells via surface receptors for these moieties, and will then be internalized via coated pits with subsequent ligand degradation, or recycling, respectively (9). Thereafter, the liposomes may gain access to subcellular compartments (e.g. Golgi) other than lysosomes.

2.2 **Anatomical considerations**

In addition to the interactions described above, liposomes administered *in vivo* are subject to additional interactions which can determine the rate of clearance and degree of organ uptake (see *Figure 4*), and which are also controlled by factors such as size and membrane composition.

Under normal circumstances, when lecithin/cholesterol liposomes are injected into the bloodstream, they are taken up to a large extent by organs rich in cells of the RES. Because of this passive targeting to the phagocytic cells of the RES, liposomes are especially suitable as drug carriers for treating diseases of the RES such as leishmaniasis and fungal infections (10−15). Small liposomes are taken up less rapidly than large ones, and when those SUVs do go to the liver there is a higher proportion entering the hepatocytes rather than Kupffer cells, since they are small enough to pass through the fenestrae of the sinusoids and into the liver parenchyma. Uptake by the liver is dose-dependent, and is subject to saturation or blockade by the liposomes themselves: large liposomes will inhibit uptake of large liposomes, and, to a lesser extent, small liposomes will block the uptake of other small liposomes. Large liposomes can also reduce the uptake of small vesicles, but not vice versa. Cholesterol taken up by liver cells in liposomal form is metabolized differently, depending on whether it is in large or small vesicles (16). Cholesterol in SUVs is converted more to glycocholate, while (larger) MLVs give more taurocholate. It is known that SUVs take the cholesterol to the hepatocytes directly, while MLVs are taken up by Kupffer cells, where the liposomes are broken down, and the cholesterol is transferred to hepatocytes via lipoproteins. It is not clear whether the difference in processing is due to differences in presentation to the same population of hepatocytes, or due to processing by different subpopulations of hepatocytes which are now known to exist.

The presence of lactosyl ceramide in liposomes increases the overall rate at which SUVs are taken up by the liver, as well as increasing the total uptake, the extra increment being found in the hepatocytes (17). The mechanism responsible is thought to be interaction of the liposomes with a surface receptor for galactose on hepatocyte membranes; pre-treatment with asialofetuin (exposing galactose residues) blocks the uptake. Uptake is also dependent on the liposome composition, and the method employed to link the saccharide group to the membrane. The hepatocyte/Kupffer cell uptake ratio can be further increased by treatment with gadolinium salts, which inhibit ingestion of liposomes by phagocytic cells, but not by parenchymal cells (18). Larger liposomes will have restricted access to hepatocytes since they will be unable to pass through the fenestrae in the sinusoids (mean diameter 0.1 μm) of the hepatic blood supply. Interestingly, reports have suggested that a proportion of liposomes taken up by hepatocytes can deliver their contents to mitochondria (19). This could have important implications for targeting and subsequent expression of genetic material in cells. Very large liposomes (5−8 μm) become associated in high proportion with the lung, and this uptake (mainly by trapping in capillaries) is augmented by incorporation in the membrane of certain negatively-charged lipids, but not others. Thus PS-containing liposomes are taken up much better than those containing PA or PE (20).

Liposomes administered intramuscularly are retained initially at the site of injection, and drain slowly through the lymphatic system to local lymph nodes, where they are again taken up by macrophages, while some will come into direct contact with lymphocytes. Liposomes containing antigenic material, co-incorporated with an immunostimulant, are capable of generating a strong immune response when administered either locally or parenterally, either as a result of surface-bound antigen interacting directly with lymphocytes, or through enhancement of macrophage processing of the antigen. Evidence suggests that for the most effective presentation of antigen, use of sphingomyelin or DSPC as major membrane component is to be recommended (21). Liposomes administered at other locations tend to stay at the site of administration (e.g. aqueous humour of the eye, or knee joint) and may be rapidly broken down (e.g. after inhalation into the lung, or after peritoneal injection). In the case of peritoneal injection, macrophages passing through the cavity take up liposomes in large quantities, and may even be recruited from the general circulation. To date there is no strong evidence that large quantities of liposomes are taken into the systemic circulation after oral or topical administration.

It is not clear what factors influence uptake of liposomes by macrophages. When liposomes are injected into the bloodstream, their surface membrane rapidly becomes coated with plasma proteins (22). A range of different proteins can be found, including α and β globulins, IgG, fibronectin, clotting factors and albumin, all of which may affect the way in which the liposome is seen by phagocytic cells, and hence the rate at which liposomes are cleared from the bloodstream by the RES (23).

Liposomes which have been polymerized in various ways bind plasma proteins in higher amounts than normal liposomes, and are cleared from the bloodstream much more rapidly (24). These liposomes also deplete factor V in plasma. Increased uptake by macrophages, *in vitro* and *in vivo*, of small liposomes bearing surface-bound amino-mannose, or other amine-containing sugars, has been shown to be dependent on the participation of heat-labile serum components (25). Coating of liposomes with certain polysaccharides (particularly incorporation of GM_1 ganglioside) prior to administration can lead to a prolongation of lifetime in the circulation (and a concomitant reduction in liver uptake) with an optimum GM_1 ganglioside content of 5 mol% relative to PC (26). Another composition which is effective in increasing plasma half-life is sphingomyelin:cholesterol (4:3) (27). Although sphingomyelin has the same headgroup as PC, there may be packing differences which influence the adsorption of proteins onto the membrane. Inclusion of GM_1 into sphingomyelin-containing liposomes gives a yet higher retention in the bloodstream (26). Liposomes of 'intermediate' fluidity are best at avoiding the RES, except for SUVs, where more rigid vesicles are taken up less avidly.

Coating of liposomes with antibodies can also increase uptake by the RES unless the F_c receptor of the antibody has been removed, or masked in some way. Using the 'long-lived' liposomes described in the previous paragraph, in conjunction with $F_{ab}{}'$ fragments or other receptor molecules raises the possibility of targeting liposomes for delivery to specific cellular components of the blood compartment, or to sites of altered vascular structure (e.g. ischaemia, immune complex deposition, etc.). It should be emphasized, however, that this approach will not permit localization of liposomes in areas outside the blood compartment, since liposomes are far too large to cross the blood vascular endothelium intact. The only places where extravasation may possibly occur are areas where the vasculature is damaged in some way, or has changed to become highly endocytic, as is thought to occur within certain tumours. The same mechanism may be responsible for uptake of sulphatide-containing liposomes across the blood−brain barrier (28). These liposomes are also taken up avidly by the liver. For local delivery to lymph nodes, intramuscular or subcutaneous injection of liposomes is efficacious, since the liposomes are carried through the lymphatics from the site of injection to draining lymph nodes (29,30), where they accumulate, without interference from filtering systems such as the hepatic RES.

It has been suggested that one mechanism whereby i.v. administered liposomes may exert an effect outside the bloodstream is through uptake by circulating monocytes which

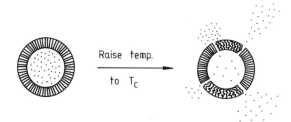

Figure 5. Temperature-mediated release of aqueous solutes. At the liposome membrane phase transition temperature (T_c) separation can occur between solid and fluid phases, leading to an increase in permeability of the membrane to aqueous solutes. In some cases, this release is enhanced by the presence of certain proteins (see *Figure 6*).

then become activated as a result of passing through a capillary bed draining wounded or infected tissue. These cells then exit the blood compartment and migrate through the tissue under the influence of chemotactic factors towards the site of the disorder. Agents which enhance the ability of these macrophages to cope with localized pathogens may usefully be incorporated into blood-circulating liposomes, as was first proposed by New and colleagues for the treatment of cutaneous leishmaniasis (31).

2.3 Site-selective delivery of liposomal contents—determination by liposome composition

In addition to influencing the *in vivo* distribution of the liposome itself, different membrane compositions can alter the way in which the liposomal contents are released. In particular, the composition can be arranged so that the release is site-specific, in response to environmental conditions, either at the gross anatomical level, at the cellular, or at the subcellular level. Mechanisms aimed at targeting at each of these levels are described in the following paragraphs.

2.3.1 *Temperature-sensitive liposomes*

This type of liposome takes advantage of the fact that liposomes leak much more readily at the phase transition temperature (T_c) of their membrane lipids (see *Figure 5*). *In vivo* or *in vitro*, this leakage is augmented by the presence of certain plasma proteins, which bind to the membrane and help to destabilize it at the T_c (32). For *in vivo* delivery, the liposomes (a mixture of DSPC and DPPC with no cholesterol) are designed to be stable at 37°C, but will break down as they pass through an area of the body in which the temperature is raised to 40°C (e.g. a tumour, or an area subjected to local external heating).

Phase transition release (PTR) induced by proteins such as HDL differs in several respects from that observed in the absence of protein. In the latter case the mechanism involved is an increased permeabilization at the T_c of an intact membrane, leading to all vesicles releasing a proportion of their low molecular weight contents. In the presence of HDL, however, [particularly the high density lipoprotein (HDL) apoprotein], the extra release at the T_c is an all-or-nothing phenomenon, in which a proportion of the vesicles are sufficiently disturbed to release all their contents (both high and low

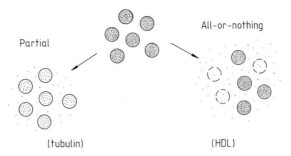

Figure 6. Modes of solute release. Protein-mediated 'phase transition release' of solutes can take place in one of two ways—either complete rupture of a proportion of vesicles, as with high density lipoproteins (HDL), or by partial permeabilization of all vesicles in the population (as with tubulin).

molecular weight) while others retain theirs in their entirety (*Figure 6*). The process is a very rapid one (in the order of a few seconds). This type of release is much more extensive if the T_c is approached from temperatures below rather than above, probably because the energy barrier for protein binding is lower for the gel than for the fluid phase. The protein, once taken up, is bound irreversibly, and is unavailable for induction of further release from other vesicles.

Using carboxyfluorescein as an indicator of release (33), freshly prepared SUVs have been found to give the most clear-cut demonstration of PTR, since they show only a small amount of spontaneous leakage in the absence of protein, and relinquish close to 100% of their contents with appropriate quantities of HDL. MLVs do not give 100% release, presumably because the protein does not gain access to internal compartments of the MLV. Furthermore, both MLVs and LUVs have a significant level of spontaneous release at the T_c in the absence of protein. SUVs also differ from other types of liposome in their ability to withstand osmotic pressure differences across the bilayer. Thus SUVs, even in hypo-osmolar buffers give only a small level of spontaneous phase transition release, whereas MLVs, while stable below the T_c (but not above) release the majority of their contents under the same conditions.

All the plasma lipoproteins can bind to liposomes to enhance PTR, as well as some non-lipoprotein plasma proteins. In addition, certain proteins such as tubulin (and actin) can interact specifically with neutral lecithin bilayers. The behaviour of tubulin contrasts strongly with that of HDL, in that it only binds at the T_c and causes partial leakage from each vesicle, rather than all-or-none release (34).

2.3.2 *Target-sensitive liposomes*

The underlying principle of target-sensitive liposomes is that they are composed of lipids which form a membrane which is intrinsically unstable (e.g. unsaturated PE, which naturally adopts the hexagonal-inverted micelle structure in preference to bilayer sheets). This membrane is stabilized by the presence of proteins anchored in the membrane by covalent attachment to fatty acid chains or other lipid molecules. Typically the protein employed is a specific antibody; if the antibody-bearing vesicle meets its 'target' (a polyvalent ligand or a cell) the antibodies will bind, and will be cross-linked by the antigen. In the process the antibody molecules will be drawn together and form a cap,

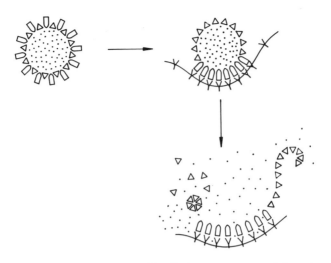

Figure 7. Receptor-mediated release. 'Target-sensitive' liposomes are constructed of membrane-bound ligands (☐) which are capable of stabilizing a membrane composed of otherwise inherently unstable lipids (△). When these ligands (e.g. fatty acid-conjugated antibodies) are aggregated by binding to a target (Y) which they recognize specifically, their ability to stabilize the membrane is reduced, and the liposome disintegrates with release of contents.

leaving a large expanse of free membrane exposed and highly susceptible to destabilization. Release of entrapped materials can thus take place in the vicinity of target cells and can be made available to these cells without the need for ingestion of the liposomes by the cell (*Figure 7*). The method is particularly appropriate for delivery of materials which exert an effect on the cell by binding to cell surface receptors (e.g. γ-interferon?) or for delivery to cells whose endocytic activity is reduced (e.g. in response to viral infections). This system is very sensitive to parameters such as protein:fatty acid coupling stoichiometry and protein:lipid ratio, and needs to be carefully optimized (35).

2.3.3 *pH-sensitive liposomes*

A similar approach to that described in the previous section has been adopted for the construction of liposomes which are sensitive to pH (*Figure 8*). Again using PE as the major lipid component, the membrane is stabilized by addition of materials which are charged at neutral pH, but which lose their charge at low pH along with their ability to stabilize the membrane. Such compounds include fatty acids, palmitoyl homocysteine, cholesterol hemisuccinate (CHEMS) and *N*-succinyl PE. Judicious mixing of these materials with PE (or better, '*trans*'-PE; which is more stable in serum) in a proportion of 10−30%, can result in liposomes which remain intact at pH 7 but will fuse when taken down to pH 5. In the case of CHEMS, such fusion requires the presence of calcium or magnesium ions; in some cases a pH gradient across the membrane is also important (36).

This type of liposome has application in delivery of bioactive materials to cells in that it provides a means for enabling the liposomal contents to escape the lysosome. As mentioned earlier, the first step in intracellular processing of liposomes is uptake into an endosome, which has an intravesicular pH of pH 5−5.5. If liposomes can be

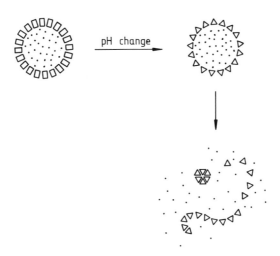

Figure 8. pH-mediated release. 'pH-sensitive' liposomes are composed of lipids with ionizable headgroups which, in the event of a pH change, are converted to a form which prefers to adopt an inverted hexagonal phase rather than the conventional bilayer sheet of the liposome membrane. Raising or lowering the pH thus gives rise to disintegration of the membrane.

induced to become fusogenic upon exposure to this pH, then they may be able to fuse with the wall of the endosome and release their contents into the cytoplasm of the cell before they become broken down and degraded in the lysosome. Such a strategy has been shown to work with plasmids containing genes which need to be delivered to the cytoplasm, and thence to the nucleus, for expression (37).

3. USE OF MARKERS TO DETERMINE FATE OF LIPOSOME COMPONENTS

In order to distinguish between the many different types of interaction which may take place between liposomes and cells, as outlined previously, care must be taken in the choice of markers, so that misinterpretations due to artefacts are avoided. In each of the following sections, the experimental differentiation of two contrasting, but easily confusable mechanisms has been highlighted.

3.1 Differentiation between lipid exchange and association

The use of radiolabelled lipid markers is one of the simplest ways of monitoring uptake of liposomes by cells. However, many common membrane phospholipids are able to transfer from one membrane to another by intermembrane contact, without any strong or long term interaction between the two membranes. Thus incubation of liposomes with cells *in vitro* may well overestimate the percentage uptake if phospholipids or cholesterol are used as markers. Cholesteryl oleate can also give spurious results since transport of this, and other cholesteryl esters, is catalysed by a specific exchange protein present in plasma. (However, note that this protein is absent from rat plasma.) The most satisfactory group of markers is that of the cholesteryl alkyl ethers, which do not exchange spontaneously, even in the presence of plasma (38) and are essentially non-metabolizable (39), so that there is no danger of underestimation of uptake as a result

231

of degradation and subsequent redistribution in culture or tissue. Because, at low concentrations, cholesteryl ethers are incorporated with 100% efficiency into all types of liposome during preparation, they are very useful markers for general use. Amersham International is a source of [^{14}C]cholesteryl hexadecyl ether, and a general method for preparation of cholesteryl ethers is given in Appendix I. Recently, cholesteryl aniline has been used as a marker. This compound also appears to be essentially non-metabolizable, and has the advantage that it can be labelled very easily with iodine, permitting detection by γ-emission. Labelling of this compound is also described in Appendix I.

Other lipid markers which could be used are *N*-conjugated fluorescent derivatives of PE (e.g. N-NBD-PE—ref. 40). These are non-exchangeable lipids which can be visualized under UV illumination, and compared with fluorescent acyl PE derivatives (see Section 2.1.1) which are subject to exchange. For long-term experiments, these markers suffer from the disadvantage that they are likely to undergo degradation by cleavage with phospholipases.

3.2 Differentiation between adsorption and internalization

One way of comparing adsorption versus internalization as methods of uptake is by direct visualization using electron microscopy of thin sections. Appropriate markers for inclusion in liposomes are ferritin, horseradish peroxidase (41,42) which can be stained cytochemically, or colloidal gold (43). Horseradish peroxidase has the slight disadvantage in that the membrane structure tends to be obscured after staining. Liposomal disposition inside cells can also be inferred using transmission electron microscopy in conjunction with X-ray micro-analysis, looking at the distribution of metal atoms such as Cr^{3+} entrapped in liposomes (44). Identification of liposomes which have not been labelled with a marker is fraught with difficulty, since treatment and processing of cells and tissues can lead to the formation of numerous artefacts and vesicle-like structures—for example 'myelin figures', which are identical to liposomes in appearance. One problem with looking at adsorbed particles under the electron microscope is that only a small fraction of the plasma membrane of a cell can be visualized. Hence, electron microscopy may considerably underestimate the extent of adsorption, unless careful calculations are carried out to allow for this. Examination by scanning electron microscopy can give a more realistic picture, although SUVs can sometimes be too small to distinguish clearly using this method. Indeed, small liposomes can present problems for all techniques of measurement, since they contain relatively very small amounts of marker.

An alternative approach for assessing the degree of internalization of liposomes is to measure the extent of degradation of metabolizable markers incorporated in liposomes. For example, sphingomyelin can be labelled at the methyl groups of the choline moiety, and in liposomes with high levels of cholesterol it does not participate in phospholipid exchange. When sphingomyelin is metabolized in cells, the choline headgroup is transferred to PC, so the degree of internalization can be monitored very easily by measuring the proportion of label that appears in PC after separation from other cellular phospholipids by TLC. Comparison of the fates of different labels in metabolizable cholesterol [^{14}C]oleate and [^{3}H]cholesterol ether will yield the same information.

Metabolizable aqueous phase markers can be used in the same way. Proteins in particular are very readily hydrolysable in lysosomes, and are degraded to amino acids which are then exported to different sites. Thus, labelled [^{125}I]BSA is broken down to amino acids including iodotyrosine, which may then either be released from the cell or remain associated. The intact protein is distinguished from hydrolysis products by its acid precipitability, and the soluble, low molecular weight products can be washed off and their radioactivity measured separately. Parallel experiments may be performed using a non-metabolizable marker—for example [^{125}I]polyvinylpyrrolidone (PVP)— so that an independent estimate of total uptake can be obtained. These two markers are particularly convenient since they employ a γ-emitting isotope, so there is no need for scintillants and laborious extraction procedures to avoid quenching conditions. Alternatively, a non-metabolizable tritiated marker (e.g. [^{3}H]inulin) could be co-entrapped with [^{125}I]BSA in the same liposome. Care must be taken if inulin is used, since it has been found that, over a period of hours, significant quantities of the intact molecule can be released from cells by exocytosis. (Inulin has often been used as a marker for *in vivo* studies, since in the free form it has an elimination rate equal to the glomerular filtration rate. Any material remaining in the blood after a long period of time must therefore still be in liposomal form.)

The disadvantage of using the metabolism of entrapped markers as an indicator of internalization is that there may be routes of internalization which do not involve lysosomal or other degradation, in which case measurement of the degree of metabolism may give an underestimate of the amount of material ingested. Also, not everything which is taken up is necessarily metabolized immediately. The different intracellular fates are discussed further in the following sections.

3.3 Differentiation between fusion and phagocytosis

The classic method of monitoring fusion of liposomes with cells is that of fluorescence dequenching of carboxyfluorescein (CF) (32). As described in Chapter 3, this technique relies on the fact that at high concentration CF displays very little fluorescence, being self-quenched, until the solution is diluted, with a consequent increase in fluorescence. Thus, upon fusion of CF-containing liposomes with the cell plasma membrane, the liposomal contents are delivered to the cytoplasm, and are diluted many hundred-fold. The cell will thus display a strong diffuse fluorescence throughout the whole body of the cell, with a dark area in the region of the nucleus (*Figure 9*). In contrast, liposomes which have been endocytosed will display punctate fluorescence, restricted to the secondary lysosomal and endocytic vacuoles. Adsorbed liposomes will not fluoresce at all, unless the CF concentration has been reduced by extensive leakage, in which case the appearance of the cell after washing should theoretically be dully fluorescent, with a bright rim. Assuming that the intracellular CF is already maximally diluted, the percentage uptake (relative to that adsorbed) may be calculated by measuring the increase in fluorescence of a suspension of cells after treatment with Triton X-100. Care must be taken in interpretation of results using this method, since some uptake of free dye may occur after it has leaked out of the liposomes. Also, cells incubated for periods of time (several hours) can leak carboxyfluorescein into the surrounding medium, and over shorter timespans, CF can leak from the lysosome into the cytoplasm. Use of calcein (which is more strongly charged) instead of CF could overcome some of these problems (see Chapter 3), although the fluorescence of calcein is very sensitive to the presence of trace metal ions.

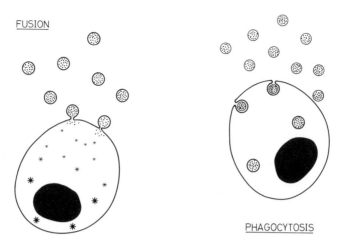

Fig. 9. Cellular uptake of liposomal carboxyfluorescein. The use of carboxyfluorescein (CF) to distinguish between fusion and phagocytosis relies on the fact that upon fusion with the cell membrane, the aqueous liposomal contents are diluted many-fold in the cell cytoplasm. Thus the CF, which is self-quenched when contained at high concentrations inside the liposome, displays a marked increase in fluorescence as it spreads throughout the cytoplasm and becomes unquenched. In contrast, uptake of liposomes by phagocytosis results in practically no dilution of the CF (and hence no fluorescence increase) since the liposomal contents are initially retained inside secondary lysosomes.

Other evidence indicative of one mechanism acting in preference to another is the influence on uptake of treatment of cells with metabolic inhibitors such as cytochalasin B or sodium azide and deoxyglucose, or with ammonium chloride or chloroquine, which are known to inhibit fusion of lysosomes with the phagosome. These agents would be expected to interfere with processing by phagocytosis but not with uptake by fusion with the plasma membrane.

Distinction could also be made by use of fluorescent phospholipid analogues, where punctate lysosomal localization can be differentiated visually from diffuse plasma membrane fluorescence. In this case, however, the possibility of adsorption of liposomes would be an added complication, which is difficult to distinguish from fusion. Although, here, such complications could be resolved by photobleaching studies, where the mobility of adsorbed lipids is lower than that of lipids incorporated into the membrane by fusion; this example is a good illustration of the difficulties in separating one phenomenon from another using a single marker, and of the advisability of conducting studies using several markers in the same liposome.

3.4 Differentiation between lysosomal and cytoplasmic localization

A cytochemical marker has been developed which is a very sensitive indicator of lysosomal delivery (45). The compound is 5-bromo,4-chloro,3-indolyl phosphate (BCIP) and is a colourless substrate for lysosomal alkaline phosphatase, being converted to the free indole which rapidly self-condenses to form an insoluble and strongly coloured precipitate which remains localized within the lysosomes. The precipitate can be seen easily in thin sections in the electron microscope. At the present time, poor solubility of the dye in solvents prevents accurate quantitation of uptake. Formation of the dye is extremely specific to lysosomes, so that its presence in sections is strongly indicative of lysosomal processing, even after exocytosis, or subsequent extrusion of lysosomal contents into the cytoplasm.

3.5 Cellular availability (differentiation between intact and degraded liposomes)

Both *in vitro* and *in vivo*, although uptake of liposomal components can be measured, it is often difficult to know whether the internal contents of these liposomes can gain access to the cell and vice versa. Indeed, some recent reports suggest that certain types of liposome can remain intact even after uptake into lysosomes (46). Elucidation of the availability of liposomal contents is particularly difficult *in vivo*. The first technique used to investigate the integrity of liposomes in whole tissues was γ-perturbed-angle correlation spectroscopy (47), in which changes in rotational motion of a liposome-entrapped γ-emitter (e.g. [$^{111}In^{3+}$] indium salts) are observed when the liposome is broken down, and the environment of the tracer is altered. Recently, similar changes have been studied using NMR probes (48).

A simpler methodology has been worked out by Szoka and colleagues (49) using a double γ-radiolabel technique. The two labels are ^{22}Na and ^{51}Cr/EDTA, and the principle of the method is that as long as the liposomes remain intact (whether inside or outside a cell) the ratio of the two labels will remain the same. If the liposomes leak their contents inside a cell, however, then the fates of the two labels will be very different, since sodium is readily excreted from most cells into the extracellular fluid via the Na^+/K^+ pump, while Cr/EDTA has no such method of exit, and remains trapped within the cell. Disruption of liposomes inside cells is therefore signalled by a decrease in the ratio of $^{22}Na/^{51}Cr$ activity which can be measured easily in a dual channel γ-counter (*Figure 10*). The 'cellular availability', C_{AV} is expressed as a percentage as:

$$[1 - (Na/Cr)_t/(Na/Cr)_0] \times 100$$

where $(Na/Cr)_t$ is the ratio at time t, and $(Na/Cr)_0$ is the ratio at time zero.

It will be seen that the expression is independent of the absolute quantity of label administered or the amounts present in the tissue. A second parameter, the 'tissue availability', T_{AV}, represents the proportion of the total dose administered *in vivo* which is available to the cells of a given tissue in the form of internalized and disrupted liposomes, and is simply the product of the C_{AV} with the proportion of activity (^{51}Cr) taken up in that tissue. The important feature of this method is that no cellular processing or breakdown of label itself is required for the differential behaviour of the labels to be manifested, so that the technique is applicable to any form of delivery into the cell.

In using this technique, it is assumed that any leakage which occurs outside the cells (before liposome uptake) will occur to the same extent for both markers, and that leaked contents will not be able to enter cells or tissues on their own. In this respect, it is important to ensure that the Cr/EDTA is not contaminated with any high molecular weight complexes, which do not leak so rapidly, but which are taken up avidly by liver cells in the free form. Such complexes can be removed easily by Biogel P-2 column

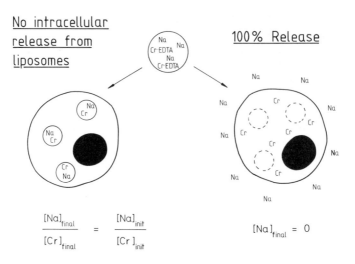

No intracellular release from liposomes

100% Release

$$\frac{[Na]_{final}}{[Cr]_{final}} = \frac{[Na]_{init}}{[Cr]_{init}}$$

$$[Na]_{final} = 0$$

Fig. 10. Double-label cellular availability assay. This assay is based on the fact that sodium and chromium ions are processed differently by the cell, and the aim of the assay is to discover to what extent the liposomal contents have access to these cellular processing mechanisms after the liposome has been internalized. Sodium ions are rapidly excreted from the cell by Na^+/K^+ pumps, so any radioactive sodium ions ($^{22}Na^+$) which are released from liposomes into the cytoplasm are rapidly cleared. In contrast, $^{51}Cr^{3+}$ ions remain trapped within the cell regardless of whether or not they have been released from liposomes. Thus measurement of the ratio of the two isotopes retained within the cell will give an indication of the extent to which liposomes have been broken down. If liposomes remain intact inside the cell, the ratio of isotopes will be identical to that for the original liposome suspension. If the liposomes have been broken down, the concentration of sodium will be less than that of chromium.

chromatography, where they elute in the void volume. The double label is prepared in phosphate-buffered saline for incorporation into liposomes, with a Na/Cr activity ratio of 6:1 to allow for preferential loss of sodium from the cells. A similar approach to the above can be employed, using [^{125}I]BSA and ^{51}Cr/EDTA as the double label, to determine 'lysosomal availability', L_{AV}, where conversion of [^{125}I]BSA into excretable units requires prior processing by the lysosomes, as described previously.

3.6 Whole body distribution

The tissue distribution of liposomes throughout the whole body in experimental systems can clearly be determined by measuring the concentration of markers (preferably γ-radiolabelled) in each of the individual organs. However, this has the disadvantage that only one or a few time points are obtained, and in clinical situations this approach is out of the question. Continuous monitoring of liposomal contents can be carried out by viewing the distribution of high energy gamma emitters by scintigraphy under a γ-camera. The most commonly used isotope in nuclear medicine imaging is technetium ^{99m}Tc, which has a half-life of 6 h. In its chelated form, [^{99m}Tc]DTPA, it is very water soluble and can be entrapped inside the aqueous compartment of liposomes by conventional methods. Pre-formed liposomes, on the other hand, can be labelled effectively by exposure to reduced sodium pertechnetate (30), which is very reactive, binds avidly to polyphosphates, and forms a fairly strong attachment (probably by the same mechanism) to phospholipid membranes. Under optimum conditions, the binding is 100% efficient. A protocol for the labelling procedure is given in Appendix I.

Recently, a third labelling method has been reported, in which ^{67}Ga is introduced into pre-formed liposomes which contain the chelating agent desferrioxamine entrapped inside, but not present outside the liposomes. The gallium is presented to the liposomes in the form of a quinolate complex, which is highly lipid-soluble, and passes across the liposome membrane very easily. Once inside the liposome, the gallium is taken up by the desferrioxamine, which competes very effectively with the quinoline for the metal ion, and is held inside the liposome as a water-soluble membrane-impermeable complex (50). A similar procedure may be performed with ^{111}In, where the indium is drawn into chelate containing-liposomes through a membrane into which an ionophore is incorporated (51).

4. RETRIEVAL OF LIPOSOME COMPONENTS

In this last section, methods are given for isolating liposomes, or their breakdown products, from biological systems in order to analyse their distribution in more detail. In the following discussion the *in vivo* case is taken, since this usually presents greater problems, but the methods for the most part will apply equally well to *in vitro* situations also.

4.1 **Isolation from plasma**

At the present time, no method of physical separation is capable of purifying plasma-derived liposomes free of VLDL and LDL. These lipoproteins are about the same size as small liposomes, and overlap in density. Liposomes themselves, after incubation in plasma, may band over a range of different densities as a result of adsorption of plasma proteins on their surface. In practice, contamination with LDL and VLDL is not a great problem, since they are present in plasma at much lower concentrations than HDL (in rat though not in man), and in both species their interaction with liposomes is not as strong. It is worth noting that plasma may be obtained if necessary that is depleted of lipoproteins (52) by treating animals with the adenine anti-metabolite 4-aminopyrazolo-(3,4-*d*) pyrimidine (Sigma Chemical Co.) in 10 mM sodium phosphate (pH 7.0) (60 mg/kg^{-1} day^{-1} for three days).

In the following methods, the aim will be to obtain a liposomal fraction from plasma that is free of any unentrapped liposomal contents which may have been released, and that is separated from HDL, albumin and/or non-lipid associated plasma proteins. It is recommended that plasma be employed rather than serum, and that it should be obtained by treatment of blood with Na$_2$-EDTA (1 mg per 10 ml final volume). Enzyme activity may be inhibited by addition of 0.3 mg of thiomersal per 10 ml. Samples should be kept at room temperature unless otherwise stated.

4.1.1 *Precipitation with manganese salts*

Scherphof and co-workers (53) have developed a method for the quantitative precipitation of VLDL, LDL, and liposomes, leaving HDL and any liposomal lipid transferred to HDL in the supernatant. The method depends on the use of MnCl$_2$ and heparin in the presence of high NaCl concentrations. The high ionic strength is necessary to prevent partial precipitation of HDL.

(i)   Prepare the precipitation solution (500 mM NaCl/215 mM MnCl$_2$) by dissolving 58.4 mg of NaCl, and 85.1 mg of MnCl$_2$.2H$_2$O in 2 ml of distilled water, and to this solution add 1000 Units of concentrated heparin solution (5000 U ml^{-1}).

(ii)   Make 300 μl of plasma (containing 0.1 mg of liposomal phospholipid) up to 0.4 ml with Tris $-$ saline buffer (10 mM Tris/150 mM NaCl). Place the mixture in a polypropylene microcentrifuge tube.

(iii)   With vigorous mixing, add 0.3 ml of precipitation solution to the plasma mixture.

(iv)   Allow to stand at room temperature for 10 min, then spin down at top speed in a microcentrifuge (10 min at 12 000 r.p.m.).

(v)   Separate the supernatant from the pellet and analyse both fractions for markers etc.

The above method is very satisfactory for recovery of the lipid fraction of liposomes, and is a useful method for monitoring lipid exchange between liposomes and HDL. However, it cannot be used in conjunction with fluorescence studies, because of interference by divalent ions. Also, it is possible that at the high concentration of divalent ions used, some leakage of aqueous contents of liposomes may occur. In this case, the protamine method described in Section 3.1.2 of Chapter 3 may give better results.

4.1.2 *Protamine aggregation*

(i)   Dilute 100 μl of the plasma sample 2-fold with buffered normal saline, and add an equal volume of protamine solution (10 mg ml^{-1}) with mixing.

(ii)   Allow to stand at room temperature for 3 min.

(iii)   Add 1.0 ml of phosphate-buffered saline, and spin down in a microcentrifuge at 12 000 r.p.m. for 10 min.

This method may be useful in estimating the extent of leakage of contents from liposomes in plasma.

4.1.3 *Column chromatography*

Bio-Gel A 1.5m (200-4 mesh: Bio-Rad Laboratories) is recommended since it gives the highest recovery of samples. It is supplied fully hydrated.

(i)   Degas the gel (about 75 ml) under a vacuum for 30 min.

(ii)   Fill a column 1.5 cm $\times$ 50 cm with the gel, and wash with at least two volumes of buffer.

(iii)   Pass the sample (maximum volume 1 ml) through, keeping the flow rate at about 5 ml h^{-1} using a peristaltic pump.

(iv)   Run the column overnight at 4$°$C.

Liposomes, LDL and VLDL will elute in the void volume (25 $-$ 30 ml). HDL will elute between 35 ml and 45 ml. Follow the protein elution by UV absorption at 280 nm.

4.1.4 *Ultracentrifugation*

For the reasons mentioned above, differential density centrifugation gives a very poor separation of liposomes from lipoproteins. However, large liposomes can be separated from other plasma proteins by a short (approx. 60 min) centrifugation at 100 000 g at 20$°$C. The liposomes form a pellet, while the plasma proteins stay in suspension since, although they are denser than the liposomes, they are much smaller, and consequently take a much longer time to sediment than the duration of spin used here. Small liposomes cannot be separated at this speed. In this case it is advisable to carry out up to four spins at 100 000 *g* to wash the liposome pellet (in 10 mM Tris−saline) completely free of unassociated, but contaminating protein. The first spin should be at room temperature (or the temperature of the assay). Subsequent spins are performed at 4°C.

Purified liposomes prepared by any of the above methods may be analysed for the presence of associated protein (24), by electrophoresis on polyacrylamide gels according to standard procedures, in the presence of 0.1% Triton X-100 (final concentration) to dissolve the liposomes.

4.2 Tissue fractionation

Before finishing with methods for subcellular fractionation of cells into different organelles, to try to assess the intracellular fate of liposomes, attention will be given to methods for separating the different cell subpopulations of the liver. This is because *in vivo* a large proportion of the liposomes administered to animals end up in this organ, and the same methods can be used for preparation of purified cell populations for use in *in vitro* studies. The procedure described here is a slight modification after Ghosh (54) of the original method of Berry and Friend (55), and involves separation of the different cells by enzymic digestion of connective tissue.

4.2.1 *Materials*

(i) *Animals*

75−200 g male Sprague−Dawley rats.

(ii) *Chemicals and enzymes*

All chemicals should be of analytical grade. Collagenase (type IV for hepatocyte isolation), protease (type XIV, pronase E from *Streptomyces griseus*), heparin 5000 IU ml^{-1} and 3,3′-diaminobenzidine tetrahydrochloride (DAB) can all be obtained from Sigma Chemical Co.

(iii) *Major equipment*

LKB peristaltic pump and plastic tubing (siliconized if necessary) to connect the pump to the plastic hub of the hypodermic needles.
LKB multitemperature control water bath set at 37°C.
Low speed refrigerated bench centrifuge (e.g. Beckman TJ-6).
If available, a temperature-controlled operating table for small animals.

4.2.2 *Solutions for perfusion*

Prepare the buffers in advance and warm them to 37°C in the water bath. As an added refinement, the buffers may be gassed with 95% air−5% CO_2 from a cylinder if desired.

(i) *Buffer A*—perfusion without collagenase

Volume required is about 1000 ml per liver.
    125 mM sodium chloride (7.305 g/l^{-1})

15 mM disodium hydrogen phosphate (2.13 g l^{-1})
2.47 mM potassium dihydrogen phosphate (336.14 mg l^{-1})
5 mM potassium chloride (372.8 mg l^{-1})
5 mM glucose (900 mg l^{-1})

Adjust pH to 7.4 with HCl or NaOH if necessary.

(ii) *Buffer B*—perfusion with collagenase

Volume required is about 200 ml per liver.
Same contents as buffer A, plus:
1.53 mM CaCl$_2$ (169.81 mg l^{-1})
0.05% collagenase (0.5 gl^{-1})

Adjust final pH to 7.4 with 1 M NaOH.

(iii) *Buffer C*—buffer with pronase

Volume required is about 100 ml per liver.
Same contents as buffer A, plus:
1.53 mM CaCl$_2$ (169.81 mg l^{-1})
Pronase (2.5 g l^{-1})

Adjust final pH to 7.4 with 1 M NaOH.

(iv) *DAB stain*

Volume required is about 5 ml per liver. Take care in weighing out the DAB—it is a suspect carcinogen.
0.05 M Tris−HCl (6.057 g l^{-1}) pH 7.6
3,3'-diaminobenzidine tetrahydrochloride (DAB) (2 g l^{-1})
Hydrogen peroxide (0.03%)

4.2.3 *Preparation of the animal for perfusion*

(i) Inject each rat in a tail vein with 0.5 ml (2500 U) of heparin to prevent blood clotting during perfusion. Leave for at least 5 min.

(ii) Rinse out the pump and tubing with buffer A. Fill the tubing with the same buffer and set the peristaltic pump to deliver 10−15 ml min^{-1}.

(iii) Anaesthetize the animals with halothane. Alternatively, in cases where halothane might interfere with the biodistribution or measurement of the drug, inject the animals with nembutal (50 mg kg^{-1} body weight) intraperitoneally 15−20 min before use.

(iv) Shave the abdominal area. Place the anaesthetized animal on a heated operating surface (if available) or on a cork board.

(v) Open the abdomen with a V-shaped cut out of the skin in the middle of the abdomen. Fold the abdominal flap up onto the chest wall. Scoop the gut to the right side of the animal and, if necessary, displace the lobes of the liver very gently upward to permit the identification of the portal vein. Never use forceps to handle the liver, and, in general, keep handling of the liver to a minimum. Make sure that undue pressure is not placed on the diaphragm as this will hinder breathing.

(vi) Using a one inch, 23 gauge hypodermic needle, cannulate the portal vein by gently introducing the needle (needle #1) into and along the vein (pointing towards the liver). Make sure that the needle does not puncture the far side of the vessel. Tie the needle in place with 4-0 braided silk using two or three ligatures (3−4 mm between ties), and again make sure that the ties are not so tight as to cut into the wall of the vein.

(vii) Ligate (two ties) the posterior vena cava below the liver at the level of the renal vein.

(viii) Locate the hepatic artery (runs beside the hepatic portal vein) and ligate it with one tie close to the abdominal aorta.

(ix) Restrain the animal by the limbs so that further movement does not disturb the needle.

(x) Discontinue inhalation anaesthetic and open the chest by cutting the rib cage up each side. Leave the diaphragm and the lowest ribs in place.

(xi) Identify the vena cava where it comes through the diaphragm. Carefully insert another $23 \times 1''$ hypodermic needle (needle #2) into the vena cava between the heart and the diaphragm pointing away from the heart. Tie the needle #2 into place with two or three ligatures of 4/0 cotton.

(xii) Ligate the vena cava (two ties) between the needle and the heart.

(xiii) Connect the tubing of the peristaltic pump from the perfusion medium to needle #1. Take care not to disturb the needle from its position in the hepatic portal vein.

(xiv) Connect needle #2 to the tubing which in turn is connected to a reservoir for waste perfusate.

4.2.4 *Liver perfusion*

(i) Perfuse the liver by allowing the peristaltic pump to pump 50 ml of buffer A at $10-15$ ml min^{-1}. Maintain the temperature of the buffer at 37°C. As the blood is displaced from the liver by the perfusate, the liver will become progressively paler (tan) in colour.

(ii) After 50 ml of perfusion medium has passed, stop the pump and wait for 30 sec. Do not allow air bubbles into the liver. Discard the perfusate and continue perfusion with a further 50 ml of fresh buffer A to ensure that all the blood is removed. Keep the surface of the liver moist with warm buffer A. Discard the perfusate. Repeat these two steps with a further 50 ml of buffer A.

(iii) Now perfuse the liver with 50 ml of buffer B and collect the perfusate into a reservoir. The pH of the perfusate drops to around pH 5.5 after passage through the liver because of the metabolic activity of the cells. Readjust the pH of the collected perfusate with 1 M NaOH to pH 7.4. This perfusate can be reused or preferably saved at 37°C.

(iv) Repeat this step twice more (total 150 ml of buffer B). Keep the liver surface moistened throughout the whole operation.

(v) Remove the liver by cutting round the diaphragm and subsequently cutting the liver free from the remnants of diaphragm. At this stage the liver should be a fragile soft bag of cells indicating a successful perfusion.

4.2.5 *Preparation of parenchymal cells*

(i) Wash the perfused liver surface with a small amount of buffer A. Chop the liver into small pieces and transfer it to the reservoir with the perfusate.

(ii)   Incubate the preparation for 20 min at 37°C. During incubation the pH will again fall to about pH 5.5, readjust the pH to pH 7.4 intermittently. Disperse the cells by stirring very slowly with a magnetic stirrer or with a blunt spatula.

(iii)   Filter the cells through a Nylon mesh (0.5 × 0.3 mm) four times. Transfer the cells which pass through the mesh into 50-ml plastic centrifuge tubes and place them in a refrigerator at 4°C for 5 – 10 min so that the cells sediment by gravity. Cooling the cells too fast (e.g.directly on ice) leads to loss of cell viability in the parenchymal fraction. Note that calcium-free buffer must be used here, since calcium tends to cause macrophages to clump, and would lead to contamination of the sediment with non-parenchymal cells.

(iv)   Remove and save the supernatant from the thick pellet of parenchymal cells. The supernatant should contain only a few parenchymal cells and be rich in non-parenchymal cells [see Section 4.2.6].

(v)   Slowly centrifuge (100 g) the parenchymal cell pellet and wash three times with buffer A. Remove and pool the supernatants with supernatant from Step (iv). Resuspend the cell pellet in 10 ml of buffer A and count the number of cells which exclude 0.25% Trypan blue in buffer A in a haemocytometer.

(vi)   Measure the total protein content of the cells, and assay aliquots of the cells for the liposomal components. Express the results as units per 10^6 cells or total units per total parenchymal cells.

4.2.6 *Preparation of non-parenchymal cells*

(i)   Centrifuge the supernatant from above at 500 g at 4°C to sediment the non-parenchymal cells. Wash with buffer three times. Resuspend the final pellet in buffer C.

(ii)   To remove contaminating hepatocytes, incubate the pellet with pronase (buffer C) at 37°C for 60 min in a slowly shaking water bath.

(iii)   Centrifuge the non-parenchymal cells from the incubation mixture at 500 g for 4 min at 4°C and wash them three times with buffer A.

(iv)   Resuspend the cells in 5 ml of ice-cold buffer A and determine the number of cells which exclude 0.25% Trypan blue in buffer A in a haemocytometer.

(v)   Assay aliquots of the cells as before.

4.2.7 *Histochemical staining of cells*

The large size of parenchymal cells allows them to be distinguished easily from the small non-parenchymal cells. Histochemical staining for peroxidase using DAB stain can be used to further identify and quantitate the cell types (56).

(i)   The large-sized parenchymal cells have big nuclei and stain positive for peroxidase. The approximate yield should be about 6 × 10^7 cells/gram liver.

(ii)   Non-parenchymal cells (approximate yield 2 × 10^7/gram liver) are small and contain two populations:

     (a)   non-staining cells; predominantly endothelial cells (60%) with some fat-storing cells,

     (b)   vacuolated Kupffer cells (20%) and leukocytes (20%).

4.3 **Subcellular fractionation**

The method described below for subcellular fraction by differential centrifugation was developed by de Duve in 1955 for rat liver (57). With more recent improvements this is by far the simplest of the many available methods, and has the advantage that standard centrifuges, rotors and solutions are used. The theory and choice of methodology for subcellular fractionation have been reviewed by Beaufay and Amar-Costesec (58).

Before the methodology is described a word of caution must be given. Differential centrifugation, especially in a homogeneous medium, produces incomplete separation of components. The pellets are inevitably contaminated with smaller materials which were present in or near the pellet zone before sedimentation. Washing of the pellet fractions only partially overcomes this problem. The finding of liposomes or their contents 'associated' with a subcellular fraction should, therefore, be taken as an indication rather than substantial proof of localization. Readers who wish to use more extensive methods for evaluation of individual subcellular fractions should consider using density gradient techniques with vertical rotors. Since lysosomes make up only 1% of liver volume, zonal rotors, which require larger volumes, are usually unsuitable for fractionation of small amounts of tissue or livers from small animals. Relatively large differences exist among the various subcellular organelles in rat liver (average radii are: nuclei 4 μm, mitochondria 0.4 μm, lysosomes 0.1 $-$0.7 μm, microbodies 0.07 μm) but these values may not hold true for other tissues or in liver from other species. In order to confirm that adequate separation has occurred, it is necessary to establish that fractions have become enriched with enzymes known to be exclusively localized within a particular organelle. Once again, the presumption that enzyme localization is the same for all species and tissues cannot be assured. For example, the enzyme G-6-P (see below) is restricted to liver and kidney, and is absent from lung and heart tissue.

4.3.1 *Materials*

The flow diagram in *Figure 11* gives an overall view of the method for separating rat liver homogenates into five fractions: nuclear (N), heavy mitochondrial (M), light mitochondrial (L), small granule (P) and soluble (S). In order to facilitate the use of this method with most centrifuges, the values for the speed and time of centrifugation will be given both in terms of g, mins and time-accumulated angular velocity ($\omega^2 t$, in $rad^2 sec^{-1}$). The latter is preferred as it is available as a programmable parameter in newer centrifuges.

(a) *Homogenization media*

(i)   0.25 M sucrose containing 5 mM Tris$-$HCl, pH 7.4
(ii)   0.25 sucrose containing 5 mM EDTA and 5 mM Tris$-$HCl, pH 7.4.

Keep and use these media at 4°C.

(b) *Animals*

Sprague$-$Dawley rats (100$-$200 g), previously injected with liposomes. Starve the animals for 12$-$18 h before taking the livers to lower the levels of hepatic glycogen.

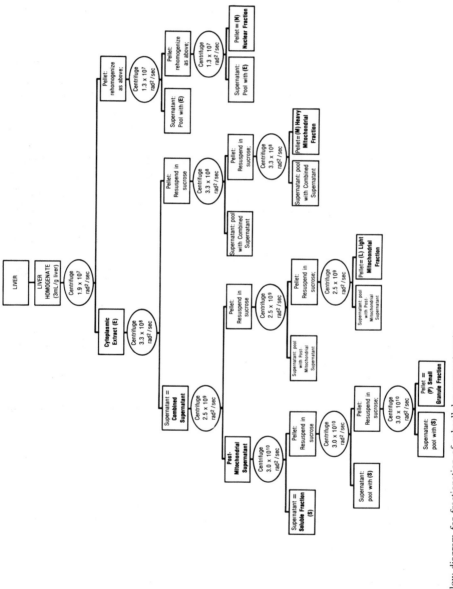

Fig. 11. Flow diagram for fractionation of subcellular compartments.

(c) *Major equipment*

(i)   Potter-Elvehejm homogenizer with a Teflon pestle (clearance 0.15 mm) driven by an overhead electric motor capable of at least 2000 r.p.m.

(ii) Refrigerated benchtop centrifuge (e.g. Beckman TJ-6 centrifuge with refrigeration attachment).

(iii) High speed rotor and a centrifuge capable of at least 100 000 g (e.g. Beckman 50 Ti fixed-angle rotor operating in a Beckman L8-M ultracentrifuge).

4.3.2 *Homogenization of liver*

(i)   Take fresh liver, as free as possible from blood contamination (or preferably liver perfused briefly with the contents of a syringe), and place in a cold, weighed plastic or glass beaker containing 5.0 ml of ice-cold 0.25 M sucrose/5 mM Tris−HCl (pH 7.4). Henceforth carry out all procedures at 4°C.

(ii) Weigh the liver and keep it on ice; quickly cut it into 10−15 small pieces with scissors.

(iii) Transfer the whole liver (or a weighed amount) to the ice-cold glass vessel of the Potter-Elvehjem homogenizer and add more fractionation medium to bring the total concentration to 3 ml sucrose/Tris per gram of liver. Keep the glass outer vessel in a plastic beaker of crushed ice throughout the homogenization.

(iv) Set up the Teflon pestle of the homogenizer to run at between 1200 and 1500 r.p.m. and then force the pestle once past the liver to the bottom of the glass outer vessel. This should take about 15 sec.

(v)   Retain a sample of known size (1−2 ml) of the whole homogenate for future tests.

4.3.3 *Differential centrifugation of liver homogenates*

(a) *Preparation of nuclear (N) fraction*

(i)   Decant the homogenized liver (about 15−30 ml/tube) into a 50-ml conical plastic centrifuge tube and balance with other tubes.

(ii) Centrifuge the homogenate for 6000 g min (1.9 × 10^7 rad^2 sec^{-1}) at 4°C in a benchtop centrifuge.

(iii) Decant the supernatant, without removing any of the fluffy aggregates, and keep on ice.

(iv) Resuspend the pellet, which contains intact cells, cell debris and nuclei, with a further 3 ml of sucrose/Tris per gram of original liver weight. Rehomogenize the pellets as in Step (iv) above.

(v)   Repeat Steps (ii)−(iv) twice.

(vi) The final pellet is N, the nuclear fraction (keep on ice).

(vii) Pool the three supernatants from above (cytoplasmic extract, E), and adjust the volume to be 10 ml of sucrose/Tris per gram of original liver weight. Save a sample of cytoplasmic extract for use in enzyme analyses.

(b) *Preparation of heavy (M) and light (L) mitochondrial fractions*

The remaining centrifugations can be performed at 4°C using a single rotor. In order to save time, the heavy (M) and light (L) mitochondrial fractions can be centrifuged

245

down as a single 'ML' pellet. To do this use the supernatant from the last step above but go directly to Step (viii) below.

(i)   Cool the rotor to 4°C before use and keep it cool between spins.
(ii)   Centrifuge three 11-ml aliquots of the supernatant (1.1 g liver per tube) in capped 16 mm × 76 mm Ultra-clear^{TM} tubes for 3.7×10^4 g min (3.3×10^8 rad^2 sec^{-1}). Weigh and balance pairs of tubes before loading the rotor.
(iii)   Remove and save the supernatant. Resuspend the three dark brown pellets in 11 ml of sucrose/Tris and add this to a single tube (3.3 g/liver/tube).
(iv)   Balance and cap the tubes, centrifuge as in Step (ii).
(v)   Pool supernatants with those from Step (iii). Resuspend the pellet from each tube in 11 ml of sucrose/Tris and repeat Step (iv).
(vi)   Repeat Step (v).
(vii)   Resuspend the final pellet (M) in 3 ml of sucrose/Tris.
(viii) Aliquot three 11-ml portions of each pooled supernatant from Steps (iii)−(v) into 16 × 76 mm tubes. Balance and cap the tubes and centrifuge for 2.4×10^5 g min (2.5×10^9 rad^2 sec^{-1}).
(ix)   Remove the supernatant along with any pinkish material (microsomes) which overlies the darker brown pellets. Save the supernatant and resuspend the three brown pellets from each liver in 11 ml of sucrose/EDTA/Tris in a single tube (2.0 g liver per tube). The presence of EDTA reduces contamination of the pellet with elements of the endoplasmic reticulum.
(x)   Balance and cap the tubes; centrifuge as in Step (viii).
(xi)   Pool the supernatants with those from Step (ix). Resuspend the pellet from each tube in 11 ml of sucrose/EDTA/Tris and repeat Step (x).
(xii)   Repeat Step (xi).
(xiii) Resuspend the final pellet (L) in 3 ml of sucrose/Tris.

(c) *Preparation of small granule fraction (P) and soluble fraction (S)*

(i)   Aliquot three 11-ml portions of the pooled supernatants from Steps (ix)−(xii) above into the same type of 16 × 76 mm tubes. Balance and cap the tubes and centrifuge for 3.54×10^6 g min (3.0×10^{10} rad^2 sec^{-1}).
(ii)   Remove and save the supernatant. Resuspend the three pinkish pellets from each liver in 11 ml of sucrose/Tris in a single tube (1.2 g liver per tube).
(iii)   Balance and cap the tubes and centrifuge as in Step (i).
(iv)   Pool the supernatants (S) with those from Step (ii). Resuspend the pellet from each tube in 11 ml of sucrose/Tris and repeat Step (iii).
(v)   Repeat Step (iv).
(vi)   Resuspend the final pellet (P) in 1 ml of sucrose/Tris.

(d) *Control fractionation*

During the course of the homogenization and fractionation procedures, liposomes in the blood, liver sinusoids, and whole liposomes from within the cell have the chance to mix freely with the cell organelles. In addition, entrapped drugs may leak from liposomes and contaminate the soluble fraction. In order to estimate the amount of contamination occurring during the procedure, carry out a fractionation using a liver from a normal untreated rat which has been spiked with whole liposomes as follows:

(i)   Prepare the liver as described in the very first step [Section 4.3.2].
(ii) Immediately before homogenization of the liver, add to the glass outer vessel an aliquot (0.1−0.25 ml) of the liposomes or other materials administered to the test animals.
(iii) Adjust the final homogenate volume with sucrose to take into account the added aliquot.
(iv) Fractionate the liver as outlined above from Section 4.3.3(a) onwards.

4.3.4 *Tests on subcellular fractions*

(i) *Protein estimation*

Sucrose interferes with some of the standard methods for protein estimation (e.g. Lowry). Use the fluorescamine (Roche) method instead (59) in the presence of 0.1% (final concentration) Triton X-100. Typically, whole rat liver contains 200−250 mg of protein per gram of liver with individual fractions as follows:

(a) Cytoplasmic extract 160−220 mg protein/gram liver
(b) Nuclear   30−50 mg protein/gram liver
(c) Heavy mitochondria 35−50 mg protein/gram liver
(d) Light mitochondria 25−35 mg protein/gram liver
(e) Microsomes   40−55 mg protein/gram liver
(f) Soluble   75−100 mg protein/gram liver

(ii) *Enzyme assays*

For the following assays dilute the subcellular fractions to 1 mg protein ml^{-1} in the same buffer as is used in the enzyme test. In this way later calculations can be made more easily. Each enzyme test normally requires 20−50 μl of material. Run the assays in (at least) duplicate and, if time permits, use more than one volume of sample.

DNA (nuclear marker, N)

Estimate the quantity of DNA using the fluorescence (emission 580 nm, excitation 540 nm) of ethidium bromide (Sigma) in the presence of 0.1% Triton X-100 (60). Use 10−40 μl for N and about 100 μl for M and L; the other fractions should contain none. Construct a standard curve using pure double-stranded DNA (Sigma). Digest the DNA in the samples with deoxyribonuclease 1 (EC 3.1.21.1) (Type IV, Sigma) and reread fluorescence.

Cytochrome oxidase [EC 1.9.3.1] (mitochondrial marker, M)

Measure the activity of this enzyme using reduced horse heart cytochrome *c* (Sigma) (60). The activity can be found from the slope of a log plot of reduced cytochrome *c* concentration (at 550 nm) over a period of 3−4 min following the addition of the enzyme. Larger volumes of the fractions may be required because of low levels of the enzyme in N and P (absent in S).

N-acetyl-β-glucosaminidase [EC 3.2.1.30] (Lysosomal marker, L)

Incubate fractions with *p*-nitrophenyl-2-acetamido-D-glucopyranoside (Koch-Light) in the presence of 0.1 % Triton X-100 in citrate buffer (pH 4.5) (61) but stop the reaction with 0.2 M borate buffer (pH 9.8) which stabilizes the colour. Measure the amount of *p*-nitrophenol liberated in a spectrophotometer at 400 nm against a blank where the enzyme was added after the borate buffer. Construct a standard curve $(0-10 \, \mu g \, ml^{-1})$ of *p*-nitrophenol (BDH) in buffers with Triton X-100.

Glucose 6-phosphatase [EC 3.1.3.9] (microsomal marker, P)

Incubate samples of the fractions with glucose-6-phosphate in the presence of EDTA and 0.1 % Triton X-100 (61). Stop the reaction by adding 8 % trichloroacetic acid. Centrifuge (10^4 g min) and assay the supernatant for inorganic phosphate (62).

(iii) Radiolabelled drugs

If a radiolabelled drug is entrapped and/or radiolabelled liposomes have been used, count samples (up to 100 μl) of each fraction for radioactivity (see Appendix 1). Samples which are highly coloured or very turbid (e.g. the whole liver homogenate, or cytoplasmic extract, E) may have to be decolorized or incubated with solubilizing agents before being assayed for β activity.

Analysis of the quotient 'drug d.p.m./liposome d.p.m.' in the injected material and in the subcellular fractions will show if the liposomes have leaked during their sojourn in the body and if the leakage is greater in one fraction than another.

4.3.5 Analysis of results

Analysis of the subcellular distribution of liposomes and their content has to include a comparison of the activity in each fraction of a number of marker enzymes as well as protein and DNA. These quantities are compared with the quantities found in the whole homogenate [or more usually, the quantities found in the sum of the cytoplasmic extract (E) plus the nuclear fraction (N), since the whole homogenate is often so thick and heterogeneous that errors in sampling are common]. If the recovery of markers is high, more confidence can be placed in the method and on the results. Recoveries are most often too high or too low when the enzyme in question remains partially latent within the fraction due to incomplete lysis with detergent. Low recoveries often indicate that protein denaturation has occurred, that a cofactor has been removed or that other enzymes in a multienzyme reaction are absent. Excess recovery can mean the presence of another enzyme that adds to or further metabolizes the substrate, or that an inhibitor has been removed. To a lesser degree all of these factors could be taking place even within a perfectly recovered experiment.

Table 1. Typical values for constituent markers following differential centrifugation

Marker	Value (/g fresh liver) in reconstituted homogenate (E + N)	Distribution (% of total) in fractions					% recovery
		N	M	L	P	S	
Protein	223.71 mg	15.6	20.3	7.7	19.2	34.8	97.6
DNA	2.14 mg	77.3	7.2	1.1	0.8	4.6	91.0
Cytochrome oxidase	19.4 U	18.5	58.3	12.4	5.7	0	94.8
N-acetyl-β-glucosaminidase	5.6 U	12.9	23.7	44.8	9.3	8.1	98.8
Glucose-6-phosphatase	18.4 U	13.4	7.9	8.3	68.3	5.2	103.2

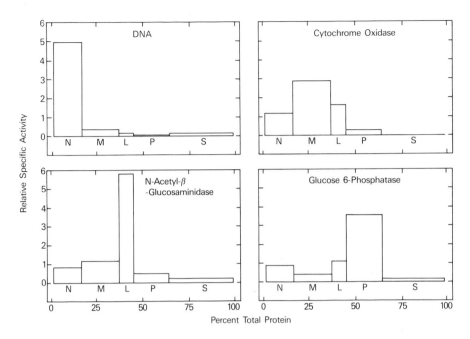

Figure 12. Distribution of enzyme markers within subcellular compartments.

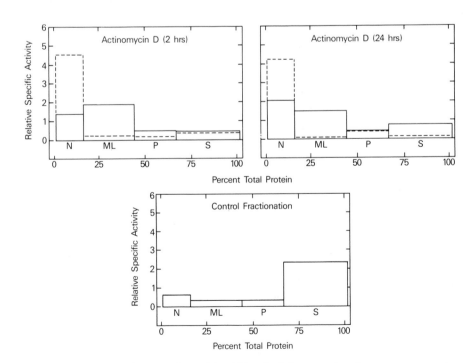

Figure 13. Distribution of entrapped materials within subcellular compartments.

Table 2. Typical values for entrapped and unentrapped drugs following differential centrifugation

Marker	Value (/g fresh liver) in reconstituted homogenate (E + N)	Distribution (% of total) in fractions*				% recovery
		N	M + L	P	S	
Protein	218.5 mg	15.2	27.8	22.6	35.2	100.8
Entrapped [³H]Actinomycin D						
2 h	0.46 µg	22.5	53.3	11.1	16.7	103.6
24 h	0.15 µg	31.2	41.8	9.6	27.2	109.8
Free [³H]Actinomycin D						
2 h	0.14 µg	68.8	5.7	3.9	12.8	91.2
24 h	0.13 µg	64.3	2.1	9.0	4.9	80.4
Control fractionation entrapped bleomycin	0.25 µg	9.9	9.3	7.5	83.2	109.9

*Values taken from (63).

In Table 2 rat livers were fractionated into large granule (M + L) and small granule (P) instead as in Table 1. In order to simplify the table the amount of protein/gm liver is the mean from 7 separate animals. Values for entrapped and free actinomycin D are taken from animals injected intravenously 2 and 24 h previously with 10 µg of drug. Actinomycin D binds to DNA if free in solution or released from liposomes so another liposome preparation (containing ¹²⁵I-labelled bleomycin) is shown to illustrate the position of liposomes in a control fractionation (*Figure 13*). All the liposomes preparations were sonicated MLVs.

The easiest method of presenting subcellular distribution results is to use the block diagram method introduced by de Duve *et al.* (57). The data from *Table 1* (from ref. 63) has been plotted by this method in *Figure 12*. The relative specific activity (RSA) is the ratio of the absolute specific activity of the marker recovered in each fraction (i.e. units per mg protein per g fresh liver) to the absolute specific activity recovered in all the fractions (N + M + L + P + S i.e. Σunits/Σmg protein/g fresh liver). Normally, the diagram is plotted with percentage of total protein in each fraction on the abscissa against the fractional RSA on the ordinate. The area of the fractional block then gives the portion of the marker found in the fraction; the total area is equal to 100%. The figure shows the expected subcellular localization of the various markers. Cytochrome oxidase from mitochondria contaminates the nuclear fraction because this fraction contains cell debris and whole cells in addition to nuclei. This contamination is seen with all the markers. The L fraction is well enriched by N-acetyl-β-glucose-aminidase, but is also contaminated by smaller mitochondria (cytochrome oxidase) and by microsomes (glucose-6-phosphatase). The microsomes (P) are well separated from most of the other granular fractions. The soluble fraction contains little of any of the enzymes suggesting that lysis of organelles during the procedure was low. The amount of lysis (or latency) can be confirmed by measuring the markers in the presence and absence of Triton X-100.

Figure 13 is the block-diagram representation of the data in *Table 2*. The differences in subcellular distribution of actinomycin D entrapped in liposomes (solid lines) and given as a free drug (dotted lines) can be clearly seen. However, the recovery of the free drug at 24 h after injection is not as high as in the other cases (see *Table 2*) so some caution should be employed in interpreting these data. Overall, free (unentrapped) actinomycin D is found in the nucleus with almost no other sites of localization. Therefore the presence of entrapped actinomycin D in the ML fraction can be taken as a strong indication of localization of liposomes in this compartment. After 24 h, the quantity of entrapped drug in the ML has decreased while that of the nuclear fraction has increased, suggesting that actinomycin D is being released intracellularly. The modest increase of entrapped actinomycin D in S over 24 h may reflect the passage of the drug through the cytoplasm on the way to the nucleus, although it should be noted that liposomes themselves tend to be associated with this fraction (see control fractionation).

5. REFERENCES

1. Sendra,A. and Pagano,R.E. (1979) *J. Biol. Chem.*, **254**, 2244.
2. Scherphof,G., Roerdink,F., Waite,M. and Parks,J. (1978) *Biochim. Biophys. Acta*, **542**, 296.
3. Alderson,J.C.E. and Green,C. (1975) *FEBS Lett.*, **52**, 208.
4. Van Renswoude,A.J.B.M. and Hoekstra,D. (1981) *Biochemistry*, **20**, 540.
5. Pagano,R.E. and Takeichi,M. (1977) *J. Cell Biol.*, **74**, 431.
6. Leserman,L.D., Barbet,J., Kourilsky,F. and Weinstein,J.N. (1981) *Nature*, **288**, 602.
7. Felgner,P.L., Gadek,T.R., Holm,M., Roman,R., Chan,H.W., Wenz,M., Northrop,J.P., Ringold,G.M. and Danielson,M. (1987) *Proc. Natl. Acad. Sci. USA*, **84**, 7413.
8. Lüllmann-Rauch,R. (1979) In *Lysosomes in Applied Biology and Therapeutics*, Dingle,J.T., Jacques,P.J. and Shaw,I.H. (eds), North Holland Publishing Co, Amsterdam, Volume 6, p .49.
9. Philippot,J.R.—personal communication.
10. Black,C.D.V., Watson,G.J. and Ward,R.J. (1977) *Trans. Roy. Soc. Trop. Med. Hyg.*, **76**, 550.
11. New,R.R.C., Chance,M.L., Thomas,S.C. and Peters,W. (1978) *Nature*, **272**, 55.
12. Alving,C.R., Steck,E.A., Hanson,W.L., Loizeaus,P.S., Chapman,W.J.,Jr and Waits,V.B. (1978) *Life Sci.*, **22**, 1021.
13. New,R.R.C. and Chance,M.L. (1980) *Acta Trop.*, **37**, 253.

14. New,R.R.C., Chance,M.L. and Heath,S. (1981) *J. Antimicrob. Chemother.*, **8**, 371.
15. Graybill,J.R., Craven,P.C., Taylor,R.L., Williams,D.M. and Magee,W.E. (1982) *J. Inf. Dis.*, **145**, 748.
16. Scherphof,G.L., Kuipers,F., Derksen,J.T., Spanjer,H.H. and Vonk,R.J. (1987) *Biochem. Soc. Trans.*, Suppl., 62S.
17. Spanjer,H.H. and Scherphof,G. (1983) *Biochim. Biophys. Acta*, **734**, 40.
18. Lazar,G., van Galen,M. and Scherphof,G.L. (1989) *Biochim. Biophys. Acta*, **1011**, 97.
19. Cudd,A. and Nicolau,C. (1985) *Biochim. Biophys. Acta*, **845**, 477.
20. Fidler,I.J., Raz,A., Fogler,W.E., Kirsh,R., Bugelski,P. and Poste,G. (1980) *Cancer Res.*, **40**, 4460.
21. Dancey,G.F., Yasuda,T. and Kinsky,S.C. (1978) *J. Immunol.*, **120**, 1109.
22. Black,C.D.V. and Gregoriadis,G. (1976) *Biochem. Soc. Trans.*, **4**, 253.
23. Juliano,R.L. and Lin,G. (1980) In Liposomes and Immunobiology, Tom,B. and Six,H. (eds), Elsevier, North Holland, New York, p. 49.
24. Bonté,F., Hsu,M.J., Papp,A., Wu,K., Regan,S.L. and Juliano,R.L. (1987) *Biochim. Biophys. Acta*, **900**, 1.
25. Stirk,A.H. and Baldeschwieler,J.D. (1986) In *Medical Applications of Liposomes*, Yagi,K. (ed.), Japan Scientific Societies Press, Tokyo/Karger, Basel, p. 31.
26. Allen,T.M. and Chonn,A. (1987) *FEBS Lett.*, **223**, 42.
27. Gregoriadis,G. and Senior,J. (1980) *FEBS Lett.*, **119**, 43.
28. Yagi,K. and Naoi,M. (1986) In *Medical Applications of Liposomes*, Yagi,K. (ed.), Japan Scientific Societies Press, Tokyo/Kargar Press, Basel, p. 91.
29. Segal,A., Gregoriadis,G. and Black,C.D.V. (1975) *Clin. Sci. and Mol. Med.*, **49**, 99.
30. Osborne,M.P., Richardson,V.J., Jeyasingh,K. and Ryman,B.E. (1979) *Int. J. Nucl. Med. Biol.*, **6**, 75.
31. New,R.R.C., Chance,M.L. and Heath,S. (1983) *Biol. Cell.*, **47**, 59.
32. Weinstein,J.N., Magin,R.L., Cysyk,R.L. and Zaharko,D.S. (1980) *Cancer Res.*, **40**, 1388.
33. Weinstein,J.N., Klausner,R.D., Innerarity,T., Ralston,E. and Blumenthal,R. (1981) *Biochim. Biophys. Acta*, **647**, 270.
34. Klausner,R.D., Kumar,N., Weinstein,J.N., Blumenthal,R. and Flavin,M. (1981) *J. Biol. Chem.*, **256**, 5879.
35. Ho,R.J., Rouse,B.T. and Huang,L. (1986) *Biochem.*, **25**, 5500.
36. Bentz,J., Ellens,H. and Szoka,F.C. (1987) *Biochem.*, **26**, 2105.
37. Wang,C.Y. and Huang,L. (1987) *Proc. Natl. Acad. Sci. USA*, **84**, 7851.
38. Huang,L. (1983) In *Liposomes*, Ostro,M.J. (ed.), Marcel Dekker Inc New York and Basel, p. 87.
39. Stein,Y., Halperin,G. and Stein,O. (1980) *FEBS Lett.*, **111**, 104.
40. Struck,D.K. and Pagano,R.E. (1980) *J. Biol. Chem.*, **255**, 5404.
41. Magee,W.E., Goff,C.W., Schoknecht,J., Smith,M.D. and Cherian,K. (1974) *J. Cell Biol.*, **74**, 531.
42. Weissman,G., Cohen,C. and Hoffstein,S. (1977) *Biochim. Biophys. Acta*, **498**, 375.
43. Hong,K., Friend,D.S., Glabe,C.G. and Papahadjopoulos,D. (1983) *Biochim. Biophys. Acta*, **732**, 320.
44. Heath,S., Chance,M.L. and New,R.R.C. (1984) *Molec. and Biochem. Parasitol.*, **12**, 49.
45. Ho,S.C. and Huang,L. (1983) *J. Histochem. Cytochem.*, **31**, 404.
46. Wu,P., Tin,G.W. and Baldeschwieler,J.D. (1981) *Proc. Natl. Acad. Sci. USA*, **87**, 2033.
47. Hwang,K.J. and Mauk,M.R (1977) *Proc. Natl. Acad. Sci. USA*, **77**, 4430.
48. Navon,G., Panigel,R. and Valensin,G. (1986) *Magn. Reson. Med.*, **3**, 876.
49. Szoka,F.C. (1986) In *Medical Applications of Liposomes*, Yagi,K. (ed.), Japan Scientific Societies Press, Tokyo/Karger Press, Basel, p. 21.
50. Gabizon,A., Huberty,J., Straubinger,R.M., Price,D.C. and Papahadjopoulos,D. (1988) *J. Liposome Res.*, **1**, 123–35.
51. Mauk,M.R. and Gamble,R.C. (1979) *Anal. Chem.*, **94**, 302.
52. van't Hooft,F.M. and van Tol,A. (1986) *Biochim. Biophys. Acta*, **876**, 333.
53. Damen,J., Regts,J. and Scherphof,G. (1982) *Biochim. Biophys. Acta*, **712**, 444.
54. Ghosh,P., Kantidas,P. and Bachhawat,B.K. (1982) *Arch. Biochem. Biophys.*, **213**, 266.
55. Berry,M.N. and Friends,D.S. (1969) *J. Cell. Biol.*, **43**, 506.
56. Rahman,Y.E., Cerny,E.A., Patel,K.R., Lau,E.H. and Wright,B.J. (1982) *Life Sci.*, **31**, 2061.
57. de Duve,C., Pressman,B.C., Gianetto,R., Wattiaux, and Appelmans,F. (1955) *Biochem. J.*, **60**, 604.
58. Beaufay,H. and Amar-Costesec,A. (1976) In *Methods in Membrane Biology*, Korn,E.D. (ed.), Plenum Press Inc, New York and London, Vol. 6, p. 1.
59. Böhlen,P., Stein,S., Dairman,W. and Udenfriend,S. (1973) *Arch. Biochem. Biophys.*, **155**, 213.
60. Blackburn,M.J., Andrews,T.M. and Watts,R.W.E. (1973) *Anal. Biochem.*, **51**, 1.
61. Beaufay,H., Amar-Costesec,A., Feytmans,E., Thinès-Sempoux,D., Wibo,M., Robbi,M. and Berthet,J. (1974) *J. Cell. Biol.*, **61**, 188.
62. Fiske,C.H. and Subbarrow,Y. (1925) *J. Biol. Chem.*, **66**, 375.
63. Black,C.D.V. (1983) PhD Thesis, London University.

Miscellaneous methods

1. PURIFICATION OF SOLVENTS

Organic solvents can be freed from traces of water by treatment with molecular sieves such as those obtainable from BDH (UK). These consist of various types of porous clay (aluminosilicates) which contain cavities permitting only very small molecules to enter. In addition, their exposed surface has a very high affinity for water molecules, and the material can absorb between 10 and 20% of its own weight of water. To prepare bone dry methanol, chloroform etc., add about 10 g of Molecular Sieve pellets, Type 3A, to 100 ml of solvent and allow it to stand with occasional mixing for an hour, then decant before use. It is often convenient to store solvents over Molecular Sieve pellets ready for use.

Build-up of phosgene in commercial supplies of chlorinated solvents (chloroform or dichloromethane) can be usually suppressed by means of prior addition of 1% ethanol or methanol. Aldehydes may be removed from alcohols by distillation over potassium hydroxide pellets, and peroxides can be removed by extraction with acid ferrous sulphate solution (2 g $FeSO_4$ in 100 ml of 0.5 M H_2SO_4), followed by a further wash with distilled water, drying over calcium chloride, and then distillation.

Other solvents commonly used—acetone, toluene, and ethyl acetate require no special purification procedure other than redistillation after a few months. For specialized applications, and for detailed descriptions of methodology, see Perrin,D.D., Armarego,W.L.F. and Perrin,D.R. (1980) *Purification of Laboratory Chemicals*, 2nd edn, Pergamon Press, Oxford, and Riddick,J.A. and Bunger,W.B. (1970) *Organic Solvents: Physical Properties and Methods of Purification*. 3rd edn, Wiley Interscience, NY.

2. PURIFICATION OF EGG YOLK PHOSPHATIDYL CHOLINE

This procedure is based on methods published by Pangborn,M.C. (1980) *J. Biol. Chem.*, **471**, 476, and Singleton,W.S., Gray,M.S, Brown,M.L. and White,J.L. (1965) *J. Am. Oil. Chem. Soc.*, **42**, 53.

2.1 Preparation of crude lecithin

(i)   Separate the yolks of 12 fresh hens eggs free from the whites, and collect in a 2-litre glass beaker.

(ii) Add 50 ml of chloroform:methanol (2:1 v/v), and homogenize for 30 sec using a Polytron homogenizer or other type of blender.

(iii) Add an extra 250 ml of chloroform followed by 250 ml of distilled water, and homogenize again.

(iv) Centrifuge the mixture in a preparative centrifuge at 3000 g for 15 min at 4°C.

(v)   Discard the upper aqueous layer, and remove the chloroform through a hole in the interfacial protein layer.

(vi)   Clarify the organic solution by passing through a sintered glass funnel.

(vii)   Dry down the chloroform solution in a rotary evaporator, and redissolve the yellow oil in 800 ml of ethanol. Remove any undissolved material by centrifugation as in Step (iv).

(viii) To the ethanolic solution, add 30 ml of 50% (w/v) aqueous cadmium chloride solution.

(ix)   Allow to stand for 60 min at 4°C.

(x)   Separate the precipitate from solution by centrifugation as in Step (iv).

(xi)   Redissolve the pellet in 100 ml of chloroform, then add the lipid solution to 700 ml of ethanol to which 5 ml of 50% (w/v) aqueous $CdCl_2$ solution has been added.

(xii)   Allow to stand for 60 min at 4°C as before, then separate the precipitate by centrifugation as in Step (iv).

(xiii) Repeat Steps (xi) and (xii) until the precipitate is white and the supernatant is colourless.

(xiv)   Redissolve the final precipitate in 150 ml of chloroform.

(xv)   Shake the chloroform solution well with 150 ml of 30% (v/v) ethanol in water in a separating tunnel.

(xvi)   Remove the lower chloroform layer and discard the upper aqueous layer.

(xvii) Repeat Steps (xv) and (xvi) until $CdCl_2$ is shown to be completely removed from the system by absence of a precipitate when one drop of silver nitrate is added to 1 ml of the upper phase.

(xviii) Evaporate the solution to dryness on a rotary evaporator, in a pre-weighed glass Quickfit round-bottomed flask. Weigh the residual lipid, then redissolve in chloroform to give a final concentration of 50 mg ml^{-1}.

2.2 Purification of crude lecithin

(i)   Suspend 125 g of neutral alumina, activity grade I (Woelm Pharma), in 150 ml of chloroform and pour the slurry into a glass column (60 cm $\times$ 5 cm, fitted with a stopcock) plugged at the bottom with glass wool.

(ii)   After the bed of the column has settled, and excess chloroform has been removed, pour 100 ml of fresh chloroform onto the column, without disturbing the bed, and allow it to run through at a flow rate of about 10 ml min^{-1}.

(iii)   After the chloroform has passed through, introduce 100 ml of the 5% solution of crude lecithin onto the column. Run this into the column, and wash in with an extra 80 ml of chloroform.

(iv)   Slowly run in a litre of chloroform:methanol (9:1) into the column, and at the same time take off the first 100 ml eluting from the column and discard.

(v)   Collect 25-ml fractions and retain.

(vi)   Assay each fraction by TLC, to ascertain which contains pure lecithin, then pool and take to dryness on a rotary evaporator.

(vii)   Redissolve in chloroform to give the desired concentration (between $20-100$ mg ml^{-1} for a stock solution).

3. RECRYSTALLIZATION OF CHOLESTEROL

(i)   Fold a piece of 20-cm filter paper (e.g. Whatman No. 1) into flutes and place in a glass filter funnel.

(ii)   Dissolve 20 g of cholesterol in 100 ml of hot methanol in a 250-ml conical flask in a water bath at a temperature of $60-70°C$. While the solid is dissolving, place the filter funnel with the fluted paper in the neck of the flask, so that the funnel is heated up to the same temperature as the solution.

(iii) When all the cholesterol is almost all in solution, filter it rapidly through the prepared funnel into a fresh conical flask.

(iv)   Allow it to stand overnight at room temperature in order to crystallize.

(v)   Remove the methanol by filtration through a Buchner funnel.

(vi)   Weigh the dry solid and analyse for purity by GLC. If necessary, repeat the whole procedure several times until impurities are no longer present.

4. PURIFICATION OF CARBOXYFLUORESCEIN

Method adapted from Weinstein,J.N., *et al.* (Chapter 3, ref. 16) and Lelkes, P.

4.1 Recrystallization (for removal of polar contaminants)

(i)   Dissolve 35 g of carboxyfluorescein (acid form from Eastman Kodak) in 200 ml of ethanol in a conical flask.

(ii)   Add 2 g of activated charcoal, and boil for several minutes in a water bath.

(iii) Filter the mixture through a Buchner funnel, and collect the filtrate in a one-litre conical flask.

(iv)   Add cold distilled water slowly with stirring, until the solution is no longer clear (about 400 ml will be required).

(v)   Cool slowly to $+4°C$, then leave overnight in a freezer at $-20°C$.

(vi)   The following day, collect the precipitated carboxyfluorescein on a Buchner funnel (Whatman filter paper No. 50) and wash thoroughly with ice-cold distilled water.

(vii) Dry the solid carboxyfluorescein in the funnel, then in a desiccator.

4.2 Column Chromatography (for removal of hydrophobic contaminants)

(i)   Place 30 g of carboxyfluorescein in a 100 ml beaker, and add 40 ml of 6 M NaOH. With stirring, bring the pH to pH 7.5 by slow dropwise addition of 6 M HCl. Take care not to overshoot. Warm the solution to $50°C$ if necessary to achieve complete dissolution. Concentration of the resultant solution will be approximately 2 M.

(ii)   Apply 10 ml of concentrated CF solution to the top of a Sephadex LH20 hydrophobic column (2.5 × 40 cm).

(iii) Elute the sample with 10 mM Tris−HCl buffer (pH 7.5) and collect it in 2-ml fractions.

(iv)   Monitor the purity of fractions by silica gel TLC, using a solvent system composed of $CHCl_3:CH_3OH:H_2O$ in the ratio 65:25:4 (by vol).

(v)   Pool the fractions containing pure CF, and adjust to the desired concentration in Tris−HCl buffer. Store in the dark at 4°C until required for use. (Stable for many months.)

The first peak to elute off the column is a non-fluorescent polar contaminant, closely followed by carboxyfluorescein itself. Hydrophobic impurities are retained on the top of the column as a brownish non-fluorescent residue. The most suitable concentration for use in standard liposome experiments is that which is iso-osmolar with physiological buffers (275 mOsm), which works out at about 100 mM carboxyfluorescein in 10 mM Tris−HCl buffer. The concentration of the solution eluted off the column may be determined by measuring its optical density at 492 nm and relating that to its molar extinction coefficient at 492 nm, which is around 75 000 M^{-1} cm^{-1}. An alternative method is to measure the osmolarity directly using an osmometer, and then adjust the solution to 275 mOsm by addition of 10 mM Tris−HCl Buffer.

Several workers have reported that good purification can be achieved using the column chromatography method alone, and that the recrystallization step can be omitted. It should be noted, however, that even after purification, most commercially-derived carboxyfluorescein is a mixture of two isomers—5-carboxy-and 6-carboxyfluorescein. Preparations using mixtures of both these isomers give good results in the types of application described here. The same basic methods as described above may be used for the purification of calcein.

5. LIPOSOME EXTRACTION METHODS

5.1 Bligh-Dyer two-phase extraction

(See *Figure 1*).

(a) *Reagent preparation*

Upper Phase

Mix together: 20 ml of methanol, 20 ml of chloroform and 20 ml of saline. Allow to stand until the two layers separate completely from each other. Remove the top layer and retain for further use.

(b) *Procedure*

(i)   Add 0.1 ml of the liposome sample to 0.9 ml of saline in a 10 ml centrifuge

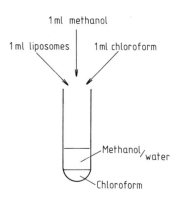

Figure 1. Bligh-Dyer two-phase extraction.

tube. Add 1 ml of methanol, mix well, then add 1 ml of chloroform to the sample. Vortex for 10 sec.

(ii) Centrifuge the tubes at 700 *g* for 5 min at 4°C to separate the two phases.

(iii) Remove all the lower layer using a glass syringe with a long flat-ended needle, and transfer it to a clean tube.

(iv) Add 1 ml of upper phase, vortex and spin as before.

(v) Repeat Steps (iii) and (iv) for a total of up to four times, retaining the lower phase each time.

Note: Perform all manipulations at 4°C. Place sample tubes in an ice bucket while being stored. The final lower phase contains the total lipid of the sample, separated from water-soluble liposomal contents, and may be used as the starting material for further lipid assays. In cases where there is a strong ionic interaction between aqueous solutes and charged lipids, which makes separation of the components difficult, acid or alkali (0.1 M) may be used instead of saline.

5.2 Sep-Pak minicolumn extraction

(See *Figure 2*).

(a) *Reagent preparation*

Aqueous elution medium:
Mix 40 ml of methanol with 20 ml of 0.1 N HCl.

(b) *Procedure*

Separation of entrapped solute from lipids:

(i) Draw up 0.1 ml of liposomes into a 1 ml disposable plastic syringe.

(ii) Fix a syringe on to the inlet port of a Waters Sep-Pak C18 minicolumn, and introduce the liposomes on to the minicolumn. Make sure all the material is on the column.

(iii) Replace the empty 1 ml syringe with a 10-ml glass syringe containing 10 ml of the aqueous elution medium, and flush the column through with the contents of the syringe. Keep the syringe and column pointing vertically downwards each time elution is performed. Collect aqueous eluate in beaker. Lipids will remain on the column.

Separation of cholesterol and α-tocopherol from phospholipids:

(iv) Attach a 10 ml glass syringe containing chloroform to the inlet port of the minicolumn. Allow 0.5 ml of chloroform to enter the column, and discard the

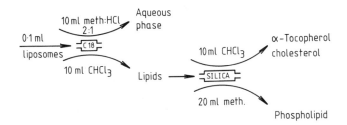

Figure 2. Sep-Pak separation of lipids.

eluate. Then fix the outlet port of the C18 column to the inlet port of a silica gel Sep-Pak minicolumn using a short male-to-male luer adaptor. Flush both columns with the remainder of the chloroform in the syringe. Disconnect the two columns, and pass a further 10 ml of chloroform through the second column. Retain the eluate, which will contain cholesterol and α-tocopherol, while phospholipids will remain on the silica gel column.

(v)   Remove phospholipids from the silica column by flushing with 20 ml of methanol.

Note: Acid or alkali may be employed in the aqueous elution medium in cases where separation is difficult because of ionic interactions between solute and lipids.

5.3 Ultrasonic disruption

The following method, taken from Bakouche,O. and Gerlier,G. (1983) *Anal. Biochem.*, **130**, 379, permits the separation of the aqueous compartment from the surrounding membrane without complete disintegration of the membrane, so that an indication can be obtained of the extent to which proteins and other molecules partition between the two compartments in liposomes. Exposure to ultrasonic sound below the T_c converts sealed vesicles into bilayer sheets devoid of any enclosed volume.

(i)   Adjust the dilution of the MLV suspension so that the phospholipid concentration is approximately 5 mg ml^{-1}.

(ii)   Cool the suspension to a temperature at least 10°C below the transition temperature of the lowest melting membrane lipid.

(iii)   Expose the suspension to six bursts of probe sonication of 30 sec duration, with 30 sec rest between each burst.

(iv)   Separate the disrupted liposomal membranes from released aqueous compartment by ultracentrifugation for 5 min at 132 000 g at the same temperature as for sonication. Alternatively, if working with SUVs, separate the lipid from the aqueous medium by column centrifugation.

To bring egg lecithin liposomes to below their T_c (around -10°C) without freezing, add to 1 ml of liposomes (phospholipid 12.5 mg ml^{-1}) 1.5 ml of ethylene glycol, mix well, and then cool in an isopropanol-dry ice bath.

6. PREPARATION OF SENDAI VIRUS

(i)   Prepare Na solution consisting of:
         100 mM NaCl 6 g l^{-1}
         50 mM Tricine 8.96 g l^{-1}
         adjusted to pH 7.4.

(ii)   Prepare a suspension of filter-sterilized Sendai virus particles diluted in Na solution to a concentration of 10 haemagglutinating units (HAU) per ml.

(iii)   Obtain 10-day-old fertilized large hens eggs incubated at 37°C in a rocking incubator.

(iv)   Locate the air sac by candelling the eggs, and mark the rim with a pencil. At this stage, any eggs which are non-fertile can be rejected.

(v)   Drill a small hole in the shell over the air space, about $2-3$ mm from the rim of the air sac, using a high speed dentist's drill, or a model maker's drill.

(vi)   Inoculate 0.2 ml of diluted Sendai virus through the drilled hole into the chorioallantoic space of each egg using a 1 ml sterile disposable syringe, fitted with a 25 gauge needle.

(vii)   Reseal the hole with nail varnish, or hot wax, and place the eggs in the incubator for a further three days. The eggs are rocked gently, or may be turned frequently by hand.

(viii)   At the end of three days, place the eggs in a cold room for 1 h at 4°C to kill the embryos.

(ix)   Remove the shell over the air sac, and aspirate the chorioallantoic fluid (about $8-10$ ml per egg) containing approx 1500 HAU ml^{-1}. Eggs showing obvious bacterial contamination (cloudiness of fluid) should be discarded.

(x)   Pool the fluid from all the eggs and centrifuge at 1400 g for 10 min at 4°C in a bench centrifuge to remove cell debris and erythrocytes. Collect the supernatant.

(xi)   Centrifuge the supernatant at 37 000 g for 60 min at 4°C, and collect the pellet.

(xii)   Resuspend the pellet in solution Na, homogenize and centrifuge at 6000 g for 5 min to remove large clumps of virus particles.

(xiii)   Load the supernatant onto a 15% (w/w) sucrose cushion, centrifuge it for 10 min at 100 000 g and collect the pellet.

(xiv)   Resuspend the pellet in solution 'Na' to give a concentration of about 1 mg ml^{-1} $(1 \times 10^4$ HAU $ml^{-1})$. Approximately 1×10^5 HAU of Sendai virions are obtained from 100 ml of allantoic fluid.

6.1 Measurement of haemagglutinating activity

(i)   Prepare High K medium as follows:
         140 mM KCl 10.7 g 1^{-1}
         5.4 mM NaCl 0.324 g 1^{-1}
         20 mM Tricine 2.8 g 1^{-1}
         adjusted to pH 7.8 with KOH.

(ii)   Wash a suspension of erythrocytes (either chicken, human or sheep) at least three times, and prepare a 0.5% v/v suspension in 'High K' medium.

(iii)   To one row of wells in a round-bottomed microtitre plate add 50 μl of 'High K' medium.

(iv)   To well No. 1 add Sendai virus suspension (diluted to 50 μl in High K medium) and perform doubling dilutions along one row. (Leave one well free of virus suspension as a negative control.)

(v)   To each row add 50 μl of the RBC suspension, mix well and place in a cold room at 4°C (to prevent haemolysis) with as little disturbance as possible.

(vi)   Read the plates after $1-2$ h. The last well showing complete haemagglutination (diffuse carpet) is taken to contain one haemagglutinating unit.

7. HYDROGENATION OF LECITHIN

Method contributed by D.M..Phillips.

   Lecithin in methanol is stirred under hydrogen in the presence of a palladium catalyst at room temperature and atmospheric pressure. The method describes the hydrogenation of 60 g of lecithin, but it can be easily scaled down.

(a) *Materials*

60 g of egg yolk lecithin ($\sim$78 mmol) in chloroform solution. 1 litre of redistilled methanol.

Catalyst: 1.5 g of 5% palladium on carbon (this is equivalent to $\sim$3 g of damp material as purchased).

(b) *Apparatus*

1 litre round-bottomed flask with three necks. (This will be used as the reaction flask.)
Three-way stopcock.
1-litre measuring cylinder for use as a gas holder.
500-ml separating funnel to fit the centre neck of the reaction flask.
Magnetic stirrer and 2-cm follower.
Gas inlet tube and exit stopcock for reaction flask.

It is important to remove most of the chloroform from the lecithin, as it is inhibitory, and to have efficient stirring. The progress of the reaction is followed by plotting the hydrogen uptake with time.

(i)   Set up the reaction flask and stirrer bar over the magnetic stirrer. Fit the hydrogen inlet to one side-neck and a stopcock (outlet) to the other side.

(ii) Dry down the lecithin solution by rotary evaporation with the bath at 35°C, and repeat the process after dissolving in methanol. Finally, dissolve the lecithin in about 400 ml of methanol and transfer it to the separating funnel.

(iii) Weigh out the catalyst, stir up with 50 ml of methanol and centrifuge it. Discard the supernatant and transfer the catalyst in more methanol to the reaction flask; place the separating funnel containing the lecithin in the centre neck.

(iv) Fit the gas holder (graduated cylinder) with two tubes: one reaching to the bottom and connected at its top to a 2-litre reservoir of water on an adjustable stand. The other tube is short and is connected to the 3-way stopcock, whose other arms go to the hydrogen or nitrogen supply and to one side-neck of the reaction flask.

(v)   Flush the whole system with nitrogen three times to remove air, using the water-levelling reservoir to drive the gas out of the measuring cylinder. Repeat this procedure twice with hydrogen; close the outlet stopcock and leave the system full of hydrogen. Run the lecithin solution into the flask. Adjust the water levels in the reservoir and cylinder so that they are equal; read the water level in the cylinder and start the stirrer.

(vi) Ensure that the stirrer is producing a marked vortex in the black liquid, otherwise the gas uptake will be very slow. As the liquid thickens with the progression of the reaction, run in two lots of 100 ml of methanol via the separating funnel after 3 h and 6 h of reaction time so that stirring can continue vigorously. As gas is taken up, the water level in the gas holder rises. Read the volume of gas consumed after making the inner and outer water levels equal. Then refill the gas holder from the hydrogen supply via the three-way stopcock. Altogether about 4 litres of hydrogen will be absorbed and the reaction may take 10 h before the rate of uptake falls to zero.

(vii) When the gas uptake virtually ceases, add 400 ml of chloroform to the reactants and filter through Whatman GF/F glass-fibre filter paper. Recover the product from the clear colourless filtrate by rotary evaporation, adding more chloroform repeatedly during the process to prevent the relatively insoluble hydrogenated lecithin from coming out of solution prematurely as a large lump.

(viii) The hydrogenated lecithin is tested in the usual way by TLC in chloroform−methanol−water (65:25:4 by vol.) for the presence of any artefacts such as lysolecithin. Gas chromatography should show the absence of oleic and linoleic acids.

8. PREPARATION OF N-HYDROXYSUCCINIMIDE ESTERS OF FATTY ACIDS

Method taken from Lapidot,Y., Rappoport,S. and Wolman,Y. (1967) *J. Lipid Res.*, **8**, 142.

(i)   Prepare a 230 mM solution of N-hydroxysuccinimide by dissolving 3.45 g (30 mmol) in 30 ml of ethyl acetate dried over Molecular Sieve pellets in a stoppered 250-ml glass conical flask.

(ii)   To the above solution add 30 mmol of the desired fatty acid.

(iii)   Prepare a solution containing 30 mmol (6.18 g) of dicyclohexyl carbodiimide in 10 ml of ethyl acetate, and add it to the solution of fatty acid.

(iv)   Allow the reaction to proceed overnight at room temperature.

(v)   Remove the precipitated dicyclohexyl urea by filtration using a suction tap.

(vi)   Evaporate the filtrate to dryness on a rotary evaporator, and purify it by recrystallization from ethanol (see method for cholesterol above).

(vii)   Check the purity by TLC, using solvent systems:
        (a) Chloroform, (b) petroleum (b.p. 40−60°C)−diethyl ether, 8:2.

Stain for N-hydroxysuccinimide and ester (red colour) by spraying with 10% hydroxylamine in 0.1 M NaOH, followed after 2 min by 5% $FeCl_3$ in 1.2 M HCl. Yields of 80−90% can be obtained.

9. PREPARATION OF CHOLESTEROL ALKYL ETHERS

Method taken from Halperin,G. and Gatt,S. (1980) *Steroids,* p. 39.

(i)   Introduce 200 mg of cholesterol *p*-toluene sulphonate and 500 mg of fatty alcohol into a 15-ml screw-capped test tube, flush with nitrogen and screw the cap on tightly.

(ii)   Heat at 110°C for 150 min in an oven or a heating block.

(iii)   Allow to cool, then add 0.5 g of $NaHCO_3$ and 10 ml of hexane.

(iv)   Mix by vortexing, and clarify by centrifugation.

(v)   Draw off the upper hexane phase and retain.

(vi)   Extract the residue with three further 10-ml portions of hexane, and pool all four portions.

(vii)   Make up a silicic acid column containing 15 g of silicic acid in hexane.

(viii) Apply all the solution containing the reaction products on to the column, and after this has run in, elute with each of the following:
        (a) 50 ml of hexane,
        (b) 50 ml of 5% benzene in hexane,
        (c) 50 ml of 10% benzene in hexane,

(d) 100 ml of 125% benzene in hexane,

(e) 50 ml of chloroform,

(f) 50 ml of methanol.

Toluene may be used as a substitute for benzene if necessary.

(ix) Collect 20-ml fractions and analyse by TLC (petroleum ether:ether, 98:2).

Most of the cholesterol ether will come off at the beginning of the 15% benzene/hexane fraction. Recrystallize from acetone (cooled to $-40°C$ before filtration).

Radiolabel may be incorporated into the final product by using labelled starting reagents. A method for production of cholesterol *p*-toluene sulphonate from (labelled) cholesterol is given below.

(i) To 15 mg of dry cholesterol (spiked with radiolabelled cholesterol) add 1 ml of dry pyridine and 0.3 g of *p*-toluene sulphonyl chloride in a glass, screw-capped 15-ml tube.

(ii) Screw the cap on tightly and incubate at 37°C for 24 h.

(iii) Add 0.5 g of ice, wait for 30 min, then add 5 ml of hexane, vortex, centrifuge at 1000 *g* for 10 min at 4°C and remove the upper layer.

(iv) Wash the organic solution by mixing with an equal volume of water.

(v) After the two phases have been separated (by centrifugation if necessary) remove the upper organic layer and wash again, this time with a saturated solution of sodium bicarbonate.

(vi) Dry over solid Na_2SO_4 for $1-2$ h, decant off the supernatant, and evaporate this solution to dryness under a stream of nitrogen.

10. PREPARATION OF SAMPLES FOR LIQUID SCINTILLATION COUNTING
(Contributed by C.D.V. Black)

Small volumes of aqueous liposomes or tissue homogenates containing liposomal lipids (0.1 ml) can be added directly to scintillant providing they contain adequate numbers of counts and do not exhibit excessive colour quenching when counted.

(i) Place 0.1 ml of homogenate in a 20-ml counting vial and add $10-15$ ml of scintillation cocktail that has been formulated for aqueous samples (e.g. Aquasol—New England Nuclear).

(ii) Suspend the homogenate in the scintillant by adding a thixotropic gelling agent (e.g. Carb-O-Sil—Packard Instrument Company) 0.3% (w/v) to each vial, shake vigorously to mix the contents and count.

If weighed tissue samples or larger volumes of tissue homogenates are to be counted, a tissue solubilization or combustion procedure is required prior to counting. The three methods by which this may be accomplished are outlined below.

10.1 **Quaternary ammonium hydroxide-based tissue solubilizers**

The method described below is based on the use of protosol (New England Nuclear) which is a 0.5 M solution of quaternary ammonium hydroxide in toluene. If other solubilizers are to be used, follow the manufacturer's recommendations with regard to the optimum tissue:solublizer ratio since their value may differ from that given here.

(i) Place weighed tissue samples into scintillation vials. Use a maximum of 200 mg of wet tissue weight or 0.35 ml of a 4:1 (w/v) aqueous tissue homogenate or 0.35 ml of whole blood or plasma.

(ii)   Add solubilizer (0.5 ml per 100 mg of tissue) to each vial and cap tightly to minimize sample loss by evaporation. Vials may be left to stand at room temperature, or if faster solubilization is desired, place the vials in a water bath or oven at 55°C (3−6 h).

(iii)   When the samples appear relatively clear with no visible solids, solubilization is complete. If samples are only moderately coloured at this point, the next step may be omitted.

(iv)   If samples appear highly coloured, allow the vials to cool to room temperature and add 0.1 ml of 30% (w/v) hydrogen peroxide in a dropwise fashion. Cap the vials loosely and incubate at 55°C for 30 min.

(v)   Allow the vials to cool and add to each vial 10−15 ml of scintillation containing chemiluminescence inhibitors such as Dimilume-30 (Packard Instrument Company) and count. If chemiluminescence persists, add 0.5 ml of 0.5 M HCl to each vial, shake well and place in the dark for several days before counting.

10.2 Wet oxidation

This technique of tissue solubilization is faster and less prone to chemiluminescence than methods based on quaternary ammonium hydroxide solubilizers. However, it should be stated at the outset that this method carries a modest explosion hazard and the entire procedure must be carried out in a fume hood. Since the risk of explosion increases with the amount of tissue, on no account should the sample size exceed 250 mg of whole tissue.

(i)   Place a weighed sample (maximum 0.2 ml of whole blood or plasma or 250 mg of tissue) in a glass scintillation vial and add 0.2 ml of perchloric acid (60%) dropwise while swirling the vial contents.

(ii)   Next add dropwise 0.4 ml of 30% (w/v) hydrogen peroxide, again swirling the vial contents.

(iii)   Incubate the samples at 70−80°C in an oven or water bath placed in the fume hood for 30−60 min. Agitate the vials occasionally.

(iv)   When the samples are cool, add 5 ml of ethylene glycol monoethyl ether, 10 ml of scintillant containing a chemiluminescence inhibitor (Dimilume-30—Packard Instrument Company) and count.

10.3 Tissue combustion

In addition to the chemical methods of solubilization discussed above, tissue samples can be prepared for liquid scintillation counting by combustion. Semi-automated devices commonly known as sample oxidizers are used for tissue combustion. Counting efficiencies for dual-labelled samples prepared by combustion tend to be higher than for samples prepared by other methods since ^{14}C and ^{3}H are collected and counted separately. Since procedures may vary depending on the particular device used, only a brief discussion of the methodology will be given.

(i)   Place weighed samples of major organs and tissues (maximum 300 mg) in the paper cones supplied by the manufacturer of the combustion device (e.g.

Combusto-Cone—Packard). Samples of tissues containing a high proportion of fat burn poorly and must be limited to about 50 mg. Large fatty samples give severe colour quenching on counting.

(ii) Liquid samples such as whole blood, plasma and tissue homogenate (maximum 0.5 ml) are placed in cones containing a combustible absorbent pad (Combusto-Pad—Packard).

(iii) Leave the samples to dry at room temperature for 24−48 h or heat them at 60°C for 3−6 h in a drying oven. If wet samples are to be combusted, add 1−2 g of cellulose powder to each sample to aid combustion and increase the 'burning' time relative to that for dry samples.

(iv) Before combustion of any tissue samples, cones containing standard amounts of the particular isotope of interest and blank samples should be burnt and counted according to the manufacturer's instructions. This ensures that the instrument is functioning correctly and that the recovery and sample spillover is acceptable.

(v) Since the time required for complete combustion will vary with the sample size and water content, blank tissue samples should be tested to establish optimal burn time.

11. LABELLING OF PRE-FORMED LIPOSOMES WITH ^{99m}TECHNETIUM

Based on the method reported in Richardson,V.J., Ryman,B.E., Jewkes,R.F., Tattersall,M.H. and Newlands,E.S. (1978) *Int. J. Nucl. Med. Biol.*, **5**, 118.

(i) Purge 10 ml of normal saline with nitrogen for 10 min to remove any dissolved oxygen.

(ii) Weigh out 7 mg of stannous chloride ($SnCl_2.2H_2O$) and dissolve this in 10 ml of deoxygenated normal saline.

(iii) Dispense approximately 1 ml of ^{99m}Tc sodium pertechnetate solution (10−100 mCi, 370−3700 MBq, from a technetium generator) into an injection vial with a rubber seal, and degas by evacuating for 10 min through a 21 gauge needle (introduced through the rubber cap) attached to a vacuum pump. (Contain the vial in a lead pot to protect the operator from radiation.)

(iv) Dispense 1 ml of the liposome suspension (in phosphate-free saline) into a fresh injection vial, and degas as in Step (iii).

(v) Draw up 0.5 ml of stannous chloride solution into a 1-ml syringe. Fit the loaded syringe with a disposable 0.2-μm filter, connected to a clean 19 gauge needle.

(vi) Draw up 0.5 ml of pertechnetate solution into a 1-ml syringe and fit it with a disposable filter connected to a 25 gauge needle. Store until required in a lead pot.

(vii) Set up a vortex mixer behind a sheet of lead glass.

(viii) Place the vial containing liposomes (still under vacuum) on the mixer, and introduce the stannous chloride solution rapidly from the 1-ml syringe with vortexing.

(ix) Take the vial off the mixer, position the needle of the pertechnetate syringe on the rubber seal, then introduce the contents of the syringe slowly (5−10 sec) with vortexing.

(x) Release the vacuum in the vial, and store the labelled liposome suspension in a lead pot until ready for use. Check the extent of labelling (which should be virtually 100% by this method) by paper chromatography as described below.

The above procedure may be performed aseptically by the use of sterile glassware and disposable syringes and needles.

12. PAPER CHROMATOGRAPHY OF LABELLED LIPOSOMES

All manipulations should be carried out in a fume hood specially designated for work with radioactive materials.

(i)   Put approximately 1 ml of PBS into a plastic clear-sided 120-ml screw-capped centrifuge tube.

(ii)   Spin the tube in a centrifuge at low speed for 5 min so that there are no water droplets remaining on the sides of the tube.

(iii)   Cut a strip of filter paper (e.g. Whatman 44 or No. 1) 11 cm by 1.5 cm, and fold over down the centre line along the whole length.

(iv)   With a pencil, mark a point on the centreline 3 cm from one end. Use a pair of scissors to taper the end to a point from a distance of 1 cm along.

(v)   Use a pencil to write the name of the sample to be run at the other end of the paper.

(vi)   Put the paper down on a flat surface so that the centreline fold is uppermost, and not touching anything.

(vii)   With a micropipette, spot approximately 2 μl of the liposome sample on to the paper at the point where the pencil mark has been made.

(viii) Drop the paper, tapered end down, into the centrifuge tube, so that the liquid rises up the paper from the saline solution at the bottom.

(ix)   Cap the tube and leave for approximately 25 min until the solvent front has moved almost to the top.

(x)   With a blue plastic disposable tube clamp, grasp the dry end of the paper and withdraw it from the tube.

(xi)   Rest the clamp on a horizontal surface in the fume hood (e.g. test-tube rack) so that the paper is not touching anything.

(xii)   Allow the paper to dry for 30 min in the draught of the hood.

(xiii) Cut the paper in half and measure the activity in the two pieces. Liposomes will stay at the origin (pencil spot) while free iodide or pertechnetate will run with the solvent front.

(xiv)   Put pieces of paper in counting tubes and measure the activity of each in a γ-counter. Alternatively, the paper strip may be cut while still wet, and placed into counting tubes.

This method may be used for analysis of free water-soluble iodine label in liposomes prepared incorporating markers such as cholesterol aniline, labelling of which is described below.

13. RADIO-IODINATION OF CHOLESTEROL ANILINE: IODINE MONOCHLORIDE METHOD

Method devised by R.R.C.New.

13.1 Reagent preparation

2.5 mM Iodine monochloride solution

Dissolve 0.0277 g of KI, 0.091 g of KIO_3 and 2.0 g of NaCl in 100 ml of distilled water containing 0.5 ml of conc. HCl. The pH should be around pH 1.37.

Buffer (NH₄Cl/NH₄OH 0.3 M)

Weigh out 1.7 g of NH_4Cl and dissolve it in 100 ml of distilled water. Adjust the pH to pH 8.9 by dropwise addition of 0.880 NH_4OH (approx. 1 ml).

Na^{125}I solution

To 50 μl of Na^{125}I solution from Amersham International (5 mCi, 200 MBq), add 200 μl of ammonia buffer. Store at 44°C.

Lipid solution

Weigh out 10 mg of cholesteryl aniline (CA) into a 5-ml glass, screw-capped bottle and dissolve in 2 ml of chloroform. Add 2 ml of methanol to the solution and mix well. Store below 0°C until needed.

13.2 Labelling procedure

(i)   Transfer 200 μl of the lipid solution (0.5 mg CA) to a Nunc vial (plastic screw-capped) and add 10 μl of Na^{125}I solution (200 μCi, 7 MBq) and 10 μl of 2.5 mM ICl solution. Mix well and allow to stand at room temperature for an hour.

(ii)   Check the extent of labelling by spotting 5 μl of reaction mixture on to a silica gel chromatography plate, and develop for an hour in $CHCl_3$:MeOH:H_2O 65:25:4 by vol.

(iii)   Stain with iodine. The spot should appear about 1 cm behind the solvent front. Scrape off the silica gel around the spot and collect in a counting tube. Scrape off a 3-cm band beginning 1 cm from the origin and collect the silica gel in a counting tube.

(iv)   Measure the activity in a γ-counter. 95% of the activity should be in the fast-running cholesteryl aniline spot.

(v)   To the remainder of the solution, add 400 μl of saline and 800 μl of chloroform. Shake well, then spin at 2000 g in a bench centrifuge for 10 min at room temperature.

(vi)   Remove the bottom layer with an Eppendorf pipette, and keep in a 1-dram glass screw-capped vial.

(vii)   Add an extra 800 μl of chloroform to the reaction mixture, shake and spin as before, remove the bottom layer and pool with the first sample obtained so that 0.5 mg of pure labelled cholesteryl aniline is collected in a total of 1.6 ml of chloroform. Store at 4°C until required for use.

Manufacturers and suppliers

General biochemicals

BDH Ltd

UK	Broom Road		
	Poole	Tel:	0202 745520
	Dorset	Telex:	41186
	BH12 4NN	Fax:	0202 738299

Boehringer

UK	Boehringer Mannheim House		
	Bell Lane		
	Lewes	Tel:	0273 480444
	East Sussex	Telex:	877487
	BN7 1LG	Fax:	0273 480266

USA	PO Box 5086	Tel:	317 849 9350
	Indianapolis		800 262 1640
	IN 46250	Telex:	6711626
		Fax:	317 576 7527

Calbiochem

UK	Calbiochem BioScience		
	Newton House		
	42 Devonshire Road	Tel:	0223 31 6855
	Cambridge	Telex:	81304
	CB1 2BL	Fax:	0223 460396

USA	PO Box 12087	Tel:	619 450 9600
	San Diego	Telex:	697934
	CA 92112-4180	Fax:	619 453 3552

Fluka

UK	Peakdale Road		
	Glossop	Tel:	04 574 62518
	Derbyshire	Telex:	669960
	SK13 9XE	Fax:	04 574 4307

USA	980 South Second Street	Tel:	516 467 0980
	Ronkonkoma	Telex:	967807
	NY 11779	Fax:	516 467 0663

Europe	Ch-9470 Buchs	Tel:	085 76 02 75
	Switzerland	Telex:	855282
		Fax:	085 65 449

ICN Biomedicals

UK	Free Press House		
	Castle Street		
	High Wycombe	Tel:	0494 443826
	Bucks	Telex:	837969
	HP13 6RN	Fax:	0494 436048

USA	PO Box 19436	Tel:	714 545 0113
	Irvine	Telex:	685580
	CA 92713	Fax:	714 557 4872

KochLight

UK	Rookwood Way		
	Haverhill	Tel:	0440 702436
	Suffolk	Telex:	817756
	CB9 8PU	Tax:	0440 61507

USA	New Brunswick Scientific		
	44 Talmadge Road	Tel:	201 287 1200
	Edison	Telex:	275083
	NJ 08818−4005	Fax:	201 287 4222

Mallinkrodt Inc.

USA	Science Products Div		
	675 McDonnell Blvd	Tel:	314 895 2333
	PO Box 5840	Telex:	209897
	St Louis	Fax:	ITT 4990141
	MO 63134		

UK	c/o Camlab Ltd	Tel:	0223 62222
	Nuffield Road	Telex:	817664
	Cambridge, CB 4 1TH	Fax:	0223 460865

Merck E, AG

Europe	Frankfurter Strasse 250		
	Postfach 4119	Tel:	06151 72 2868
	6100 Darmstadt	Telex:	4193280
	FRG	Fax:	06151 72 3521

Molecular Probes Inc.

USA	PO Box 22010	Tel:	503 344 3007
	Eugene	Telex:	858721
			MOLECULAR
	OR 97402	Fax:	503 344 6504

Pierce

UK	Pierce Warriner UK Ltd		
	44 Upper Northgate St	Tel:	0244 382525
	Chester	Telex:	617057
	Cheshire, CH1 4EF	Fax:	0244 373212
Europe	Pierce Europe BV		
	PO Box 1512	Tel:	31 1860 19277
	3260 BA Oud-Beijerland	Telex:	21676
	The Netherlands	Fax:	31 1860 19179
USA	PO Box 117	Tel:	815 968 0747
	Rockford	Telex:	239912413
	IL 61105	Fax:	815 968 7316

Sigma Chemical Company Ltd

UK	Fancy Road	Tel:	0202 733114
	Poole		800 373731
	Dorset	Telex:	418242
	BH17 7NH	Fax:	0202 715460
USA	PO Box 14508	Tel:	314 771 5750
	St Louis		800 325 3010
	MO 63178	Telex:	2559107610593
		Fax:	314 771 5757

Woelm Pharma Gmbh & Co.

Germany	Max Woelm Strasse	Tel:	056 51 80101
	D 3440	Telex:	680566
	Eschwege	Fax:	056 51 801384

General chemicals

Aldrich Chemical Co. Ltd.

UK	The Old Brickyard		
	New Road		
	Gillingham	Tel:	07476 4414
	Dorset	Telex:	417238
	SP8 4BR	Fax:	07476 3779

Carlo Erba—Farmitalia

Europe	Analytical Div	Tel:	392 699 51
	via C Imbonati 24	Telex:	330314
	I 20159 Milan	Fax:	392 699 51
	Italy		

Eastman Kodak Co

USA	Eastman Organic Chemicals	Tel:	716 588 4817
	RPD Technical Services	Telex:	97 8481
	Rochester 14607	Fax:	716 722 3179
	NY, USA		

UK	Kodak Ltd		
	Acornfield Road	Tel:	051 548 6560
	Knowsley Industrial Park Nth	Telex:	629640
	Liverpool, L33 7UF	Fax:	051 547 2404

Lipids

Asahi Org Chem Ind Co Ltd.

Japan	2-5955 Nakanose-Machi	Tel:	81 982 333311
	Nobeoka City	Telex:	
	Miyazaki Prefectine	Fax:	81 982 322868
	Japan		

Avanti Polar Lipids Inc.

USA	5001 A Whitling Drive	Tel:	205 663 2494
	Pelham	Telex:	910250652
	AL 35214−1955	Fax:	205 663 0756

Genzyme Corporation

USA	75 Kneeland Street	Tel:	617 451 1923
	Boston	Telex:	201223
	MA 02111	Fax:	617 451 2454

UK	Genzyme Fine Chemicals		
	Hollands Road	Tel:	0440 703522
	Haverhill	Telex:	81333
	Suffolk	Fax:	0440 707783
	CB9 8PU		

Japan	Genzyme Japan		
	Jitsugetsukan GF		
	1-3-7 Kojimachi	Tel:	03 230 1541
	Chiyoda-Ku, Tokyo	Fax:	03 230 1548

Lipid Products

UK	Nutfield Nurseries		
	Crabhill Lane		
	South Nutfield	Tel:	0737 823277
	near Redhill	Telex:	
	Surrey, RH1 5PG	Fax:	0737 822561

Lipoid KG

Europe	Frigenstrasse 4	Tel:	0621 55 3018
	D-6700	Telex:	464 783
	Ludwigshafen 24	Fax:	0621 55 3559

Lucas Meyer

Europe	Ausschlager Elbdeich 62	Tel:	4940 789 550
	2000 Hamburg 28	Telex:	2163220
	FRG	Fax:	4940 789 8329

UK	Honeywill & Stein Ltd.		
	Times House		
	Throwley Way	Tel:	01 770 7090
	Sutton	Telex:	946560 BPCLGMG
	Surrey, SM1 4AF	Fax:	01 770 7295

USA	765 East Pythian Avenue	Tel:	217 875 3660
	Decatur	Telex:	404387
	IL 62526	Fax:	217 877 5046

Radiochemicals

Amersham

UK	UK Sales Office		
	Lincoln Place		
	Great End		
	Aylesbury	Tel:	0296 39522
	Buckinghamshire	Telex:	837660
	HP20 2TP	Fax:	0296 85910

USA	2636 S Clearbrook Drive	Tel:	312 364 7100
	Arlington Heights		800 323 9750
	IL 60005		

ICN Radiochemicals

USA	PO Box 19536	Tel:	714 545 0113
	Irvine	Telex:	685580
	CA 92713	Fax:	714 557 4872

New England Nuclear

UK	Du Pont UK Ltd		
	Biotechnology Systems		
	Division		
	NEN Research Products		
	Wedgewood Way		
	Stevenage	Tel:	0438 734026
	Hertfordshire	Telex:	825591
	SG1 4QN	Fax:	0438 734 154

Europe	Du Pont de Nemours		
	(Deutschland) GmbH		
	Biotechnology Systems		
	Division		
	NEN Research Products		
	Dupont Strasse 1	Tel:	06172 870
	D-6380	Telex:	410676
	FRG	Fax:	06172 871500

USA	EI du Pont de Nemours & Co		
	(Inc)		
	NEN Products		
	331 Treble Cove Road	Tel:	617 671 9531
	No Billerica	Telex:	940996
	MA 01862	Fax:	617 663 7315

Packard Instrument Co.

USA	2200 Warrenville Road	Tel:	312 969 6000
	Downers Grove	Telex:	21 0031
	IL 60515	Fax:	312 969 6511

UK	Eskdale Road		
	Winnersh Triangle		
	Wokingham	Tel:	0734 696622
	Berkshire	Telex:	848884
	RG11 5DZ	Fax:	0734 699 609

Chromatography materials

Bio-Rad

UK	Bio-Rad House		
	Maylands Avenue		
	Hemel Hempstead	Tel:	0442 232552
	Hertfordshire	Telex:	827770
	HP2 7TD	Fax:	0442 59118

Pharmacia

UK	Pharmacia Ltd		
	Pharmacia House		
	Midsummer Boulevard		
	Central Milton Keynes	Tel:	0908 661101
	Buckinghamshire	Telex:	826778
	MK9 3HP	Fax:	0908 690091

USA	800 Centennial Avenue	Tel:	201 457 8000
	Piscataway	Telex:	6858125
	NJ 08854	Fax:	201 457 9022

Sweden	Pharmacia Biotechnology		
	International AB	Tel:	46 18 17 3000
	S-751 82	Telex:	76974
	Uppsala	Fax:	46 18 12 6077

USA	1414 Harbour Way South	Tel:	415 232 7000
	Richmond	Telex:	3720184
	CA 94804	Fax:	415 232 4257

Whatman

UK	Springfield Mill	Tel:	0622 692022
	Maidstone	Telex:	96113
	Kent	Fax:	0622 691425
	ME14 2LE		

USA	9 Bridewell Place	Tel:	201 773 5800
	Clifton	Telex:	133426
	NJ 07014	Fax:	201 472 6949

Membranes

Nuclepore Corporation

USA	7035 Commerce Circle	Tel:	415 463 2530
	Pleasanton		800 882 7711
	CA 94566-3294	Telex:	3719645
		Fax:	415 463 2029

UK	c/o Sterilin Ltd		
	Lampton House		
	Lampton Road		
	Hounslow	Tel:	01 572 2468
	Middlesex	Telex:	889124
	TW3 4EE	Fax:	01 572 7301

Dialysis equipment

Dianorm

Europe	PO Box 126	Tel:	089 811 44 47
	D-8000 Meunchen 65	Telex:	
	FRG	Fax:	

Fresenius

UK	67 Christleton Court		
	Stuart Road		
	Manor Park	Tel:	092 858 0058
	Runcorn	Telex:	627678
	Cheshire, WA7 1PG	Fax:	0928 581424

Europe	Bad Homburg v dH		
	Borkenberg 14	Tel:	06171 60-0
	6370 Oberursel/Ts 1	Telex:	003418120
	FRG	Fax:	06172 24011

Minitech

USA	14905 28th Avenue North	Tel:	612 553 3300
	Minneapolis	Telex:	290825 RSIMPLS
	MN 5541		PLOH
		Fax:	612 553 3387

Travenol Laboratories Ltd

UK	Baxter Health Care		
	Caxton Way		
	Thetford	Tel:	0842 4581
	Norfolk	Telex:	81319
	IP24 3SE	Fax:	0842 2219

USA	1 Parkway North		
	Suite 300 PO 784	Tel:	312 940 5600
	Deerfield	Telex:	23724497
	IL 60015-0784	Fax:	312 940 4294

Terumo Corporation

Japan	44-1, 2 chome Hatagaya	Tel:	813 374 8111
	Shibuya-ku	Telex:	02324597
	Tokyo	Fax:	813 374 8390

Genotec

Europe	Gesellschaft fur Mess- und		
	Regeltechnik mbH		
	Eisenacher Strasse 56	Tel:	030 784 60 27
	D-1000 Berlin 62	Telex:	186218
	FRG	Fax:	030 788 12 01

UK	c/o Clandon Scientific Ltd		
	Lysons Avenue		
	Ash Vale		
	Aldershot	Tel:	025 251 4711
	Hampshire	Telex:	858210
	GU12 5QF	Fax:	025 251 1855

Sonicators

Branson Ultrasonics Corp

USA	Eagle Road	Tel:	203 796 0400
	Danbury	Telex:	643743
	CT 06810	Fax:	203 796 0450

Crest Ultrasonics

USA	Scotch Road	Tel:	609 883 4000
	Mercer County Airport	Telex:	510 685 9577
	PO Box 7266	Fax:	609 883 6452
	Trenton NJ 08628		

Kerry Ultrasonics Ltd

UK	Hunting Gate		
	Wilbury Way	Tel:	0462 50761
	Hitchin	Telex:	825068
	Hertfordshire SG4 0TQ	Fax:	0462 420712

MSE

UK	Fisons		
	Sussex Manor Park		
	Gatwick Road	Tel:	0293 31100
	Crawley	Telex:	878851
	RH10 2QQ	Fax;	0293 561980

Sonics & Materials Inc.

UK	c/o Roth Scientific Co Ltd		
	Alpha House		
	Alexandra Road		
	Farnborough	Tel:	0252 513131
	Hampshire	Telex:	858650
	GU14 6BU	Fax:	0252 543609
USA	Kenosia Avenue	Tel:	203 744 4400
	Danbury	Telex:	969639
	CT 06801	Fax:	203 798 8350

Microfluidizers

Biotechnology Development Corporation

USA	Medicontrol Corp		
	44 Mechanic Street	Tel:	617 965 7255
	Newton	Telex:	755709
	MA 02164	Fax:	617 965 1213

UK	Christison Ltd		
	Albany Road		
	East Grinstead Industrial		
	Estate	Tel:	091 477 4261
	Gateshead	Telex:	537426
	NE8 3AT	Fax:	091 490 0549

Extrusion equipment

Amicon

UK	Upper Mill		
	Stonehouse	Tel:	045 382 5181
	Gloucester	Telex:	43532
	GL10 2BJ	Fax:	045 382 6686

USA	W R Grace & Co		
	24 Cherry Hill Drive	Tel:	617 777 3622
	Danvers	Telex:	23275192
	MA 01923	Fax:	617 777 6204

Lipex Biomembranes Inc.

Canada	3550 West 11th Avenue	Tel:	604 734 8263
	Vancouver BC	Fax:	604 734 2390
	V6R 2K2		

Millipore

UK	Millipore House		
	The Boulevard		
	Ascot Road		
	Croxley Green		
	Watford	Tel:	0923 816375
	Hertfordshire	Telex:	24191
	WD1 8YW	Fax:	0923 818297

USA	80 Ashby Road	Tel:	617 275 9200
	PO Box 255		800 225 1380
	Bedford	Telex:	4430066
	MA 01730	Fax:	617 875 2051

Nuclepore Corporation
    see **Membranes**

Schleicher & Schuell GmBH

Europe	Postfach 4	Tel:	055 61 79 10
	D-3354 Dassel	Telex:	0965632
	FRG	Fax:	055 61 72 743
USA	10 Optical Avenue	Tel:	603 352 3810
	Keene	Telex:	757901
	NH 03431	Fax:	603 357 3627
UK	c/o Anderman		
	145 London Road		
	Kingston-upon-Thames	Tel:	01 541 0035
	Surrey	Telex:	947359
	KT2 6NH	Fax:	01 541 0623

Osmometers

Advanced Instruments Inc.

USA	100 Highland Avenue	Tel:	617 449 3000
	Needham Heights	Telex:	200177
	MA 02194	Fax:	617 455 8468
UK	c/o Vital Scientific Ltd		
	Huffwood Trading Estate		
	Partridge Green	Tel:	0403 710479
	West Sussex	Telex:	04012126
	RH13 8AU	Fax:	0403 710 382

Gonotec

Europe	Gesellschaft für Mess Und	Tel:	030 784 6024
	Regeltechnik mbH	Telex:	186218
	Eisenacher Strasse 56	Fax:	030 881 201
	D-1000 Berlin 62		
	FRG		
UK	c/o Clandon Scientific		
	(see under dialysis equipment)		

Appendix II

Homogenizers

Brinkmann Instruments Inc.

USA	Cantiague Road Westbury NY 11590	Tel: Telex: (No fax)	516 334 7500 6853174
UK	c/o Chemlab Scientific Products Construction House Greyfell Avenue Hornchurch Essex RM12 4EH	Tel: Telex: (No fax)	04024 76162 9419785

Photon correlation spectroscopy

Coulter Electronics Ltd

UK	Northwell Drive Luton Bedfordshire LU3 3RH	Tel: Telex: Fax:	0582 491414 825074 0582 490390
USA	PO Box 2145 Hialeah Florida	Tel: Telex: Fax:	305 883 0131 305 883 6820

Malvern Instruments

UK	Spring Lane South Malvern Worcestershire WR14 1AQ	Tel: Telex: Fax:	068 489 2456 339679 068 489 2789

Pacific Scientific

USA	Instrument Division 2431 Linden Lane Silver Spring MD 20910	Tel: Telex: Fax:	301 495 7000 197634 301 495 0478

Standard texts on liposomes and phospholipid membranes

Books

An Introduction to the Chemistry and Biochemistry of Fatty Acids and Their Glycerides. Gunstone,F.D. Chapman and Hall Ltd, London (1967)

Techniques of Lipidology. Kates,M. North Holland, Amsterdam/American Elsevier Publ Co. Inc, New York (1972)

Form and Function of Phospholipids. Ansell,G.E., Hawthorne,J.N. and Dawson,M. (eds), Elsevier Scientific Publishing Co., Amsterdam (1973) BBA Library Vol. 3.

Fatty Acids and Glycerides. Kuksis,A. (ed.), Plenum Press, New York (1978) Handbook of Lipid Research vol. 1.

Liposomes and Their Uses in Biology and Medicine. Papahadjopoulos,D. (ed.), New York Acad. Sci, New York (1978) Vol. 308.

Drug Carriers in Biology and Medicine Gregoriadis,G. (ed.), Academic Press, London, (1979).

Lysosomes in Applied Biology and Therapeutics. Dingle,J,T. and Jacques,P.I.J. (eds), North Holland, Amsterdam (1979).

Liposomes and Immunobiology. Tom,B.H. and Six,H.R. (eds), Elsevier North Holland, New York (1980).

Liposomes in Biological Systems. Gregoriadis,G. and Allison,A.C. (eds), Wiley, New York (1980).

Targeting of Drugs. Gregoriadis,G., Senior,J. and Trouet,A. (eds), Plenum Press, New York and London (1981).

Liposomes: From Physical Structure to Therapeutic Applications Knight,G. (ed.). Elsevier/North Holland Biomedical Press, Amsterdam (1981).

Liposome Technology Leserman,L. and Barbet,J. (eds), Les Editions INSERM, Paris (1982).

Liposomes in the Study of Drug Activity and Immunocompetent Cell Functions. Nicolau,C. and Paraf,A. (eds), Proc. Int. Symp. Grignon, France (1982).

Liposomes. Ostro,M. (ed.) Marcel Dekker, Inc., New York and Basel (1983).

Liposome Letters. Bangham,A.D. (ed.) Academic Press, London (1983).

Membrane Fluidity in Biology. Aloia,R.C. (ed.), Academic Press, New York and London (1983) Vol. 1 and 2.

Biomembrane Structure and Function. Chapman,D. (ed.), Macmillan Press, London (1983).

Liposome Technology. Gregoriadis,G. (ed.), CRC Press, Boca Raton, Florida (1984) Vol. 1, 2, and 3.

Liposomes as Drug Carriers. Schmidt,K.H. (ed.), Georg Thieme, Verlag, Stuttgart and New York (1986).

Chemistry and Physics of Lipids: Vol. 40, Special Issue: *Liposomes.* Cullis,P.R. and Hope,M.J. (eds) Elsevier Scientific Publishers Ireland Ltd., Shannon (1986).
Medical Applications of Liposomes. Yagi,K. (ed.), Japan Scientific Soc. Press, Tokyo, and Karger, Basel (1986).
Methods in Enzymology, Drug and Enzyme Targeting. Green,P. and Widder,K.J. (eds), Academic Press, New York (1987) Vol. 149.

Chapters

Bangham,A.D., Hill,M.W. and Miller,N.G.A. (1984) In *Methods in Membrane Biology.* Korn,E.B. (ed.), Plenum Press, New York, London, p. 1.
Juliano,R.I. and Layton,D. (1980) In *Drug Delivery Systems* Juliano,R.L. (ed.), OUP Inc, New York and Oxford, p. 189.
Papahadjopoulos,D. and Kimelberg,H.N. (1983) In *Progress in Surface Science.* Davison,S.G. (ed.), Pergamon Press, Oxford, Vol. 4, p. 141.

Key references for applications of liposomes

DRUG TARGETING

Enzyme therapy

Sessa,G. and Weissmann,G. (1970) *J. Biol. Chem.*, **245**, 3295–301.
Incorporation of lysozyme into liposomes. A model for structure-linked latency.
Gregoriadis,G., Leathwood,P.D. and Ryman,B.E. (1971) *FEBS Lett.*, **14**, 95–9.
Enzyme entrapment in liposomes.
Weissmann,G., Bloomgarden,D., Kaplan,R., Cohen,C., Hoffstein,S., Collins,T.,
Gottlieb,A. and Nagle,D. (1975) *Proc. Natl. Acad. Sci. USA*, **72**, 88–92.
A general method for the introduction of enzymes, by means of immunoglobulin-coated liposomes, into lysosomes of deficient cells.
Cohen,C.M., Weissmann,G., Hoffstein,S., Awasthi,Y.C. and Srivastava,S.K. (1976)
Biochemistry, **15**, 452–60.
Introduction of purified hexosaminidase A into Tay–Sachs leukocytes by means of immunoglobulin-coated liposomes.
Finkelstein,M. and Weissmann,G. (1978) *J. Lipid Res.*, **19**, 289–303.
The introduction of enzymes into cells by means of liposomes.
(*see also* **Administration in humans**)

Chelating agents

Rahman,Y.-E. and Wright,B.J. (1975) *J. Cell. Biol.*, **65**, 112–22.
Liposomes containing chelating agents. Cellular penetration and a possible mechanism of metal removal.
Rosenthal,M.W., Rahman,Y.-E., Moretti,E.S. and Cerny,E.A. (1975) *Radiat. Res.*,
63, 262–74.
Removal of polymeric plutonium by DPTA directed into cells by liposome encapsulation.

Cancer

Ara-C

Mayhew,E., Papahadjopoulos,D., Rustum,Y.M. and Dave,C. (1976) *Cancer Res.*, **96**,
4406–11.
Inhibition of tumor cell growth *in vitro* and *in vivo* by 1-β-D-arabinofuranosylcytosine entrapped within phospholipid vesicles.
Rustum,Y.M., Dave,C., Mayhew,E. and Papahadjopoulos,D. (1979) *Cancer Res.*, **39**,
1390–5.

Role of liposome type and route of administration in the antitumour activity of liposome-entrapped 1-β-D-arabinofuranosylcytosine against mouse L1210 leukaemia.
Mayhew,E., Papahadjopoulos,D., Rustum,Y.M. and Dave,C. (1978) *Ann. NY Acad. Sci.*, **303**, 371−86.
Use of liposomes for the enhancement of the cytotoxic effects of cytosine arabinoside.
Ellens,H., Rustum,Y.M., Mayhew,E. and Ledesma,E. (1982) *J. Pharm. Exp. Therap.*, **222**, 324−30.
Distribution and metabolism of liposome-encapsulated and free 1-β-D-arabinofuranosylcytosine (Ara-C) in dog and mouse tissues.
(*see also* **Administration in humans**)

Doxorubicin

Rahman,A., Kessler,A., More,N., Sikic,B., Rowden,G., Woolley,P. and Schein,P.S. (1980) *Cancer Res.*, **40**, 1532−7.
Liposomal protection of adriamycin-induced cardiotoxicity in mice.
Forssen,E.A. and Tokes,Z.A. (1981) *Proc. Natl. Acad. Sci. USA*, **78**, 1873−.7
Use of anionic liposomes for the reduction of chronic doxorubicin-induced cardiotoxicity.
Olsen,F., Mayhew,E., Maslow,D., Rustum,Y.M. and Szoka,F. (1982) *Eur. J. Cancer Clin. Oncol.*, **18**, 167−76.
Characterisation, toxicity and therapeutic efficacy of adriamycin encapsulated in liposomes.
Gabizon,A., Dagan,A., Goren,D., Barenholz,Y. and Fuks,Z. (1982) *Cancer Res.*, **42**, 4734−9.
Liposomes as *in vivo* carriers of adiamycin: reduced cardiac uptake and preserved antitumour activity in mice.
Mayhew,E., Rustum,Y.M. and Vail,W.J. (1983) *Cancer Drug Deliv.* **1**, 43−58.
Inhibition of liver metastases of M5076 tumour by liposome-entrapped adriamycin.
Gabizon,A., Goren,D., Fuks,Z., Barenholz,Y., Dagan,A. and Meshorer,A. (1983) *Cancer Res.*, **43**, 4730−5.
Enhancement of adriamycin delivery to liver metastatic cells with increased tumoricidal effect using liposomes as drug carriers.
(*see also* **Administration in humans**)

MDP-PE

Fidler,I.J. and Poste,G. (1982) *Immnunopathol.*, **5**, 161−74.
Macrophage-mediated destruction of malignant tumor cells and new strategies for the therapy of metastatic disease.
Schroit,A.J. and Fidler,I.J. (1982) *Cancer Res.*, **42**, 161−7.
Effects of liposome structure and lipid composition on the activation of the tumoricidal properties of macrophages by liposomes containing muramyl dipeptide.
Schroit,A.J., Hart,I.R., Madsen,J. and Fidler,I.J. (1983) *J. Biol. Response Modif.*, **2**, 97−100.
Selective delivery of drugs encapsulated in liposomes: natural targeting to macrophages involved in various disease states.

Schroit,A.J., Galligioni,E. and Fidler,I.J. (1983) *Biol. Cell.*, **47**, 87−94.
Factors influencing the *in situ* activation of macrophages by liposomes containing muramyl dipeptide.
Fidler,I.J. and Schroit,A.J. (1984) *J. Immunol.*, **133**, 515−18.
Synergism between lymphokines and muramyl dipeptide encapsulated in liposomes: *in situ* activation of macrophages and therapy of spontaneous cancer metastases.

Infectious diseases

Black,C.D.V., Watson,G.J. and Ward,R.J. (1977) *Trans. R. Soc. Trop. Med. Hyg.*, **71**, 550−2.
The use of pentostam liposomes in the chemotherapy of experimental leishmaniasis.
New,R.R.C., Chance,M.L., Thomas,S.C. and Peters,W. (1978) *Nature, Lond.*, **272**, 55−6.
Antileishmanial activity of antimonial agents entrapped in liposomes.
Alving,C.R., Steck,E.A., Hanson,W.L., Loizeaux,P.S., Chapman,W.L.,Jr. and Waits,V.B. (1978) *Life Sci.*, **22**, 1021−6.
Improved therapy of experimental leishmaniasis by use of a liposome-encapsulated antimonial drug.
New,R.R.C., Chance,M.L. and Heath,S. (1981) *Parasitology*, **83**, 519−27.
The treatment of experimental cutaneous leishmaniasis with liposome-entrapped pentostam.
New,R.R.C., Chance,M.L. and Heath,S. (1981) *J. Antimicrob. Chemother.*, **8**, 371−81.
Antileishmanial activity of amphotericin and other antifungal agents in liposomes.
Taylor,R.L., Williams,D.M., Craven,P.C., Graybill,D.J., Drutz,D.J. and Magee,W.E. (1982) *Am. Rev. Respir. Dis.*, **125**, 610−11.
Amphotericin B in liposomes: a novel therapy of histoplasmosis.
Graybill,J.R., Craven,P.C., Taylor,R.L., Williams,D.M. and Magee,W.E. (1982) *J. Infect. Dis.*, **145**, 748−52.
Treatment of murine cryptococcosis with liposome-associated amphotericin B.
Lopez-Berenstein,G., Mehta,R., Hopfer,R.L., Mills,K., Kasi,L., Mehta,K., Fainstein,V., Luna,M., Hersh,E.M. and Juliano,R.L. (1983) *J. Infect. Dis.*, **147**, 939−44.
Treatment and prophylaxis of disseminated infection due to *Candida albicans* in mice with liposome-encapsulated amphotericin B.
Lopez-Berenstein,G., Hopfer,R.L., Mehta,R., Mehta,K., Hersh,E.M. and Juliano,R.L. (1984) *J. Infect. Dis.*, **150**, 278−83.
Liposome-encapsulated amphotericin B for the treatment of disseminated candidiasis in neutropenic mice.
Koff,W.C., Fidler,I.J., Showalter,S.D., Chakrabaty,M.K., Hampar,B., Ceccorulli, L.M. and Kleinerman,E. (1984) *Science*, **224**, 1007−9.
Human monocytes activated by immunomodulators in liposomes lyse herpes virus-infected but not normal cells.
Koff,W.C. and Fidler,I.J. (1985) *Antivir. Res.*, **5**, 179−90.
The potential use of liposome-mediated antiviral therapy.
Fountain,M.W., Dees,C. and Schultz,R.D. (1981) *Current Microbiology*, **6**, 373−6.
Enhanced intracellular killing of *Staphylococcus areus* by canine monocytes treated

with liposomes containing amakacin, gentamicin, kanamycin and tobramycin.

Fountain,M.W., Weiss,S.J., Fountain,A.G., Shen,A. and Lenk,R.P. (1985) *J. Infect. Dis.*, **152**, 529−35.
Treatment of *Brucella canis* and *Brucella abortus in vitro* and *in vivo* by stable plurilamellar vesicle-encapsulated aminoglycosides.

Lopez-Berenstein,G. (1987) *Antimicrob. Agents Chemother.*, **31**, 675−8.
Minireview: Liposomes as carriers of antimicrobial agents.

(*See also* **Administration in humans**)

Modes of administration

Oral administration

Insulin

Dapergolas,G., Neerunjun,E.D. and Gregoriadis,G. (1976) *FEBS Lett.*, **63**, 235−9.
Penetration of target areas in the rat by liposome-associated bleomycin, glucose oxidase and insulin.

Dapergolas,G. and Gregoriadis,G. (1976) *Lancet*, **2**, 824−7.
Hypoglycaemic effect of liposome-entrapped insulin administered intragastrically into rats.

Patel,H.M. and Ryman,B.E. (1976) *FEBS Lett.*, **62**, 60−3.
Oral administration of insulin by encapsulation within liposomes.

Gregoriadis,G., Neerunjun,D.E. and Hunt,R. (1977) *Life Sci.*, **21**, 357−69.
Fate of a liposome-associated agent injected into normal and tumor-bearing rodents. Attempts to improve localization in tumor tissues.

Patel,H.M. and Ryman,B.E. (1977) *Biochem. Soc. Trans.*, **5**, 1054−6.
The gastro-intestinal absorption of liposomally entrapped insulin in normal rats.

Dapergolas,G. anmd Gregoriadis,G. (1977) *Biochem.Soc. Trans.*, **5**, 1383−6.
The effect of liposomal lipid composition on the fate and effect of liposome-entrapped insulin and tubocurarine.

Rowland,R.N. and Woodley,J.F. (1981) *Bioscience Rep.*, **1**, 345−52.
Uptake of free and liposome-entrapped insulin by rat intestinal sacs *in vitro*.

Other materials

Rowland,R.N. and Woodley,J.F. (1981) *Bioscience Rep.*, **1**, 39−406.
Uptake of free and liposome-entrapped ^{125}I-labelled PVP by rat intestinal sacs *in vitro*: evidence for endocytosis.

Chiang,C.M. and Weiner,N. (1986) *Int. J. Pharm.*, **37**, 75−85.
Gastrointestinal uptake of liposomes. I: *In vitro* and *in situ* studies.

Ueno,M., Nakasaki,T., Adachi,I., Sato,K. and Horikoshi,I. (1987) *Yakuzaigaku*, **47**, 56−60.
Oral administration of liposomally entrapped heparin. Effect of surface charge of liposome on gastro-intestinal absorption.

Via respiratory tract

McCullough,H.N. and Juliano,R.L. (1979) *J. Natl. Cancer Inst.*, **63**, 727−30.

Organ selective action of an anti-tumour drug: pharmacologic studies of liposome-encapsulated β-cytosine arabinoside administered via the respiratory system of rats. Juliano,R.L. and McCullough,H.N. (1980) *J. Pharm. Exp. Ther.,* **214**, 381−7.
Controlled delivery of an anti-tumor drug: localised action of liposome-encapsulated cytosine arabinoside administered via the respiratory system.

Intra-articular

Dingle,J.T., Gordon,J.L., Hazleman,B.L., Knight,C.G., Page-Thomas,D.P., Phillips,N.C., Shaw,I.H., Fildes,F.J.P., Oliver,J.E., Jones,G., Turner,E.H. and Lowe,J.S. (1978) *Nature,Lond.,* **271**, 372−3.
Novel treatment for joint inflammation.

Administration in humans

Tyrrell,D.A., Ryman,B.E., Keeton,B.R. and Dubowitz,V. (1976) *Br. Med. J.,* **2**, 88.
Use of liposomes in treating type II glycogenosis.
Sells,R.A., Owen,R.R., New,R.R.C. and Gilmore,I.T. (1987) *Lancet,* **12**, 2(8559), 624−5.
Reduction in toxocity of doxorubicin by liposomal entrapment.
Sells,R.A., Gilmore,I.T., Owen,R.R., New,R.R.C and Stringer,R.E. (1987) *Cancer Treat. Rev.,* **14**, 383−7.
Reduction in doxorubicin toxicity following liposomal delivery.
Gabizon,A., Peretz,T., Ben-Yosef,R., Katane,R., Biran,S. and Barenholz,Y. (1986) *Proc. Am. Soc. Clin. Oncol.,* **5**, 43 (Abstract).
Phase I study with liposome-associated adriamycin: preliminary report.
Delgado,G., Potkul,R.K., Treat,J.A., Lewandowski,G.S., Barter,J.F., Forst,D. and Rahman,A. (1989) *Am. J. Obstet. Gynecol.,* **160**, 812−17.
A phase I/II study of intraperitoneally administered doxorubicin entrapped in cardiolipin liposomes in patients with ovarian cancer.
Price,C., Aherne,W., New,R.R.C., Mayhew,E., Stringer,R.E., Littleton,P., Adams,K., Rustum,Y. and Lister,A. (1989) *Proc. Am. Assoc. Cancer Res.,* **30**, 988.
Encapsulation of cytosine arabinoside in liposomes: a method of manipulating in vivo pharmacokinetics in man.
Lopez-Berenstein,G., Fainstein,V., Hopfer,R.L., Mehta,K., Sullivan,M.P., Keating,M., Rosenblum,M.G., Mehta,R., Luna,M., Hersh,E.M., Reuben,J., Juliano,R.L. and Bodey,G.P. (1985) *J. Infect. Dis.,* **4**, 704−10.
Liposomal-amphotericin B for the treatment of systemic fungal infections in patients with cancer: a preliminary report.
Lopez-Berenstein,G., Bodey,G.P., Frankel,L.S. and Mehta,K. (1987) *J. Clin. Oncol.,* **5**, 310−17.
Treatment of hepatosplenic fungal infection with liposomal amphotericin B.
Patel,H.M., Harding,N.G.L., Logue,F., Kesson,C., MacCuish,A.C., McKenzie,J.C., Ryman,B.E. and Scobie,I. (1978) *Biochem. Soc. Trans.,* **6**, 784−5.
Intrajejunal absorption of liposomally entrapped insulin in normal saline.
Hemker,H.C., Hermens,W.Th., Muller,A.D. and Zwaal,R.F.A. (1980) *Lancet,* **1**, 70−1.

Oral treatment of haemophilia A by gastrointestinal absorption of Factor VIII entrapped in liposomes.

IMMUNOLOGY

Almeida,J.B., Brand,C.M., Edward,D.C. and Heath,T.D. (1975) *Lancet*, 7941, 899−901.
Formation of virosomes from influenza sub-units and liposomes.
Kinsky,S.C. (1972) *Biochim. Biophys. Acta*, **265**, 1−23.
Antibody-complement interaction with lipid model membranes.
Gregoriadis,G. and Allison,A.C. (1974) *FEBS Lett.*, **45**, 71−4.
Entrapment of proteins in liposomes prevents allergic reactions in pre-immunised mice.
Allison,A.C. and Gregoriadis,G. (1974) *Nature, Lond.*, **252**, 252.
Liposomes as immunological adjuvants.
Dancey,G.F., Yasuda,T. and Kinsky,S.C. (1977) *J. Immunol.*, **119**, 1868−73.
Enhancement of liposomal model membrane immunogenicity by incorporation of lipid A.
Van Rooijen,N. and Van Nieuwmegen,R. (1980) *Immunol. Commun.*, **9**, 747−57.
Endotoxin-enhanced adjuvant effect of liposomes particularly when antigen and endotoxin are incorporated within the same liposome.
Banerji,B. and Alving,C.R. (1981) *J. Immunol.*, **126**, 1080−4.
Anti-liposome antibodies induced by lipid A. I: Influence of ceramide, glyco-sphingolipids and phosphocholine on complement damage.
New,R.R.C., Theakston,R.D.G., Zumbühl,O., Iddon,D. and Friend,J. (1984) *New Engl. J. Med.*, **311**, 56−7.
Immunisation against snake venoms.
Van Houte,A.J., Snippe,H. and Willers,J.M.N. (1979) *Immunology*, **37**, 505−14.
Characterisation of immunogenic properties of haptenated liposomal model membranes in mice. I: Thymus independence of the antigen.
Magee,W.E., Talcott,M.L., Straub,S.X. and Vriend,C.Y. (1976) *Biochim. Biophys. Acta*, **451**, 610−18.
A comparison of negatively and positively charged liposomes containing entrapped polyinosinic polycytidylic acid for interferon induction in mice.
Lachmann, P.J., Munn,E.A. and Weissmann,G. (1970) *Immunology*, **19**, 983−6.
Complement-mediated lysis of liposomes produced by the reactive lysis procedure.
Bruyere,T., Wachsmann,D., Klein,J.-P., Schöller,M. and Frank,R.M. (1987) *Vaccine*, **5**, 39−42.
Local response in rat to liposome-associated *Streptococcus mutans* polysaccharide-protein conjugate.

DIAGNOSIS

In vitro

Haxby,J.A., Kinsky,C.B. and Kinsky,S.C. (1968) *Proc. Natl. Acad. Sci. USA*, **61**, 300−7.

Immune response of a liposomal model membrane.

Six,H.R., Young,W.W., Uemura,K. and Kinsky,S.C. (1974) *Biochemistry,* **13**, 4050–8.

Effect of antibody-complement on multiple versus single compartment liposomes. Application of a fluorimetric assay for following changes in liposomal permeability.

Weinstein,J.N., Yoshikami,S., Henkart,P., Blumenthal,R. and Hagins,W.A. (1977) *Science,* **195**, 489–92.

Liposome-cell interaction: transfer and intracellular release of a trapped fluorescent marker.

Yasuda,T., Tadakuma,T., Pierce,C.W. and Kinsky,S.C. (1979) *J. Immunol.,* **123**, 1535–9.

Primary *in vitro* immunogenicity of liposomal model membranes in mouse spleen cell cultures.

Uemara,K., Yuzawa-Watanabe,M., Kitazawa,N. and Taketoni,T. (1980) *J. Biochem.,* **87**, 1641–8.

Liposome agglutination and liposomal membrane immune damage assays for the characterisation of antibodies to glycosphingolipids.

Slovick,D.I., Saida,T., Lisak,R.P. and Schreiber,A. (1980) *J. Immunol. Meth.,* **39**, 31–8.

A new assay for lytic anti-galactocerebroside antibodies employing rubidium release from galactocerebroside-labelled liposomes.

Yasuda,T., Naito,Y., Tsumita,T. and Tadakuma,T. (1981) *J. Immunol. Meth.,* **44**, 153–8.

A simple method to measure anti-glycolipid antibody by using complement-mediated immune lysis of fluorescent dye-trapped liposomes.

Axelsson,B., Eriksson,H., Borrebaeck,C., Mattiasson,B. and Sjögren,H.O. (1981) *J. Immunol. Meth.,* **41**, 351–63.

Liposome immune assay (LIA). Use of membrane antigens inserted into labelled lipid vesicles as targets in immune assays.

Janoff,A.S., Carpenter-Green,S., Weiner,A., Siebold,J., Weissmann,G. and Ostro,M. (1983) *Clin. Chem.,* **29**, 1587–92.

Novel liposome composition for a rapid colorimetric test for systemic *lupus erythematosis*.

In vivo

McDougall,I.R., Dunnick,J.K., McNamee,M.G. and Kriss,J.P. (1974) *Proc. Natl. Acad. Sci. USA,* **71**, 3487–91.

Distribution and fate of synthetic lipid vesicles in the mouse: a combined radio-nuclide and spin label study.

McDougall,I.R., Dunnick,J.K., Goris,M.L. and Kriss,J.P. (1975) *J. Nucl. Med.,* **16**, 488–91.

In vivo distribution of vesicles loaded with radio pharmaceuticals: a study of different routes of administration.

Caride,V.J., Taylor,W., Camer,J.A. and Gotschalk,A. (1976) *J. Nucl. Med.,* **17**, 1067–72.

Evaluation of liposome-entrapped radioactive tracers as scanning agents. I Organ distribution of liposome [^{99m}Tc-DTPA] in mice.

Caride,V.J. and Zaret,B.L. (1977) *Science,* **198**, 735−8.

Liposome accumulation in regions of experimental myocardial infarction.

Richardson,V.J., Ryman,B.E., Jewkes,R.F., Tattersall,M.H.N. and Newlands,E.S. (1978) *Int. J. Nucl. Med. Biol.,* 5, 118−23.

^{99m}Tc-labelled liposomes. Preparation of radiopharmaceutical and its distribution in a hepatoma patient.

Richardson,V.J., Ryman,B.E., Jeyasingh,K., Jewkes,R.F., Tattersall,M.H.N., Newlands,E.S. and Kaye,S.B. (1979) *Br. J. Cancer,* **40**, 35−43.

Tissue distribution and tumour localisation of 99m-technetium-labelled liposomes in cancer patients.

Osborne,M.P., Richardson,V.J., Jeyasingh,K. and Ryman,B.E. (1979) *Int. J. Nucl. Med. Biol.,* **6**, 75−83.

Radionuclide-labelled liposomes—a new lymph node imaging agent.

New,R.R.C., Critchley,M. and Patten,M. (1981) *Nucl. Med. Comm.,* **2**, 31.

Distribution of radionuclide-labelled liposomes draining from inflamed tissue.

Morgan,J.R., Williams,K.E., Davies,R.L., Leach,K., Thomson,M. and Williams,L.A.P. (1981) *J. Med. Microbiol.,* **14**, 213−17.

Localisation of experimental staphylococcal absesses by ^{99m}Tc-technetium-labelled liposomes.

Havron,A., Seltzer,S.E., Davis,M.A. and Shulkin,P. (1981) *Radiology,* **140**, 507−11.

Radioopaque liposomes: a promising new contrast material for computed tomography of the spleen.

Caride,V.J., Sostman,H.D., Twickler,J., Zacharis,H., Orphanoudakis,S.C. and Jaffe,C.C. (1982) *Invest. Radiol.,* **17**, 381−5.

Brominated radioopaque liposomes: contrast agent for computed tomography of liver and spleen. A preliminary report.

DELIVERY OF GENETIC MATERIAL

Straub,S.X., Garry,R.F. and Magee,W.E. (1974) *Infect. Immunol.,* **10**, 783−92.

Interferon induction by poly(I):poly(C) enclosed in phospholipid particles.

Magee,W.E., Talcott,M.L., Straub,S.X. and Vriend,C.Y. (1976) *Biochim. Biophys. Acta,* **451**, 610−18,.

A comparison of negatively and positively charged liposomes containing entrapped polyinosinic—polycytidylic acid for interferon induction in mice.

Kulpa,C.F. and Tinghitella,T.J. (1976) *Life Sci.,* **19**, 1879−88.

Encapsulation of polyuridylic acid in phospholipid vesicles.

Mayhew,E., Papahadjopoulos,D., O'Malley,J.A., Carter,W.A. and Vail,W.J. (1977) *Mol. Pharm.,* **13**, 488−95.

Cellular uptake and protection against virus infection by polyinosinic-polycytidylic acid entrapped within phospholipid vesicles.

Ostro,M.J., Giacomoni,D. and Dray,S. (1977) *Biochem. Biophys. Res. Comm.,* **76**, 836−42.

Incorporation of high molecular weight RNA into large artificial lipid vesicles.
Dimitriadis,G.J. (1978) *Nucl. Acids Res.*, **5**, 1381−6.

Introduction of ribonucleic acids into cells by means of liposomes.
Ostro,M.J., Giacomoni,D., Lavelle,D., Paxton,W. and Dray,S. (1978) *Nature, Lond.*, **247**, 921−3.

Evidence for translation of rabbit globin mRNA after liposome-mediated insertion into a human cell line.
Dimitriadis,G.J. (1978) *Nature, Lond.*, **274**, 923−4.

Translation of rabbit globin mRNA introduced by liposomes into mouse lymphocytes.
Mukherjee,A.B., Orloff,S., Butler,J.D., Triche,,T., Lalley,P. and Schulman,J.D. (1978) *Proc. Natl. Acad. Sci. USA*, **75**, 1361−5.

Entrapment of metaphase chromosomes into phospholipid vesicles (lipochromosomes): carrier potential in gene transfer.
Wilson,T., Papahadjopoulos,D. and Taber,R. (1979) *Cell*, **17**, 77−84.

The introduction of poliovirus RNA into cells via lipid vesicles (liposomes).
Fraley,R.T., Fornari,C.S. and Kaplan,S. (1979) *Proc. Natl. Acad. Sci. USA*, **76**, 3348−52.

Entrapment of a bacterial plasmid in phospholipid vesicles.
Lurquin,P.F. (1979) *Nucl. Acids Res.*, **6**, 3773−84.

Entrapment of plasmid DNA by liposomes and their interactions with plant protoplasts.
Fraley,R., Subramani,S., Berg,P. and Papahadjopoulos,D. (1980) *J. Biol. Chem.*, **255**, 10431−35.

Introduction of liposome-encapsulated SV40 DNA into cells.
Wong,T.K., Nicolau,C. and Hofschneider,P.H. (1980) *Gene*, **10**, 87−94.

Appearance of β-lactamase activity in animal cells upon liposome-mediated gene transfer.
Fukunaga,Y., Nagata,T. and Takebe,I. (1981) *Virology*, **113**, 752−60.

Liposome-mediated infection of plant protoplasts with tobacco mosaic virus RNA.
Radford,A., Pope,S., Sazci,A., Fraser,M.J. and Parish,J.A. (1981) *Mol. Gen. Genet.*, **184**, 567−9.

Liposome-mediated genetic transformation of *Neurospora crassa*.
Lurquin,P.F. and Sheehy,R.E. (1982) *Plant Sci. Lett.*, **25**, 133−46.

Binding of large liposomes to plant protoplasts and delivery of encapsulated DNA.
Nicolau,C., Le Pape,A, Soriano,P., Fargette,F. and Juhel,M.F. (1983) *Proc. Natl. Acad. Sci. USA*, **80**, 1068−72.

In vitro expression of rat insulin following intravenous administration of liposome-entrapped rat insulin I gene.

INDEX